Genetics – Research and Issues

nova
science publishers
www.novapublishers.com

Genetics – Research and Issues

Fragile X Syndrome: From Diagnosis to Treatment
Fabrizio Stasolla, PhD (Editor)
ISBN: 978-1-68507-572-9 (Softcover)
ISBN: 978-1-68507-593-4 (eBook)

Recurrence Risk in Autism Spectrum Disorders
Michael Beenstock (Author)
ISBN: 978-1-68507-129-5 (Hardcover)
ISBN: 978-1-68507-244-5 (eBook)

p53: Structure, Functions and Role in Disease
Monte Stevens (Editor)
ISBN: 978-1-53618-771-7 (Softcover)
ISBN: 978-1-53618-808-0 (eBook)

A Comprehensive Guide to Genetic Counseling
Benjamin A. Kepert (Editor)
ISBN: 978-1-53616-975-1 (Softcover)
ISBN: 978-1-53616-976-8 (eBook)

Gene Mutations: Causes and Effects
Helena M. Christoffersen (Editor)
ISBN: 978-1-53616-984-3 (Softcover)
ISBN: 978-1-53617-030-6 (eBook)

More information about this series can be found at https://novapublishers.com/product-category/series/genetics-research-and-issues/

Ekaterina S. Evstratova, Vladislav G. Petin, Andrey D. Kaprin and Petr V. Shegai

Genetic Instability and Some Unusual Radiobiological Effects

nova
science publishers
www.novapublishers.com

DOI: https://doi.org/10.52305/OYKO4114

Library of Congress Cataloging-in-Publication Data

Names: Evstratova, Ekaterina S., author. | Petin, V. G. (Vladislav Georgievich), author. | Kaprin, A. D. (Andreĭ Dmitrievich), author. | Shegai, Petr V., author.
Title: Genetic instability and some unusual radiobiological effects / Ekaterina S. Evstratova, Vladislav G. Petin, Andrey D. Kaprin and Petr V. Shegai.
Description: New York : Nova Science Publishers, [2022] | Series: Genetics - research and issues | Includes bibliographical references and index. |
Identifiers: LCCN 2022024315 (print) | LCCN 2022024316 (ebook) | ISBN 9781685078867 (hardcover) | ISBN 9798886970074 (adobe pdf)
Subjects: LCSH: Radiogenetics. | Mutagenesis. | Ionizing radiation--Physiological effect.
Classification: LCC QH465.R3 E87 2022 (print) | LCC QH465.R3 (ebook) | DDC 572.8/38--dc23/eng/20220720
LC record available at https://lccn.loc.gov/2022024315
LC ebook record available at https://lccn.loc.gov/2022024316

Published by Nova Science Publishers, Inc. † New York

Contents

Foreword

I have gone through the volume on *Genetic Instability and Some Unusual Radiobiological Effects* authored by Ekaterina S. Evstratova, Vladislav G. Petin, Andrey D. Kaprin and Petr V. Shegai. I find the book interesting and highly impressive from radiobiological viewpoints and a detailed account of genetic instability and its relevance in various biological phenomena would prove absorbing and engage the readers. The Section on unusual radiobiological effects has presented a valuable account of frequently discussed but unsettled issues in radiobiology. From the viewpoint of readers, the proposed book will fill in the gap of knowledge in classical to modern radiobiology. Indeed, biological significance of genetic instability makes an attractive reading material. The accounts of unusual effects of radiation are particularly valuable. The controversies in the literature with regard to mutagenesis, carcinogenesis and ageing in irradiated cells are adequately covered from the studies in yeast cells. I felt fascinated to read the comparisons and universal effects of ionizing radiations like gamma rays, alpha particles and UV radiation.

The Authors have suitably covered low dose radiation effects, which are of current scientific, health and environmental interests. The indications of likely adverse effects and beneficial effects are worthy of reading by young researchers as well as specialist in radiobiological research which, I am sure, will compel researchers to generate new ideas in research.

There are 8 chapters in the book and each chapter is well placed. Readers will find each chapter rich in contents covering current hot topics in radiobiology. Apart from damage to DNA by direct effects of radiation, a coverage is found of non-targeted effects of radiation on cells in vitro and in vivo. I feel it would be unrewarding to elaborate on each chapter but it will prove a sought-after book. The dependence of sensitivity of cells to radiation on the diploid state has been succinctly described. I find highly exciting the idea of some specific points raised with regard to unusual radiobiological effects, which will make the book stand out among other contemporary publications in the field of radiobiology. The volume will prove scientifically stimulating and exciting.

Dr. K. P. Mishra
Bhabha Atomic Research Center, Mumbai, India

Preface

Genetic instability leads to the accumulation of many changes necessary to transform the stable genome of a normal cell into an unstable genome. Such a transformation can be the cause of radiation mutagenesis, carcinogenesis and aging – the main long-term effects of ionizing radiation, UV light, and other harmful environmental factors. A new quantitative assessment of the genetic instability was suggested, taking into account the delay in the formation of colonies by cells surviving after exposure to harmful agents. It was demonstrated that the genetic instability of yeast cells is mainly determined by cell ploidy, but not by the sigmoid shape of the survival curve or the cell's ability of post-irradiation recovery. New experimental data confirm the participation of recovery processes in the manifestation of the relative biological effectiveness of densely ionizing radiation both for cell inactivation and for radiation-induced genetic instability.

A number of universal patterns of synergic interactions are demonstrated, which did not depend on the used biological test and objects nor the interacting physical and chemical factors. Potential significance of synergy at low intensities of physical and chemical environmental factors is proved. Original data on the influence of cyclic heating and cooling on cell survival are obtained. It is concluded that osmotic homeostasis of cells, along with DNA, is an important target responsible for cell viability under the damaging action of various agents. Data on the favorable effect of the increased natural background of ionizing radiation are discussed. The manifestations of radiation hormesis were revealed for various radiation accidents, professional and medical exposure. The results obtained demonstrate that the action of high-energy ionizing radiation is accompanied by irradiation with UV light that occurs when charged particles move at a speed exceeding the speed of light in a given medium.

Introduction

The problem of long-term consequences of exposure to ionizing radiation is one of the oldest problems in radiobiology. Such consequences include radiation-induced instability of cell genome, and in a more generalized form, genetic instability. This phenomenon leads to the accumulation of many changes necessary to transform the stable genome of a normal cell into an unstable genome. It has been demonstrated by many authors that genetic instability is one of the most important evolutionarily significant properties of a cell. Stochastic effects include radiation-induced genetic instability, radiation mutagenesis, carcinogenesis and aging – the main long-term effects of ionizing radiation. It is believed that radiation-induced genetic instability occurs in irradiated cells is a result of a disturbance of the integrity of the DNA structure. The state of genome instability often precedes tumor progression and persists for a long time after irradiation for many cell generations. Mechanisms of radiation induced genetic instability and pathogenetic significance of genetic instability is discussed. A quantitative method is proposed to assess the genetic instability, taking into account the delay in the formation of colonies on nutrient media by cells surviving after irradiation.

The interaction of ionizing radiation with matter is briefly described with an emphasis on the fact that the action of sparsely ionizing radiation is accompanied by UV light arising from the movement of electrons at a speed exceeding the speed of light in a biological environment. It was shown that the total number of primary radiation damage increases in proportion to cell ploidy. No correlation was und between cell ploidy and the recovery constant, which characterizes the probability of recovery per unit of time while the irreversible component of radiation damage decreases with increasing cell ploidy. As a quantitative test of genetic instability, the delay in the colony formation by cells surviving after irradiation was used. It was demonstrated that the genetic instability of yeast cells is mainly determined by cell ploidy, but not by the sigmoid shape of the survival curve or cell ability to post-irradiation recovery, as was repeatedly postulated earlier by many authors.

The main regularities of biological effects after exposure to ionizing radiation with high LET are presented. The mechanisms of the different shapes of the survival curves after exposure to radiation with various LET as well as the dependence of RBE on the LET of radiation are described and discussed. The differences in the change of the survival curves shape for yeast cells and cultured mammalian cells with an increase in LET are qualitatively explained. A distinctive feature of the effect of charged particles on cell genetic structures is the formation of cluster DNA damage, from which it is difficult to cell recover due to problems with the repair of severe DNA damage. New experimental data are presented that confirm the participation of recovery processes in the manifestation of the RBE of densely ionizing radiation both for cell inactivation and for radiation-induced genetic instability. The identity of the RBE of α-particles for both of these tests means that the formation of lethal damage is accompanied by the proportional formation of sub-damage responsible for the genetic cell instability.

The general characteristics of UV-induced biological effects are discussed. New experimental data on the genetic instability of haploid and diploid yeast strains of wild type and their UV sensitive mutants are obtained. The main conclusion from the data presented here is that the genetic instability of the studied haploid and diploid strains, differing in UV sensitivity and in the shape of the survival curves is mainly determined by cell ploidy. The greatest value of genetic instability was 30–50% for haploid cells while it reached 100% with an increase in UV light dose for diploid cells regardless of their sensitivity to UV light, the shape of the survival curves and cell ability to recover from UV damage. These data show that the cells of all diploid strains that survived after irradiation retained some residual sub-damage emerging as many colonies produced after control, and these sub-damages preserved in the distant descendants of cells that survived after irradiation.

The regularities of the synergistic interaction of physical and chemical environmental factors were described in several monographs based on the results obtained by the staff of Biophysical laboratory of the A.F. Tsyb Medical Radiological Research Center, Obninsk (Petin, Komarov, 1989; Petin et al., 2012, 2020; Petin, Morozov, 2015; Petin, Kim, 2016). New results are summarized in our recent publications (Petin et al., 1998; Petin et al., 2000; Komarova et al., 2002; Petin et al., 2014; Pronkevich, Evstratova et al., 2017; Evstratova, Petin, 2018 a, b, c 2020; Evstratova et al., 2018 b), on the basis of which this chapter was prepared. It begins from mathematical description and quantitative prediction of synergistic effects after combined application of two damaging agents. The model is based on the suggestion that synergism is due to the production of additional effective damage resulting from interaction of sub-damage that are non-effective after separate application of agents. Potential significance of synergy at low intensities of physical and chemical environmental factors is proved. Biophysical interpretation of synergy dependence on the intensity of the applied agents is presented. General biological and clinical significance of synergy ideas are discussed. A number of universal patterns of synergic interactions are demonstrated, which did not depend on the used biological objects and test as well as on the interacting physical and chemical factors.

Exchange with the external environment of cells is carried out using the system of osmotic homeostasis, including the membrane apparatus. For the effective exchange, it is necessary to maintain a certain pressure balance in the cell-environment system. Violations of this balance cause changes that are incompatible with the normal cell functions. For example, ionizing radiation or hyperthermia can lead to such violations, especially in conditions of a sharp change in the absolute values of the temperature of cell environment surroundings. The dependence of cell osmotic properties on the dose of ionizing radiation and temperature parameters is presented. The osmotolerance and cell ability to reproduce in environments with various osmotic pressures at different temperatures were described. New data related with the influence of tonicity and chloramphenicol on hyperthermic cytotoxicity and cell permeability are provided. Original data on cyclic heating and cooling on cell survival are also shown. The action of osmolytes on destructive effect of ionizing radiation, hyperthermia, microwave radiation and ultrasound are compared. Peculiarities of biological effects of electromagnetic non-ionizing radiation at cellular and molecular levels are discussed. It is concluded that osmotic homeostasis of cells, along with DNA, is an important target responsible for cell viability under the damaging action of various agents.

The next chapters deals with various approaches to the concept of low doses. A review of data on radiation hormesis, the positive effect of low doses of ionizing radiation, is presented. Many healthy effects induced by small doses of ionizing radiation on biological objects of various complexity are considered. Data on the favorable effect of the increased natural background of ionizing radiation are discussed. The manifestations of radiation hormesis were also revealed as a result of the Hiroshima and Nagasaki bombing, as well as various radiation accidents, professional and medical exposure. The fact that low doses of ionizing radiation are beneficial to humanity seems unlikely, but the health effects of chronic exposure to residents who have lived in ^{60}Co-contaminated apartments for over 25 years in Taipei (Taiwan). This unintentional human experiment proved that chronic exposure to humans at specific dose rates could effectively reduce the rate of human cancer. It is noted that cancer mortality among adults living in these buildings for 9–20 years was only 3% of the total cancer deaths of adult population.

The involvement of molecule excitations in the biological action of ionizing radiation is demonstrated for UV-hypersensitive pro- and eukaryotic cells. Photoreactivation after combined action of ionizing radiation and visible light is observed. New experimental data on the antagonistic interaction of ionizing radiation and visible light are presented. Parameters of bacterial photoreactivation after exposure to UV light and ionizing radiation are calculated and compared. The results obtained demonstrate that the action of high-energy ionizing radiation is accompanied by irradiation with UV light that occurs when charged particles move at a speed exceeding the speed of light in a given medium. The dose of UV light accompanying the irradiation of biological objects with high energy of sparsely ionizing radiation has been estimated. The observed plateau of the UV dose with an increase in both the energy of ionizing radiation and in the volume of the exposed cell suspension is shown to be determined by physical properties of the Čerenkov emission.

Part 1. Genetic Instability

Chapter 1

Biological Significance of Genetic Instability

Abstract

The problem of long-term consequences of exposure to ionizing radiation is one of the oldest problems in radiobiology. Such consequences include radiation-induced instability of cell genome, and in a more generalized form, – genetic instability. This phenomenon leads to the accumulation of many changes necessary to transform the stable genome of a normal cell into an unstable genome. It has been demonstrated by many authors that genetic instability is one of the most important evolutionarily significant properties of a cell. Stochastic effects include radiation-induced genetic instability, radiation mutagenesis, carcinogenesis and aging – the main long-term effects of ionizing radiation. It is believed that radiation-induced genetic instability occurs in irradiated cells is a result of a disturbance of the integrity of the DNA structure. The state of genome instability often precedes tumor progression and persists for a long time after irradiation for many cell generations. Mechanisms of radiation induced genetic instability and pathogenetic significance of genetic instability is discussed. A quantitative method is proposed to assess the genetic instability, taking into account the delay in the formation of colonies on nutrient media by cells surviving after irradiation.

1.1. The Phenomenon of Genetic Instability

Classical radiobiology and radiation genetics have firmly established two phenomena: there is no threshold of doses – and, consequentially, there are gene mutations and chromosome aberrations, as well as biological enhancement of induced damage of the reduplication mechanism in the proliferating cells. Traditionally, it is believed that the biological effects of ionizing radiation occur in irradiated cells because of a disturbance of the integrity of the DNA structure. This means that: 1) biological effects take place only in irradiated cells; 2) the passage of ionizing particles through the cell nucleus is a prerequisite for the occurrence of a biological reaction; 3) DNA is a target molecule in the cell. In recent decades, many data have been published indicating the long-term effects of radiation – induced genomic instability in the form of various disorders of the genetic apparatus in the descendants of multiple-divided cells that were irradiated both *in vitro* and *in vivo*. In many publications, along with the more common term of genetic instability, the terms of genomic instability or chromosomal instability are also used.

The problem of long-term consequences of exposure to ionizing radiation is one of the oldest problems in radiobiology (Vilenchik, 1987, 1991; Moskalev, 1991; Limoli et al., 1997). It is directly related to studies of the inconstancy of the genome, which differs in mechanisms from ordinary mutagenesis. For the first time, these problems were formulated at the beginning of the twentieth century in the form of the concept of unstable genes, which was transformed later into the concept of mobile genetic elements that make a great contribution to the basis of biological evolution (Ferguson et al., 2015). Genome stability is one of the most important evolutionarily significant properties of a cell (Timofeev-Ressovsky, Zimmer, 1947; Vorobtsova, 1988, 2006; Geraskin, Sarapultsev, 1995; Inge-Vechtomov, 2015). Genetic instability is a term used to describe a phenomenon that leads to the accumulation of many changes necessary to transform the stable genome of a normal cell into an unstable genome. Recently, there has been a serious debate about the importance of the instability of the human genome in the carcinogenic process. Genetic instability is characterized by a large collection of diverse endpoints of observation, including large-scale chromosomal rearrangements and aberrations, amplification of genetic material, aneuploidy, micronucleus formation, microsatellite instability, and gene mutation (Wang et al., 2009).

Destabilization of chromosomes has long been considered as the first and direct sign of genetic instability. Various mutations (gene, chromosomal, and genomic), as well as artificial manipulations with the genome (inbreeding, DNA transfection, inclusion of BrdU into cell DNA, etc.) cause its destabilization. Exposure to ionizing radiation induces genome instability in distant mitotic descendants of cells irradiated *in vitro* or *in vivo*. It manifests itself both in morphological criteria – an increase in the frequency of *de novo* mutations, and in functional indicators – increased radiosensitivity, changes in gene expression, disappearance of the adaptive response, and the delay in the formation of colonies by cells that survived after irradiation. Genetic instability also occurs in the meiotic progeny of irradiated cells.

Analysis of the literature data indicates that the patterns of genetic instability do not correspond to the classical concepts of the target theory, according to which during the induction of rarely occurring mutations, it is transmitted to all the offspring of an irradiated cell (Timofeeff-Ressovsky, Zimmer, 1947; Raevsky, 1959; Fritz-Neagley, 1961; Zimmer, 1962; Lea, 1946; Timofeeff-Ressovsky et al., 1981). Evidence has emerged for untargeted radiation effects, i.e., such consequences of irradiation, when biological reactions are observed in cell that was not crossed by a charged particle. One of these phenomena is called "bystander effect." The possible influence of the non-targeting radiation effect on the biological response when exposed to low doses of ionizing radiation is now carefully discussed. Non-targeting effects include radiation-induced genetic instability, when biological effects can observe in the distant descendants of irradiated cells, the nucleus of which has not been crossed by ionizing particle (Little, 2000; Morgan, 2003). It has become apparent that radiation can induce cellular instability, which is observed in the distant offspring of cells that survived the irradiation.

Genetic instability is a phenomenon known for a number of radiobiological manifestations. The essence of the phenomenon lies in the fact that in some of the cells that survived after radiation exposure, functionally altered offspring are produced, in which aberrations of chromosomes and gene mutations arise with a high frequency over many generations, in some cases leading to increased cell mortality by apoptosis (Mazurik, Mikhailov, 2001). Genetic instability refers to a series of events that can cause some heritable changes in the genome. This term covers genetic changes that can be classified in various ways (Suskov et al., 2006; Yuen et al., 2007; Zhu et al., 2012). In particular, many delayed effects, such as delayed proliferative death, reduced growth efficiency, inhibition of the ability of cells to recover from radiation damage, death among distant offspring, clonal heterogeneity, increased thermal and radiosensitivity, malignant transformation, the emergence of micronuclei and giant cells may be considered as examples of genetic instability. Genetic instability includes a number of other effects – DNA damage, structural chromosome aberrations, gene mutations, sister chromatid exchanges, changes in the number of chromosomes in a cell (aneuploidy or polyplodia) as well as the reproductive cell death, genome rearrangement, malignant transformation, decreased cloning efficiency and heterogeneity among offspring of irradiated cells (Chang, Little, 1992a, b; Morgan, Murnane, 1995; Little, 1998; Mazurik, Mikhailov, 2001; Korogodin et al., 2007; Saenko, Zamulaeva, 2007; Petin et al., 2008; Vorobtsova, 2008; Shen, 2011; Stirling et al., 2011; Aggarwal, Brosh, 2012; Kumaran et al., 2013; Lee et al., 2015).

The expression of genetic instability after chronic or acute exposure to ionizing radiation at the organismic level can be attributed to carcinogenesis, aging, and an increase in the frequency of chronic diseases (Suskov, Kuzmina, 2001; Kuzmina, Suskov, 2002; Suskov et al., 2008). Most of these effects can be observed after many divisions and even passages after irradiation (Cheng, Loeb, 1993; Morgan, Murnane, 1995; Little, 1998; Shen, 2011). The phenomenon of genetic instability is an important factor in carcinogenesis and some neurodegenerative diseases. Since endogenous DNA damage in every human cell occurs on average more than 60,000 times a day, any structural damage to the integrity of DNA, as well as a decrease in the reliability and efficiency of its repair are a source of genetic instability. Genetic instability concerns such important consequences of exposure to ionizing radiation and UV light as mutagenesis, carcinogenesis, and aging, which are the main long-term effects of radiation (Vilenchik, 1987; Moskalev, 1991; Cheng, Loeb, 1993; Shen, 2011).

A common property of various delayed effects observed after exposure to ionizing radiation is the transmission of some subdamage (sublesions), ineffective for lethal events, to distant offspring (Chang, Little, 1992 a, b; Dugan, Bedford, 2003; Petin et al., 2008). The state of genome instability often precedes tumor progression (Little, 1998; Kalb et al., 2006; Brosh, Bohr, 2007; Shen, 2011) and persists for a long time after irradiation for many cell generations (Klein et al., 1990; Little et al., 1997; Korogodin et al., 2007; Petin et al., 2008). An insufficiently studied aspect of this problem is a quantitative comparison of the survival and genetic instability of haploid and diploid cells surviving after exposure to many factors, for example, γ-rays, α-particles and UV light. The most convenient model for this purpose is yeast cells, whose radiobiological reactions are qualitatively identical to those for cultured mammalian cells (Korogodin, 1966, 2010; Korogodin et al., 1968; Kapultsevich, 1978; Critchlow, Jackson, 1998; Resnick, Cox, 2000; Feldman, 2012; Karpenshif, Bernstein. 2012; Mulla et al., 2014; Petin, Kapultcevich, 2014).

The authors of many works discussed various cell factors that can affect the stability of the genome. These include cell ploidy, the phase of the cell cycle, transcriptional activity, the ability of cells to control the repair of DNA breaks, the metabolic stage of cell growth, and the effect of various stresses (Kapultsevich et al., 1972; Zhloba, Sevankaev, 1991; Korogodin, 1993; Saenko et al., 1998; Yuen et al., 2007; Suskov et al., 2008; Zhu et al., 2012; Sankaranarayanan et al., 2013; Skoneczna et al., 2015; Koltovaya, 2017). From the data discussed here, it is clear that the term "genetic instability" encompasses many genetic changes that can be classified in various ways.

Like any biological phenomenon, genetic instability can have both negative – gene amplification, mutations, chromosome destabilization, neoplastic transformation, and a positive value – a source of evolutionary processes, the formation of hypervariable antibody regions, and the emergence of alternative ways of adaptation to changing environmental conditions. To date, kinetic patterns of the release of gene and chromosomal aberrations have been established, which obey the classical concepts of the target theory and the hit principle (Luchnik, 1968; Bochkov, 1971; Sevankaev, 1987). However, in recent decades, it has been firmly established that, in addition to direct mutagenic effects, ionizing radiation leads to various disorders of the genetic apparatus in the descendants of multiply divided cells that were irradiated both *in vitro* and *in vivo*. Little (2007) summarizes the important characteristics of radiation-induced genomic instability as follows. This phenomenon has been described for several genetic consequences, including a sustained increase in the incidence of reproductive impairment and apoptosis in the offspring of irradiated cells. These genetic changes occur in the distant offspring of the irradiated cells, sometimes 10–30 generations after the irradiation of the parental cells.

The main causes of radiation sickness and death of mammals and humans include reproductive death, apoptosis, and cell necrosis immediately after irradiation (Burlakova et al., 2001; Mazurik, Mikhailov, 2001). It has been shown that these deterministic effects are complex genetic and biochemical phenomena accompanying DNA repair. Stochastic effects include radiation-induced genetic instability, including radiation mutagenesis, carcinogenesis and aging – the main long-term effects of ionizing radiation. The phenotypic manifestation of genetic instability also includes stochastic delayed reproductive cell death, the result of which is the slow formation of colonies by cells that survived after irradiation. There is a correlation between persistent genome instability and delayed reproductive death as well as the decreased clonogenic activity (James, Werner, 1966; Tolstorukov et al., 1979, 1982; Little et al., 1997).

Genomic instability due to polyploidy, i.e., to an increase in the number of chromosome sets in cell, has become generally investigated in the last decade. Storchova (2014) notes a number of issues that need to be resolved in future studies. What are the immediate benefits of polyploidy associated with the relatively frequent appearance and retention of polyploid cells? What molecular mechanisms are involved in genetic instability, cellular metabolism, and growth of aneuploidy cell? And finally, what molecular mechanisms allow to adapt polyploid and aneuploid cells? It is assumed that the answers to these questions will ultimately improve our understanding of the evolution and origin of diseases (Storchova, Kuffer, 2008). A growing amount of information indicates that polyploid cells also arise in various pathological conditions (Storchova, Pellman, 2004). Genetic instability in these cells can provide a pathway for aneuploidy and contribute to the development of cancer.

The balance between advantages and disadvantages resulting from genome rearrangements into polyploid or aneuploid cells determines the further fate of these cells. The presence of such a balance increases the reproductive capacity of many species, influences the evolution and occurrence of malignant tumors. Changes in ploidy contribute to tumorigenesis, probably due to an increase in genomic instability (Storchova, Kuffer, 2008). Tetraploid cells have been found in several types of cancer (Storchova, Pellman, 2004; Ganem et al., 2007; Storchova, Kuffer, 2008; Davoli, de Lange, 2011). Besides, tetraploid cells of mice injected into healthy mice often initiated tumors, while isogenic diploid cells did not influence this process. These results indicate the importance role of genetic instability in tumorigenesis.

Polyploidy is a widespread phenomenon, especially in plants and fungi. In humans, polyploidy often occurs in certain tissues (Storchova, Kuffer, 2008). Tetraploidization can initiate chromosomal instability, probably due to abnormal centrosomes and double chromosome mass. Chromosomal instability, in turn, may persist or soon give way to a stably spreading but aneuploid karyotype. Both chromosomal instability and stable aneuploidy are commonly seen in cancer. There have been convincing data demonstrating that an increase in the number of sets of chromosomes can cause aneuploid tumor (Cheng, Loeb, 1993; Hanahan, Weinberg, 2000; Blasko, 2005).

It should be noted that genetic instability of cells is the subject of intensive modern research in terms of human aging and the participation of this effect in the formation of oncological diseases, which are a global problem of humanity (Tsyb, Gulidov, 2010; Davydov et al., 2013; Pankratov et al., 2014; Kaprin et al., 2017; Kaprin et al., 2019). The works of many authors (Chang, Little, 1992a, b; Morgan et al., 1996; Korogodin et al., 2007; Kumaran et al., 2013) discuss events and processes that may be involved in the initiation and maintenance of an unstable phenotype – delayed mutations, double-stranded DNA breaks, erroneous DNA repair, telomere size, signal transduction, changes in gene expression, epigenetic factors.

A distinctive feature of many tumors is the presence of additional centrosomes, which also cause genetic instability. To determine the mechanism underlying this defect, Ganem et al. (2007) designed cells that differ only in the number of centrosomes. It was demonstrated that additional centrosomes contributed to the formation of segregation errors. These findings provide a direct link between extracentrosomes and chromosomal instability, the main common characteristics of solid tumors.

1.2. Patterns of Genetic Instability Display

Cells are constantly faced with endogenous and exogenous factors that influence their genome. In the process of evolution, cells have developed a mechanism responsible for repair of radiation damage inside the genome, while at the same time mutually exclusive goals have been fulfilled – to preserve the unchanged genome and the ability to modify (mutate), which allows them to adapt to changes in the environment. Numerous studies have established that radiation-induced chromosomal instability occurs in unicellular organisms, cultured mammalian cells, as well as in various cells of both humans and rodents (Chang, Little, 1992a, b; Morgan, Murnane, 1995; Little, 1998; Mazurik, Mikhailov, 2001; Pelevina et al., 2003; Korogodin et al., 2007; Saenko, Zamulaeva, 2007; Yuen et al., 2007; Petin et al., 2008;Vorobtsova, 2008; Korolev, 2009; Shen, 2011; Stirling et al., 2011; Aggarwal, Brosh, 2012; Kumaran et al., 2013; Lee et al., 2015). Many of these studies were carried out in tumor cell lines. As a test of this phenomenon, the work (Dugan, Bedford, 2003) studied chromosomal instability in the offspring of normal diploid human fibroblasts (AG1521A) that survived after exposure both to low- and high-LET ionizing radiation. Genomic instability in the offspring of irradiated human fibroblasts (AG01522B) was observed in the form of micronuclei, apoptosis, and delayed reproductive death (Belyakov et al., 1999). The increased efficiency of alpha particles compared to X-rays has been noted.

Based on the evidence that delayed mutation events can be formed in the offspring of irradiated cells, it has been hypothesized (Little et al., 1997) that radiation can induce some type of genomic instability in cells, which leads to an increased rate of spontaneous mutations. The authors observed an increased frequency of minisatellite mutations in irradiated cells by changing the thymidine kinase locus. In their opinion, the results obtained mean that subpopulations of genetically unstable cells can arise due to radiation exposure, regardless of the quality of the radiation. The frequency of mutations in the Hprt locus of CHO cells was measured at doses up to 12 Gy of X-ray and up to 2 Gy of α-particles. Approximately 8-9% of 446 selected colonies showed a high frequency of mutations, which was about 100 times higher than background. The frequency with which such clones arose was approximately the same for irradiation with X-rays and α-particles. At the same time, there was no dose dependence of the recorded effect after exposure to X-rays in the dose range of 4-12 Gy. Little (1998), continuing these studies, demonstrated that 14 out of 60 clones mutant at the Hprt locus were characterized by slow cell growth, a high frequency of induced polyploidy (10–38%), and aberrations of the chromosomal and chromatid types. These results imply that among the clones showing a high rate of delayed mutation formation, there may be subpopulations of cells that are particularly unstable. The described results show that ionizing radiation can induce genetic instability transmitted to offspring surviving cells, which can lead to a constant increase in the frequency of new mutations occurring in offspring.

Note that mutations induced in cells of various origins, including yeast and mammalian cells, result from DNA damage. These lesions are formed as a result of direct damage of certain sections of DNA, and can also turn into mutations during subsequent DNA replications or as a result of erroneous repair of DNA damage. It is believed that such a mutagenic event in a critical gene is the first step towards genetic cell instability and possible carcinogenesis, i.e., the emergence and development of a malignant tumor from a normal cell. It is known that carcinogenesis is a multi-step process that occurs due to the accumulation of mutations in many genes required to maintain normal growth of control cells. Studies have shown that for the complete transformation of a cell into a tumor one must be realized from 6 to 8 independent sequential events (Baverstock, 2000). It remains difficult to understand how a single cell can accumulate these independent mutational events throughout its life *in vivo*. The phenomenon of genetic instability is most suitable for interpreting this multistep process.

The results of the work (Ponomarev et al., 2008) indicate a relationship between the delayed formation of radiation mutagenic effects and chromosomal instability. The authors analyzed the possible relationship between the delayed mutation and cytogenetic changes using a clone showing a high incidence of delayed mutation. In this unstable population, an increased frequency of aneuploid cells (13%) was noted compared to 1% in the initial control population of CHO cells. When Hprt mutations were isolated from this population, a significant proportion showed slow cell doubling under standard culture conditions. An increase in both chromosomal and chromatid aberrations was noted among slowly growing subclones. The frequency of aberrations was much lower in the group of 36 Hprt mutant clones with normal doubling times.

To investigate the target responsible for chromosomal instability, as well as the dependence of this effect on the dose and dose rate of ionizing radiation, control GM10115 cells were used, as well as cells in whose DNA bromodeoxyuridine was incorporated (Limoli et al., 1999). Chromosomal stability status was determined by *in situ* fluorescence hybridization approximately 20 generations after irradiation of clonal populations obtained from single progenitor cells that survived acute irradiation. In total, about 700 individual clones were analyzed, representing over 140,000 metaphases. Based on the results obtained, the authors conclude that the increase in chromosomal instability caused by the presence of bromodeoxyuridine proves that DNA is one of the critical targets important for the induction of chromosomal instability.

Chromosomal instability in cells of a human-hamster hybrid cell line after exposure to X-rays is characterized by the appearance of chromosomal rearrangements in several generations after exposure to ionizing radiation. To identify the cellular targets responsible for radiation-induced chromosomal instability, cells were treated with ^{125}I-labeled compounds and frozen (Kaplan, Morgan, 1998). The radioactive decay of ^{125}I caused damage to cells, primarily at the site of their decay, and freezing the cells made it possible to accumulate damage in the absence of other cellular processes. It has been shown that the ^{125}I-deoxyuridine decay, which is incorporated into DNA, causes chromosomal instability. Chromosomal instability can also be caused by the administration of this drug without freezing the cells. Labeling of cells with a radionuclide accumulating in the membrane or cytoplasm (^{125}I-succinyl-con A) did not lead to chromosomal instability. These results indicate that the target for radiation-induced chromosomal instability in these cells is the nucleus.

Morgan and Murnane (1995) discussed the role of genomic instability in cell radioresistance. The authors note that the main factor limiting the successful use of radiotherapy is the natural or induced radioresistance of cells. Tumors that did not fully respond to radiation therapy contained more radioresistant cells than tumors that did not receive radiation therapy. It is concluded that genomic instability, in particular chromosome aberrations, can contribute to the radioresistance of the cell population in tumors. Loss of genome stability in mammalian tumor cells is considered the most important property of cancer. Indeed, in cancer cells, there is an increased level of chromosomal aberrations, including deletions, instability of DNA sequences, gene amplification, disruption of the synthesis of daughter DNA molecules, in which instead of one copy of a part of the chromosome several copies are formed. These effects may be precursors to cancer formation. It has been shown that genetic instability and an increased rate of changes in the genome of mammalian cells can directly modulate the radiosensitivity of cells. In particular, destabilization of chromosomes due to genomic instability preceding tumorigenesis leads to a state of the tumor that allows better adaptation to changes in its local environment or increases the resistance of DNA to radiation damage. The authors conclude that it is necessary to develop new approaches in radiotherapy based on the reduction of genomic instability.

In addition to the "bystander effect" and genome instability, other "non-targeting" effects, such as the adaptive response and radiation hormesis, manifested both at the cellular and organismal levels, have been described after irradiation with ionizing radiation. According to the current theory, direct hit is not necessary to damage cells, and biological effects can occur when the track of an ionizing particle travels at a distance greater than the size of the cell. In this case, gene expression is of great importance, which can lead to malignant transformation without direct mutations. Based on the analysis of the experimental results, the following effects of chronic low-intensity irradiation were mentioned (Pelevina et al., 2006), which cannot be interpreted by the direct hit of ionizing particles into the cell nucleus. 1) Populations of cells, animals and, probably, people with special properties and a particular phenotype may appear. 2) These populations are more sensitive to additional application of damaging factors may be due to the synergistic interaction of the resulting damage. 3) These populations have a reduced ability for an adaptive response and there is also a high individual variability in radiosensitivity, adaptive response, and spontaneous level of damage in cellular structures. 4) In these populations, the phenomenon of increased radiosensitivity after exposure to low doses can be detected.

Radiation-induced chromosome instability *in vitro* was studied in the cell culture of the Chinese hamster CHO-K1 (Pyatenko et al., 2013). The aim of this work was to study the influence of the proliferative state of the cell population on the manifestation of instability according to the criterion of the delayed chromosomal aberrations (dicentrics). The cells were irradiated with ^{60}Co gamma rays in the exponential and stationary states of growth with subsequent frequent replatings, a scheme of restricted growth was additionally applied. The data obtained indicate that the stationary population of cells is more sensitive to the induction of long-term effects of radiation than the proliferating one. Resting the offspring of irradiated cells leads to a significantly higher frequency of the delayed aberrations compared to the traditionally used continuously proliferating population. The data observed show that restricted growth of the dividing offspring leads not only to the formation of the delayed chromosome aberrations, but also to their increase with time. The authors conclude that these results indicate the possibility of a genetic instability mechanism not directly related to DNA replication.

To determine whether genetic instability is transmitted *in vivo*, bone marrow cells obtained from male mice were irradiated with α-particles *in vitro* and transplanted into female recipients. In the hematopoietic system, persistent chromosomal instability was observed up to 1 year after transplantation (Watson et al., 1996). These data demonstrate the long-lasting effect of α-particle-induced genomic instability in donor stem cells. The aim of the works (Kadhim et al., 1994, 1998) was to determine whether chromosomal instability is induced in bone marrow cells and fibroblasts of human lungs by ionizing radiation, and to investigate the effect of radiation quality by comparing the effectiveness of X-rays, neutrons and α-particles. The expression of genomic instability in the form of an increased level of chromatid and chromosomal aberrations was demonstrated in clonal descendants of murine hemopoietic stem cells irradiated with α-particles at doses of 0.25–1.0 Gy. It was shown that densely ionizing radiation was more effective in inducing genetic instability than X-ray radiation, although the author was unable to quantify RBE because of tests used.

Radiations with high LET are currently considered promising for the treatment of cancer due to their higher RBE, especially in the area of low doses, usually used in fractionated irradiation of malignant tumors (Yarmonenko et al., 1992; Tsyb, Mardynskiy, 2010; Kaprin, Mardynskiy, 2018). Another advantage of using radiation with high LET for cancer treatment is the suppression of the ability of cells to recover from sublethal and potentially lethal radiation damage (Kogl, 1986; Yarmonenko et al., 1992; Hall, Giaccia, 2018). Although genetic instability is well known after exposure to ionizing radiation with low- and high-LET (Kadhim et al., 1994; Little et al., 1997; Smith et al., 2003; Ponnaiya et al., 2004), quantitatively RBE of densely ionizing radiation for radiation-induced genetic instability has never been evaluated. The authors of the paper (Dugan, Bedford, 2003) indicate that the frequency of unstable clones was higher after exposure of cultured normal human cells to radiation with high- than low-LET radiation. The frequency of unstable clones in human lymphoblasts was compared in clones formed after irradiation with ^{137}Cs γ-rays or ^{56}Fe particles (Evans et al., 2003). Most unstable clones isolated after exposure to γ-rays or ^{56}Fe-particles showed chromosomal instability. Changes in growth characteristics and induced mutations occurred much less frequently than cytogenetic changes in these unstable clones. The frequency and complexity of unstable clones were higher after exposure to ^{56}Fe-particles than γ-rays, although these authors did not quantify the RBE of densely ionizing radiation.

Increased frequencies of chromatid fragments were found in the descendants of 35 generations of HF19 human diploid fibroblast cells exposed to X-ray irradiation (1.5 Gy). After exposure to α-particles (0.4 Gy) and neutrons (0.5 Gy), an earlier expression of genomic instability was noted. When exposed to low-LET radiation, the effect of a greater remoteness of expression of chromosomal instability was found than after exposure to high-LET radiation. Both these and other data presented in the review work (Suskov, Kuzmina, 2001) indicate an increased efficiency of irradiation with α-particles. The dependence of the spectrum of chromosome aberrations in the offspring of irradiated cells of various origins on the dose and quality of radiation was noted. It is concluded that the expression of chromosomal instability is determined by both the type of cells and the quality of ionizing radiation.

As a result of the analysis of genetic instability patterns, it can be concluded that phenotypically genetic instability is manifested by a special state of the offspring of irradiated cells, in which they differ significantly from normal non-irradiated cells. This condition is characterized by radiobiological, genetic, cytological, molecular and biochemical features that are not characteristic of normal cells.

In all works devoted to the comparison of genetic instability under the action of radiation with different LETs, a high efficiency of densely ionizing radiation was demonstrated (Kadhim et al., 1994; Watson et al., 1996; Little et al., 1997; Suskov, Kuzmina, 2001; Dugan, Bedford, 2003; Evans et al., 2003; Smith et al., 2003; Ponnaiya et al., 2004). However, the indicators of genetic instability used by the authors did not allow them to quantify RBE values; none of these studies attempted to compare RBE of densely ionizing radiation for cell inactivation and genetic instability. There are no data in the literature on the comparison of genetic instability of wild-type haploid and diploid cells and their radiosensitive mutants.

1.3. The Mechanisms of Radiation-Induced Genetic Instability

It is known that physicochemical changes in molecules after exposure to ionizing radiation are carried out within a short period of time, while radiation-induced genetic instability can be detected much later after exposure (hours, days, and even more distant periods of time). The study of the molecular and cellular mechanisms underlying the long-term effects of radiation is one of the keys and most urgent problems of radiation biology. Intensive studies of the molecular mechanisms of regulation of the cell cycle, apoptosis, and malignant transformation directly indicate the relationship of genome instability with the listed phenomena (Mazurik, Mikhailov, 2001). The authors distinguished three interconnected systems that maintain the stability of the genome at the cellular level, the violation of which leads to genetic instability. The first group includes the system of redox homeostasis, which produces various cytotoxic factors, including reactive oxygen species involved in the elimination of genetically outsider material. It has been suggested that the main source of genetic instability in the offspring of irradiated cells is the production of increased amounts of reactive oxygen species. For this case, cellular components, including DNA, are damaged, as a result of which the frequency of cell death, mutations and chromosome aberrations increases. The second system, which responsible for genome stability, the authors attribute to the control of the cell cycle at checkpoints, which ensures the emergency removal of a cell with pathologically altered DNA that can disrupt genome constancy. An important component of this system is the tumor suppressor protein p53. Violation of this control can lead to genetic instability, including the formation of aberrations of cell chromosomes, increased expression of apoptosis, decreased ability of the offspring of irradiated cells to form colonies on solid nutrient medium, and other effects. The authors attribute the repair of DNA damage to the third mechanism responsible for the genetic stability of cells. For prokaryotes, the SOS-repair system is known, which is activated when DNA damage can disrupt cell viability. The best studied of SOS repair was carried out in bacterial cells of *Escherichia coli*, the main participants of which are proteins encoded by genes REC A and LEX A. A similar system of rapid response to DNA damage exists in eukaryotes. In such cells, double-stranded DNA breaks can be the result of normal metabolic processes, or the action of DNA-damaging agents – ionizing or UV radiation, etc. Unrepaired DNA double-strand breaks can lead to such consequences as genetic instability, neoplastic transformation or cell death. Mistakenly repaired DNA damage can also induce genetic instability. The repair of double-strand breaks in eukaryotic cells occurs by either homologous recombination or non-homologous joining of the broken DNA ends. Both types of repairs play an important role in maintaining chromosome stability.

Irradiation changes the state of cellular defense mechanisms: antioxidant protective systems, DNA repair regulation of the cell cycle and apoptosis (Mohr et al., 1999). It is believed that the mechanisms of changes in cellular radiosensitivity are closely related to the mechanisms of induction and implementation of radiation-induced genome instability (Pelevina et al., 2003; Schwartz et al., 2007). The currently established molecular manifestations of radiation-induced instability in the offspring of irradiated cells consist in a change in the DNA structure, impaired DNA repair, and a significant shift in the functioning of the redox system. The latter not only causes a change in the intracellular content of reactive oxygen species, but also affects the regulatory functions that they perform, influencing the intracellular reactions controlled by genetic programs and activated in response to the influence of various environmental factors (Bruskov et al., 2020). The need for the participation of reactive oxygen species in the implementation of radiation-induced genetic instability suggests that the introduction of substances affecting the state of redox homeostasis can modify the development of cells along the path of genetic instability. It should be borne in mind that free radicals accumulating in the cell, formed as a result of water radiolysis or lipid peroxidation, have a damaging effect both on the hereditary apparatus and on the membrane structures of the cell. Normally, there is some balance between the concentration of reactive oxygen species and the activity of the antioxidant system. However, under conditions of chronic exposure to low doses of radiation, this balance is disturbed. It was noted that children living in areas contaminated with radionuclides are in a state of deep peroxide stress (Durnov et al., 2001; Suskov et al., 2006). There was a negative correlation between the level of chromosomal damage and the concentration of antioxidants. In addition, it has been shown in experimental studies with animals that irradiation of both parents' leads to more aggravated consequences in their offspring than irradiation of only one of them (Nefyodov et al., 2001).

The work (Mikhailov et al., 2005) is devoted to the molecular manifestations of radiation-induced genetic instability. A number of indicators were used – changes in the structure of DNA, the efficiency of excisional DNA repair, the content of reactive oxygen species in bone marrow cells in mice of the reparative (CBA) and incapable to repair (101/H) lines in dynamics up to 185 days after exposure to ionizing radiation at a sublethal dose of 1.5 Gy. It was found that after irradiation, offspring with a reduced activity of excisional repair and an increased tendency to structural changes in DNA are formed in the cells of the bone marrow. Based on the results obtained, the authors come to the conclusion that molecular manifestations of radiation-induced instability exist in the offspring of bone marrow cells of mice for a long time. In the same work, the biological effect of phenosan, an adaptogen capable of increasing the body's nonspecific resistance to a wide range of harmful effects of physical, chemical and social nature, was studied. The administration of the drug did not protect the DNA of bone marrow cells from radiation damage caused by irradiation at a sublethal dose, but it had an effect on long-term post-radiation changes. Thanks to phenosan, the bone marrow cells of the irradiated CBA mice exhibited a greater efficiency of DNA repair from damage and maintenance of viability. Bone marrow cells from male mice of the repair-defective line 101/H, which were injected with phenosan before irradiation, remained incapable of eliminating DNA damage by repair, which, according to the authors, led to activation of the program for maintaining genome constancy by apoptosis in cells carrying structural defects and caused the death of animals.

From the modern radiobiology point of view, DNA damage is a trigger for all events leading to a biological effect, and double-stranded DNA breaks play a key role here. The initiating event of genetic instability can be DNA damage, for example, double-strand breaks, which leads to the activation of a number of protein kinases that regulate the expression of genes, the functioning of which leads to the activation of growth factors, induction of DNA repair, and regulation of the cell cycle.

When analyzing the mechanisms of genomic instability, Suskov and Kuzmina (2001) state that local changes in DNA can be considered as the initial material, which, in the process of repeated reduplication, leads to the formation of an unstable phenotype of the cell genome, which is a reflection of the principle of a biological amplifier (Timofeev-Ressovsky, 1996). It is noted that the following essential points should be taken into account. First, the cell genome is represented by a set of morphologically discrete chromosomes with a strictly defined position and constant connections with each other and with the nuclear envelope. Since in the process of cell division there is a constant disintegration of this integral structure, some violations of this process are possible, which can lead to an imbalance of the genetic apparatus, and, consequently, to destabilization of the genome. Secondly, in the process of the mechanism of convariant reduplication, rare events of crossings between maternal and daughter DNA strands can occur, which leads to the formation of sister chromatid exchanges, i.e., to conformational rearrangements of chromatin. Since the state of the genome is a fundamental property of a cell that determines its normal functioning, its destabilization leads to physiological failure, genetic instability, and in some cases to malignancy – the acquisition of the properties of a malignant tumor by cells.

It is known that the structure of DNA in a cell is regularly disturbed due to the formation of compounds in the process of metabolism that have the ability to damage DNA. These include, for example, reactive oxygen species, lipid peroxidation products, and reactive nitrogen species (Zhestyanikov, 1968, 1979; Ryabchenko, 1979; Tomilin, 1983; Bruskov et al., 2020). It has been shown that the frequency of spontaneous DNA damage can reach thousands of events per cell per a day. The action of ionizing radiation and other physical and chemical agents significantly increases the yield of DNA damage, which can initiate genetic instability and programmed cell death (apoptosis). Most single and many double breaks in DNA caused by radiation can be repaired by DNA repair. However, this process does not occur with 100% efficiency; there is an irreversible component of radiation damage (Korogodin, 1966, 1993, 2010). Double-stranded DNA breaks can be mistakenly repaired using DNA from another chromosome, resulting in asymmetric or symmetric exchange chromosomal rearrangements. It is known that large deletions or asymmetric chromosomal rearrangements lead to cell death after several cell cycles. Symmetrical chromosomal exchanges can exist in a cell for a long time, but they can influence the phenotypic manifestation of genetic instability. Ionizing radiation also triggers nuclear signal transduction genes, whose products may be responsible for long-term changes in phenotype.

The role of DNA repair and recombination in genomic instability has been discussed in many publications (Cheng, Loeb, 1993; McMurray, Goltschling, 2004; Shen, 2011). Genomic instability can affect the disfunction of enzymes, leading to erroneous restoration of spontaneously occurring or radiation-induced DNA breaks. Chromosomal damage induced by exogenous factors, including exposure to ionizing radiation, UV light, or chemical mutagens, results from DNA breaks, DNA strand and DNA-protein crosslinks. Mutations that reduce the fidelity of DNA replication and repair also affect genomic instability. The role of DNA damage in genetic instability, aging and cancer formation is being actively investigated. The authors (Mcurray, Gottschling, 2004) suggest that the accumulation of damaged proteins during cell aging leads to the loss of function of gene products that are important for maintaining the integrity of the genome. The identification of these proteins and the process of their damage represent a new direction for understanding the relationship between age and genetic instability.

The role of cell radioresistance in genetic instability has been widely discussed in many publications (Morgan, Murnane, 1995; Kaplan, Morgan, 1998; Petin et al., 2008). However, there are some contradictions on this issue. Radiosensitive mutant CHO cells (Chang, Little, 1992a, b; Dugan, Bedford, 2003), defective in the repair of double-stranded DNA breaks, do not show delayed formation of colony by surviving cells, which confirms the idea that the repair of double-stranded breaks should play important role in genetic instability. On the other hand, radiosensitive human mutant cell lines, such as xeroderma pigmentosa, Fanconi anemia, and ataxia-telangiectasia, have defects in DNA repair and demonstrate genetic instability (Gatti et al., 1991; Kalb et al., 2006; Sugasawa, 2008).

Note that the described sections of the chapter were devoted to the manifestation of genetic instability at the subcellular and cellular levels. It is well known that this effect also takes place at the organismal level of mammals. The next section will consider the pathogenetic manifestations of genetic instability in animals and humans.

1.4. Pathogenetic Significance of Genetic Instability

The consequence of radiation-induced instability of the cell genome, both in irradiated organisms and in their offspring, is: at the tissue level – morphological and functional inferiority of tissues and organs, and at the organismal level – an increase in the risk of tumor developing and, possibly, non-tumor somatic pathology (Luke et al., 1997; Baverstock, 2000; Nefedov et al., 2001). Many human diseases with high prevalence, such as neurodegenerative diseases and cancer, are associated with a violation of the genetic stability of cells. Tumor cells show chromosomal damage, which is considered as a delayed manifestation of the radiation effect (Aggarwal, Brosh, 2012). In non-irradiated cells of normal origin, such changes are not recorded. In most tumor cells that have not been irradiated, there is a tendency to aneuploidy and complex chromosomal breakdowns. The cells in these tumors are capable of proliferation and adaptation to a new environment and form secondary cancers that can differ from the original. These data indicate the adaptive capabilities of the destabilized genome. Evidence for the role of chromosomal instability in potentiating the risk of malignant formation was obtained in the study of syndromes of rare recessive diseases (ataxia telangiectasia, pigmented xeroderma, Fanconi anemia, Bloom's syndrome), characterized by unique cytogenetic characteristics, genetic instability and a tendency to malignant transformation. Some patients with such diseases show an increased yield of chromosomal aberrations, especially after exposure to ionizing radiation, but a normal level of exchange of sister chromatids (Gatti et al., 1991; Kalb et al., 2006; Sugasawa, 2008).

White BALB/c mice are significantly more susceptible to radiation-induced breast cancer than black C57BL/6 mice. Primary cultures of mammary epithelial cells from these two lines were used to assess the delayed manifestation of chromosomal aberrations as a measure of the radiation instability of the genome after exposure to ^{137}Cs gamma radiation (Ponnaiya et al., 1997). As expected, active division of cultures from both lines showed a rapid decrease in the initial asymmetric aberrations over time after irradiation. However, after 16 doublings of the population in the cells of BALB/c mice, there was a noticeable increase in the frequency of breaks and breaks of the chromatid type, which remained elevated throughout the entire period of the experiment (28 doublings). This effect was not observed for cells from C57BL/6 mice. These results demonstrate a correlation between latent expression of chromosomal damage *in vitro* and susceptibility to breast tumors, further supporting the central role of radiation-induced instability in tumorigenesis.

Ponnaiya et al. (1997) showed the expression of genomic instability in the form of an increased level of chromatid and chromosomal aberrations (mainly chromatid) in clonal descendants of mouse hematopoietic stem cells irradiated with α-particles at doses of 0.25–1.0 Gy and undergoing 13 cell divisions. In experiments on laboratory animals, a decrease in the resistance of the genetic apparatus of the cells from the offspring of irradiated parents to mutagenic factors was shown. Thus, with the help of additional testing irradiation, an increased radiosensitivity of the chromosomes of the genome of somatic cells (hepatocytes, karyocytes, embryonic fibroblasts) of the offspring of male mice and rats irradiated at doses of 4.2–6.0 Gy was revealed in comparison with the chromosomes of intact parents (Vorobtsova, 1988). In this work, an increased sensitivity to the ionizing radiation was revealed in the somatic cells of the offspring of drosophila male irradiated at doses of 10–30 Gy. The results of these studies indicate the inheritance of the genome instability of the somatic cells of the offspring through the gametes of irradiated parents, which is revealed by an additional radiation load (synergism).

The authors of the work (Fomenko et al., 2006) analyzed the genomic instability detected by the micronucleus test. In the offspring of the F1-generation BALB/c male mice exposed to chronic low-intensity γ-radiation (at doses of 0.1; 0.25, and 0.5 Gy and at a dose rate of 0.01 Gy/day), as well as in the offspring of F1-generations of females, irradiated once with X-rays (0.5, 1.0, and 2 Gy at a dose rate of 0.1 Gy/min), a significant increase in the frequency of micronuclei in erythrocytes of the bone marrow was found in comparison with the offspring of non-irradiated males and females. The increase in the frequency of micronuclei in offspring was proportional to the value of the radiation dose of the parents. The effects of increasing the frequency of micronuclei in the offspring of chronically irradiated males and in the offspring of acutely irradiated females are comparable in level at an irradiation dose of 0.5 Gy: in the offspring of irradiated female parents or male parents, the frequency of micronuclei increased by 1.5 and 1.6 times, respectively. The results obtained indicate the possibility of induction of genome instability in the somatic cells of the offspring of irradiated parents. The formation of genome instability in somatic cells of the F1-generation offspring, whose parents (males or females) were irradiated before mating in small and medium doses, was also demonstrated. The effects of increasing the frequency of micronuclei in erythrocytes of the bone marrow in the offspring of chronically irradiated males or acutely irradiated females were comparable in level at an equal radiation dose of 0.5 Gy.

The mechanism of genome instability formation in somatic cells of F1-generation descendants from irradiated parents is still unclear. It can be assumed that transgenerational transmission of genome instability into somatic cells of the offspring occurs as a result of induction of non-lethal genome damage in the germ cells of parents exposed to low and medium doses of ionizing radiation before fertilization. Experiments on transgenic mice (Luke et al., 1997) showed an increase in genome instability in bone marrow cells, determined in a certain F1-generation gene obtained from male parents exposed to γ-radiation at doses of 1-4 Gy. These results indicate the possibility of transgenerational transmission of factors leading to genome instability in the F1-generation offspring of irradiated parents. Transgenerational carcinogenic effects have also been demonstrated in CBA mice when F0-generation males were exposed to X-rays at doses of 1 and 2 Gy (Mohr et al., 1999). The frequency of mutations at two loci was studied in the germ line of the offspring of the first and second generation of inbred male CBA/H, C57BL/6, and BALB/c mice exposed to either fission neutrons or X-rays. Irradiation of CBA/H males with any of these types of radiation led to an increase in the mutation frequency in the next two generations. Such transgenerational effects were also observed in C57BL/6 mice irradiated with X-rays and in BALB/c mice irradiated with neutrons. Since radiation-induced genetic instability manifests itself in diploid cells after fertilization, the authors emphasize that the revealed pattern is important for assessing risk in humans.

It is assumed that genomic instability is an early stage in radiation-induced tumorigenesis. It follows from this hypothesis that individuals highly susceptible to tumor induction by ionizing radiation should exhibit increased radiation-induced genetic instability. Many works cite the results of long-term studies carried out on cells in tissue culture, laboratory animals (mice), blood lymphocytes in residents from contaminated areas affected by the Chernobyl accident (Pelevina et al., 2006; Shevchenko, Snigireva, 2006; Suskov et al., 2006). It is concluded that staying in contaminated regions leads to genome instability, which manifests itself in a number of effects at the cytogenetic, cellular and organismal levels. It has been shown that after exposure in the accident zone in distant generations there is an increase in the frequency of cell death, the emergence of micronuclei and giant cells, an increase in radiosensitivity, and the absence of an adaptive response. In the culture of cells of embryonic fibroblasts obtained from mice exposed in areas of radioactive contamination, an increase in the number of cells with multiple chromosomal aberrations is noted. In mice exposed in the accident zone, a decrease in the number of endothelial cells in different parts of the brain and an increase in the radiosensitivity of animals were found. All the described effects are observed in distant generations of cells and may result from the induction of genome instability after exposure to radiation.

Using a micronucleus test for blood lymphocytes, the spontaneous level of damaged lymphocytes in groups of children with different living conditions and different environmental stress was studied, as well as its change when exposed to ionizing radiation *in vitro* (Pelevina et al., 2001). 860 children were examined. It has been shown that at a higher spontaneous level of damaged lymphocytes, a higher radiosensitivity is observed and the ability to adaptive response decreased. The data described in the literature indicate the reality of genomic instability in the somatic cells of children exposed to low-intensity exposure to low doses of radiation, the severity of which is determined by the individual characteristics of the organism. This primarily refers to the levels of cytogenetic disorders of parents born after the accident to irradiated parents, as well as in children exposed to intrauterine irradiation during the accident in 1986 (Suskov et al., 2006). This fact may have the following explanation. Under conditions of prolonged exposure to low doses of radiation, radiation-induced damage is added to spontaneous DNA damage normally occurring during molecular genetic processes (replication, transcription, translation, segregation of daughter structures) in proliferating cells. Unrepaired lesions are either realized as DNA/chromosome breaks, gene/chromosome mutations, or remain as long-lived lesions. The latter can be the starting material for the induction of genomic instability.

The work (Bezlepkin, Gaziev, 2001) discusses the role of non-targeted mechanisms for fixing radiation damage and the participation of recombination events in meiosis or in the early stages of embryogenesis in the induction of an unstable genome of reproductive cells. It is assumed that quantitative determination of the level of induced instability of the genome of such cells may be an acceptable option for assessing genetic risk and an indicator of the increased likelihood of oncological and other pathologies in the offspring of parents exposed to radiation. Currently, a number of experimental data have been obtained confirming the possibility of the formation of an unstable genome in the somatic cells of the offspring of irradiated parents. These include the results of an increase in the frequency of chromosomal mutations in somatic cells, as well as an increase in the risk of tumor formation in the offspring of the F1 generation of male rodents exposed to ionizing radiation (Vorobtsova, 2006, 2008).

The frequency of lymphocytes mutant in the genes of the T-cell receptor was determined in 42 workers of nuclear chemical enterprises, who were also examined using cytogenetic methods. In 11 individuals exposed to irradiation at different times, the average frequency of TCR-mutant lymphocytes was statistically significantly higher (2.6 times) than in the control group of non-irradiated donors. At the same time, the average frequency of unstable chromosomal aberrations did not exceed the control level. External radiation doses for different individuals were in the range of 10–400 cGy, however, no dose dependence of the recorded tests was found (Smirnova et al., 2005). Comparison of the results of the analysis of structural and gene mutations showed that in the case of external irradiation, the observed disturbances may be the result of genome instability in the long term after exposure. In the case of combined exposure, the genetic changes are presumably due to exposure to radiation from radioactive plutonium in the body. The results obtained showed that the simultaneous use of several different methods for assessing somatic mutagenesis makes it possible to more fully characterize the processes in cells after irradiation.

It is known, that after the Chernobyl accident, large contingents of children and adults have been and continue to be exposed to low-intensity exposure to small doses of radiation. At the same time, there is an increase in morbidity, a tendency towards the formation of chronic diseases, an increase in the development of immunodeficiency states, anemization, malignant neoplasms and neuropsychic maladjustments. It should be noted that in the past decades in radiobiological research, including the use of cell cloning techniques, it has been firmly established that, in addition to direct mutagenic effects, ionizing radiation leads to the induction of genomic instability. It includes various disorders of the genetic apparatus (DNA breaks, structural aberrations of chromosomes, gene mutations, sister chromatid exchanges, aneuploidy, polyploidy) in the descendants of irradiated cells. There also appeared information about the increased radiosensitivity of chromosomes in the cells of the offspring of irradiated parents (Eliseeva et al., 1994; Kadhim et al., 1994).

Cytogenetic three-year monitoring showed a significant increase in the frequencies of aberrant cells and chromosome-type aberrations in the child population of two regions of Ukraine contaminated with precipitated radionuclides, compared with those in the control (Eliseeva et al., 1994). The authors conclude that over the years there is an accumulation of chromosomal aberrations (paired fragments, dicentrics, rings), an increase in the proportion of children with various chromosomal abnormalities, as well as an increase in the radiosensitivity of their lymphocytes.

The work (Suskov, Kuzmina, 2001) analyzes the results of a comprehensive genetic examination of 43 children of the liquidators of the accident and 24 of their fathers. Radiation-induced chromosome aberrations were detected in blood lymphocytes in 75% of the fathers-liquidators. In 63% of their unirradiated children, similar chromosome abnormalities were found, but with a lower frequency. Along with structural chromosomal aberrations, sporadic aneuploid and polyploid cells were revealed in both generations. The same work provides data on the study of the stability of the genetic apparatus of the somatic cells of children living in conditions of low-intensity action of small doses of radiation because of radioactive fallout after the Chernobyl accident. Both children and parents had increased levels of both radiation-induced chromosome aberrations and chromatid fragments. In both generations, deviations in the repair efficiency of genomic DNA are observed. In addition to the direct action of radiation, the authors also allow indirect action. Free radicals accumulating in the cell, formed as a result of radiolysis of water, lipid peroxidation, potentiate the direct effect of radiation, having a damaging effect on both the hereditary apparatus and the membrane structures of the cell. An experimental study of the expression of genomic instability using testing irradiation in peripheral blood lymphocytes of 15 children (born 1986–1998) living in territories with radionuclide contamination (over 15 Ci/km^2 for ^{137}Cs, Novozybkovsky district, Bryansk region) was carried out. In 5 children irradiated *in utero* in 1986, this phenomenon was studied in 3 consecutive cell generations (mitoses). The data obtained indicate the reality of the expression of the induced genomic instability in the descendants of multiply divided cells of a growing organism, exposed to prolonged exposure to low levels of radiation (Suskov, Kuzmina, 2001).

The work (Pelevina et al., 2003) presents the results of long-term studies of the effectiveness of the action of ionizing radiation in small doses (up to 0.5 Gy), performed on cells in tissue culture and blood lymphocytes of humans and children. In laboratory studies and in studies carried out in areas contaminated with radionuclides, genome instability was found in the descendants of irradiated cells, the manifestation of which is recorded in the form of distant cell death, an increase in the frequency of cells with micronuclei, a decrease in proliferative activity, a loss of the ability to adaptive response, an increase in sensitivity to additional irradiation, increasing the frequency of sister chromatid exchanges. The results of the study of the adaptive response in the population of human blood lymphocytes are also presented. It was found that in all studied populations there are individuals in whom an adaptive response is not recorded, and individuals with increased radiosensitivity of lymphocytes after adaptive irradiation (0.05 Gy). On the basis of their own data and literature data, the authors discuss possible mechanisms for the realization of the effects of ionizing radiation in low doses. It is concluded that after low-dose irradiation, firstly, a population with new properties can form, and secondly, the effects and mechanisms of their implementation can differ from those with high-dose irradiation. Thus, among normal healthy children, up to 20–25% of individuals are characterized by increased radiosensitivity of lymphocytes after exposure to low doses. The frequency of children with an adaptive response is sharply reduced when living in areas contaminated with radionuclides. Defects in DNA repair are possible explanations for this phenomenon. It was shown that an increase in radiosensitivity after adaptive irradiation of human blood lymphocytes is associated with spontaneous damage to the genome – in most individuals in whom the frequency of cells with micronuclei is increased, increased radiosensitivity is also observed after irradiation with low doses (Pelevina et al., 2003). It was shown that the adaptive response can arise not due to a decrease in the number of damaged cells, but due to an increase in the number of intact cells in the population. In this work, for the first time, the assumption is made that the effects of radiation in small doses are sometimes difficult to determine, but when combined with other agents, they can become more dangerous and be expressed in the intensification of many reactions due to synergistic interactions.

The work (Suskov et al., 2006) presents the main results of studying genomic instability in children permanently living in territories with radionuclide contamination (16–18 Ci/km^2) and exposed to low-intensity irradiation at various stages of ontogenetic development. In all examined groups of irradiated children, increased frequencies of radiation-induced chromosome aberrations were observed, due to a decrease in the activity of unscheduled genomic DNA synthesis in lymphocytes. In children living in areas contaminated with radionuclides, an increased radiosensitivity of lymphocyte genomes to testing radiation *in vitro* and features of the dynamics of the frequencies of chromosomal aberrations and sister chromatid exchanges in three cell generations were revealed. The data obtained indicate the systemic nature of the observed effects, the reality of the induction of genomic instability in a growing child's organism, which is exposed to low-dose radiation exposure, the severity of which is determined by the individual genotypic characteristics of the organism. The biological significance of the phenomenon of post-radiation genomic instability, its relationship with the state of health and its pathogenetic role in the development of somatic pathology are postulated.

As a result of a comprehensive study (Suskov et al., 2008) according to cytogenetic and immunogenetic criteria of the lymphocyte genome state in fathers who liquidated the consequences of the Chernobyl accident and their unirradiated children, increased frequencies of chromosome aberrations, gene mutations (TCR mutations) and predictors of apoptosis were found. As a result of the analysis of the conjugation between the studied indicators, the distinctive features characteristic of the induction of genomic instability in the body of unirradiated children in comparison with fathers-liquidators who underwent direct radiation exposure were revealed. For all the studied criteria, individual variability of genome destabilization was observed, which manifested itself both in the diversity of the spectrum of mutations and in their varying degrees of severity. The results obtained indicate the need for an integral assessment of the state of the genome using several genetic indicators to identify transgenerational genomic instability in children – the descendants of irradiated parents.

The change over 3 years of radiation-induced unstable chromosomal aberrations of leukocytes in people exposed to external exposure to γ-radiation in the course of professional activity was estimated. As a result of studies carried out in 2013, mosaic deletions and amplifications were found in these individuals. It was found that with an unchanged dose of external irradiation, the frequency of aberrant cells, chromosomal aberrations, paired fragments, and dicentric chromosomes in 2017 increased 2–4 times compared to 2014. It was shown that mosaic deletions and amplifications of chromosomal loci in individual workers could be detected even at external doses of over 100 mSv.

Genomic instability can initiate cancer, enhance its progression and affect the overall prognosis of a sick patient; it occurs for a variety of reasons – damage to telomeres, amplification of centrosomes, epigenetic modifications and DNA damage as a result of exposure to endogenous and exogenous damaging factors (Marder, Morgan, 1993; Mikhailov et al., 2003; Blasco, 2005; Vorobtsova, 2008; Aggarwal, Brosh, 2012; Duffy et al., 2016). It was hypothesized that the mechanism for the generation of chromosome breaks in the offspring of irradiated cells is the cycle of breaks-reunification of chromosomal bridges, increased levels of reactive oxygen species, impaired DNA replication or replicative stress, activation of fragile sites as a source of delayed DNA damage transmitted to daughter cells, telomere dysfunction under the influence of radiation and telomere-telomeric fusion of chromosomes, leading to delayed dicentrics (Cheng, Loeb, 1993; Little, 1998; Mazurik, Mikhailov, 2001; Mikhailov et al., 2005; Brosh, Bohr, 2007; Saenko, Zamulaeva, 2007; Shen, 2011; Yuen et al., 2007). Many cancer treatments related with the DNA damage of tumor cells in order to disrupt the progressive proliferation of cancer cells (Ferguson et al., 2015). The authors also consider it generally accepted that there is a need to develop tailor-made therapies for a particular patient and type of cancer. This work analyzes in detail the mechanisms by which genomic instability can lead to cancer, as well as therapies and measures that prevent genomic instability. In particular, the authors discuss five priority areas aimed at reducing genome instability: (1) prevention of DNA damage; (2) enhancement of DNA repair; (3) reduction in erroneous repair; (4) reduced clustering of centrosomes; (5) inhibition of telomerase activity. In addition, the authors attach great importance to vitamins D and B, selenium, carotenoids, and inhibitors of PARP (poly (ADP-ribose) polymerase) enzymes as priority approaches against genome instability. Possible combinations of these factors are also discussed to identify potential synergistic effects.

Analysis of a large amount of their own data and the results of other authors made it possible to formulate some generalized, but very vague definition of genomic instability, taking into account various aspects of this effect manifestation (Suskov, Kuzmina, 2001). Genomic instability is a phenomenon of structural and functional inconstancy of genetic material that occurs in the descendants of multiple divisions of cells exposed to radiation (or other genotoxic agents), manifested by DNA breaks and reconversions, chromatin recompaction, chromosome aberrations, sister chromatid exchanges, aneu/polyploidy, unplanned gene expression/suppression, gene and chromosomal mutations, which leads to disturbances in the genomic balance and is accompanied by the development of cellular dysfunctions, malignancy, induction of apoptosis and cell death.

1.5. Yeast Cells Genetic Instability

Many authors (Kapultsevich, 1978; Korogodin et al., 1991, 2007; Degtyareva et al., 2008; Petin et al., 2008; Korogodin, 2010; Botstein, Fink, 2011; Stirling et al., 2011; Zhu et al., 2012; Donigan et al., 2015; Bell, Labib, 2016; Duffy et al., 2016; Tombline et al., 2017) have studied various aspects of the genetic instability of the yeast *Saccharomyces cerevisiae*. Convincing data have been obtained that some of the cells that survived after irradiation became different from those before irradiation, and give functionally altered offspring. These delayed manifestations of the radiation effect are not clonal, i.e., they may be not identical for all cells of the colony formed later than control. This phenomenon is called radiation-induced instability (Baverstock, 2000; Mikhailov et al., 2003). The yeast *Saccharomyces cerevisiae* is a convenient object for studying the long-term effects of radiation, including genetic instability (Klein et al., 1990; McMurray, Goltschling, 2004; Korogodin et al., 2007; Yuen et al., 2007; Sheltzer et al., 2011). Yeast research has been particularly useful for studying the molecular mechanisms and pathways that affect aging at the cellular level.

The yeast *Saccharomyces cerevisiae* are unicellular fungi, which are one of the most studied eukaryotic model organisms in the field of molecular and cellular biology (James, Werner, 1965; Korogodin, 1966, 1993; Korogodin et al., 1968; Luchnik et al., 1977; Kapultsevich, 1978; Frankenberg et al., 1981; Frankenberg-Schwager et al., 1987; Petin, 1987; Feldman, 2012; Falcone, Mazzoni, 2016; Petin, Kim, 2016). As an object, yeast has a number of advantages and unique properties. At the same time, yeasts are the simplest eukaryotes with a sexual process, making them a convenient object for studying the problems of general and especially molecular genetics of eukaryotes. Haploid and diploid strains were easily isolated in saccharomycetes. Since yeast cells can exist, receiving energy both through the process of respiration and during fermentation, it is possible to isolate a variety of mitochondrial (cytoplasmic) mutations. Yeast cells have a small size and small volume of chromosomes (radiation target), a short generation time and have the ability to recover from damage caused by exposure to ionizing or UV radiation. Yeast cells are less sensitive to ionizing radiation than mammalian cells. The yeast *Saccharomyces cerevisiae* grows rapidly and forms colonies visible on a solid nutrient medium, and is also well studied in radiobiological terms. Most of the radiobiological responses of yeast cells are qualitatively similar to those of cultured mammalian cells (Kapultsevich, 1978; Korogodin, 2010; Critchlow, Jackson, 1998; Resnick, Cox, 2000; Feldman, 2012; Karpenshif, Bernstein, 2012; Mulla et al., 2014; Petin, Kapultcevich, 2014). The authors cite examples of studies in which mammalian (mainly human) DNA repair genes are expressed in some genetic mutants of yeast. Yeast has served as an excellent model for studying the mechanisms of genetic stability and repairing DNA damage. The processes that affect the lifespan of yeast cells may be similar to those in higher eukaryotes. Yeast has allowed researchers to discover and characterize the function of human genes involved in DNA repair mechanisms, maintaining genomic stability and cellular DNA replication.

Most of the results on genetic instability have been obtained with cultured mammalian cells in a diploid state. Yeast cells, as the simplest example of eukaryotic cells, are a convenient object for comparing the lethal effect and genetic instability of haploid and diploid cells, which makes it possible to obtain new fundamental data.

The mechanisms for controlling the DNA integrity and the correctness of the passage of the cell cycle are best studied in a model eukaryotic organism of the yeast *Saccharomyces cerevisiae*, in which a "checkpoint," a control that stops cell division in the event of DNA damage, was first discovered (Koltovaya, 2017). DNA damage and disruption of the cell division cycle cause genetic instability, which is the cause of many human diseases, including neoplastic cell transformation. In the process of evolution, a special protective mechanism has been developed – at all stages of the cell cycle there are checkpoints in which the correctness of the cell cycle passage and the integrity of the genome are checked. At these points, violations of the process of division or damage to DNA induce a special mechanism that stops cell division, starts DNA repair, controls telomere length, induces transcription and initiates apoptosis. Although the retardation of the passage of the cell cycle after irradiation has been known for a long time, it was shown for the first time on the yeast *Saccharomyces cerevisiae* that during the slowing down or complete cessation of division, radiation damage is repaired and that this process is genetically determined. Genomic instability resulting from defects in repair proteins causes various hereditary genetic disorders in humans and also leads to a predisposition to cancer.

Different types of yeast cells are a widespread model of eukaryotic cells for studying the role of ploidy in the manifestation of various biological effects. The authors of the work (Lettier et al., 2006) studied the role of DNA double-strand breaks in spontaneous homologous DNA recombination, which is a source of genetic instability and loss of heterozygosity in mitotic cells. The authors provide direct evidence that most spontaneous homologous recombinations in *Saccharomyces cerevisiae* are initiated by DNA damage other than double-strand breaks. In particular, a class of *rad*52 mutants has been described that are completely able to perform homologous recombination, but are unable to repair double-stranded DNA breaks. The studies were carried out on the basis of genetic analysis and assessment of the consequences of the absence of repair after the introduction of a plasmid with double-stranded DNA breaks. The results obtained led to the conclusion that single, rather than double DNA strand breaks, are the main sources of spontaneous homologous recombination in mitotic yeast cells.

The work (Kumaran et al., 2013) is devoted to the characterization of chromosomal instability of diploid, polyploid, and hybrid yeast cells. It has been shown that a number of factors affect the stability of chromosomes: replication, chromosome centromeres, attachment of the mitotic spindle to chromosomes and telomeres. Chromosomal stability also depends on chromosome size, fusion of sister chromatids, genes involved in DNA double-strand break repair, and the functioning of cell cycle checkpoints. Chromosomal instability is observed in most solid tumors and is associated with somatic mutations in genes responsible for the integrity of the genome. The spectrum of such mutations is only partially known, and it is not yet possible to predict all the disorders that can lead to genetic instability. To solve this problem, the authors of paper (Stirling et al., 2011) created a catalog of genes and pathways for the formation of genetic instability by screening about 2000 alleles, which accounted for 90% of the main genes in *Saccharomyces cerevisiae*. The strains carrying the plasmid were grown on synthetic complete media lacking the appropriate ingredient to detect various mutations. Integration of these results with previously published data for other yeast genes allowed the authors to obtain a systematic set of genes for genetic instability, consisting of 692 genes responsible for genetic instability. On their basis, the authors created a list of human genes – candidates for the control of genetic instability, which, according to the authors, will help to understand the phenotypes of chromosomal instability in human diseases.

The doubled amount of DNA in cells after doubling the ploidy should be reflected in the increased amount of DNA damage. The yeast *Saccharomyces cerevisiae* was used to study the genetic consequences of polyploidy (Mayer, Aguilera, 1990; Storchova, 2014). A 5–10-fold increase in spontaneous recombination and UV-induced recombination of unsynchronized cultures in tetraploid strains compared to diploid ones was shown. Isogenic diploid, triploid and tetraploid strains with a genetically encoded chromosome were designed, which tracked the loss of one, two or three chromosomes during mitosis. It was found that the loss of one copy of the chromosome occurred at a rate almost 30 times higher in triploid strains and 1000 times higher in tetraploid than in diploid strains. These results imply that aneuploidy increases genome instability due to the increase in DNA volume with increasing ploidy. Aneuploidy occurs as a result of a violation of chromosome segregation in mitosis or meiosis and reduces the stability of cell functioning, and is also associated with cancer, a disease of increased cell proliferative capacity. Human cells characterized by aneuploidy, when the number of chromosomes in the nucleus is not a multiple of the haploid set, can lead to various diseases and developmental disorders. While the study of the effect of aneuploidy on genomic instability in cancer cells is complicated by the fact that most cancer cells have both multiple point mutations and other types of chromosomal abnormalities, the yeast *Saccharomyces cerevisiae* is a useful biological object for evaluating the independent effects of certain types of genetic changes.

To investigate the mechanism by which aneuploidy can promote tumorigenesis, this phenomenon was investigated by the method of selective generation of aneuploid yeast strains (McMurray, Gottschling, 2004; Sheltzer et al., 2011; Zhu et al., 2012). The yeast *Saccharomyces cerevisiae* is especially suitable for studying the effect of aneuploidy on cellular physiology due to their small genome, consisting of sixteen chromosomes, therefore, in the diploid state, they tolerate aneuploidy better than in the haploid state (Mulla et al., 2014). To investigate the mechanism by which aneuploidy can promote oncogenesis, the effect of aneuploidy on genomic stability was investigated (Sheltzer et al., 2011). The authors analyzed 13 budding haploid yeast strains, when cells in the budding stage have a double set of chromosomes before the transition of daughter chromosomes into the formed bud. The authors found that all aneuploid strains exhibit one or more forms of genomic instability. Most of these strains showed an increased likelihood of chromosome loss and were characterized by a decrease in the efficiency of mitotic recombination, as well as erroneous repair of DNA damage. The presence of chromosomal aneuploidy occurs in 90% of tumors. It has been suggested that aneuploidy increases genomic instability, may accelerate genetic changes that promote unrestricted proliferation. However, although aneuploidy is the result of genomic instability, there is currently limited evidence whether genomic instability may be due to aneuploidy itself. To test this possibility directly, the accuracy of chromosome segregation was analyzed in several haploid *Saccharomyces cerevisiae* strains, which carry additional copies of single yeast chromosomes during budding (Sheltzer et al., 2011). It has been shown that aneuploid strains exhibit impaired proliferation and are characterized by increased sensitivity to conditions that disrupt protein synthesis. It is not excluded that aneuploidy and the mutagenic effects can be beneficial, increasing the likelihood of genetic instability, which contributes to the adaptation of cells to the environment and evolutionary processes.

Chromosomal instability is a key component of cancer progression and many inherited diseases. Understanding why some chromosomes are more unstable than others can provide insight into the integrity of the genome. A systematic study was carried out on the spontaneous loss of chromosomes for all sixteen chromosomes in *Saccharomyces cerevisiae* to elucidate the mechanisms underlying chromosomal instability. It was shown that the stability of different chromosomes varied by more than 100 times (Kumaran et al., 2013). Although the karyotype of a tumor can remain fairly stable over time, it is often the case that individual cells in a tumor have different karyotypes.

Homologous recombination is a source of genome instability and loss of heterozygosity in mitotic cells. Since these events can result in a serious health risk, the work (Lettier et al., 2006) analyzed in detail the role of DNA double-strand breaks in this process for *Saccharomyces cerevisiae*. In eukaryotes, a high level of homologous recombination is a normal sign of meiosis and results from the induction of a large number of DNA double-strand breaks. Correct repair of DNA double-strand breaks is critical to maintaining genome stability (Frumkin et al., 2016). This work describes the results of experimental studies and their quantitative modeling. It was identified 19 genes that cause genetic instability in the yeast *Saccharomyces cerevisiae*. To better understand the molecular mechanisms by which these genes operate, the authors investigated their interaction with the deletion. It is concluded that instability genes caused cell cycle disturbances.

The authors of the review publication (Skoneczna et al., 2015) focused their attention on the mechanisms leading to the genome instability in yeast cells. A comprehensive review of the molecular pathways for ensuring the stability of eukaryotic cells is presented. In the process of evolution, cells have adapted to mechanisms responsible for the fulfillment of conflicting goals: to preserve the integrity of the genome and the ability to initiate mutations that allow cells to adapt to changes in the environment. One of the evolutionary mechanisms that has been improved for survival is genetic instability. The authors describe in detail the mechanisms responsible for genome stability and tolerance to mutations in mitotic cells of *Saccharomyces cerevisiae* and *Schizosaccharomyces pombe*. These processes include mechanisms that ensure replication accuracy, DNA damage recognition and conditions that provide accurate chromosome segregation during cell division. Various factors that can affect genetic instability are discussed – cell ploidy, cell cycle phase, DNA transcriptional activity, reliability of DNA integrity, potential impact of endogenous or exogenous stress. The analysis of several priority directions of molecular mechanisms of the formation of genetic instability – replication errors and systems for correcting these errors, recombination and transcription, mutations induced by recombination, as well as the role of mitochondria and ploidy – is separately highlighted in their work. The authors note that the long evolutionary history of life has led to the improvement of the mechanisms of genome stability. Various mechanisms have been used to maintain constant genome content. In yeast cells, replication is very accurate, DNA errors or damage are effectively removed, DNA breaks are repaired, and abnormal DNA structures that occur during the cell cycle are eliminated. The molecular mechanism of cell division control to ensure equal segregation of chromosomes is described. Violation of these control processes can lead to death or genome instability. Both phenomena are responsible for the evolution and tolerance to mutations in parallel with the mechanisms for maintaining the stability of the genome. Optimizing replication accuracy and balance between DNA repair efficiency provides success in future generations, which must adapt to a changing environment. It is concluded that all the processes that ensure the stability of the genome are highly conservative and occur both in yeast cells and in human cells.

Genetic instability caused by the stress of DNA replication is an important factor in stimulating the development of cancer. Analysis of genomic instability in *Saccharomyces cerevisiae* induced by impaired DNA replication was carried out in paper (Zheng et al., 2016). The authors measured the rate of genomic changes in a yeast strain with reduced expression of replicative DNA polymerase. It has been shown that most of the recombinant DNA lesions were formed during the S or G2 phase, presumably as a consequence of the breakdown of the replication fork. It has been shown that cells with a low level of DNA polymerase are capable of accumulating small promoter mutations (deletions) that increase the expression of this polymerase. These deletions confer a selective growth advantage of cells, demonstrating that replication stress may be one of the factors contributing to phenotypic evolution.

Amplification of a number of somatic copy and gene overexpression are common features of many cancers. To determine the role of gene overexpression in chromosome instability, 245 budding yeast genes have been identified that cause genetic instability when overexpressed (Duffy et al., 2016). To identify these genes, the authors scanned a collection of 5,100 yeast strains, each of which was tested for genetic instability using two different methods. The first method consisted in determining the accuracy of transfer of a chromosome into which a plasmid with a unique gene responsible for the galactose-induced promoter was inserted. Using the second method, the authors measured the loss of the MATα locus, which allowed haploid cells to mate with a female mating strain. Some of the identified genes are also representative for mammalian cells, including humans; therefore, the authors believe that, since the genotypes of various cancer cells often determine their sensitivity to damaging factors, the data obtained can be useful for choosing the optimal chemotherapy regimen.

Studies using yeast cells have been especially useful for understanding the molecular mechanisms and pathways that affect cell aging (Aggarwal, Brosh, 2012). The authors of this work discuss the question of how the mechanisms of aging may be realized. For various cellular systems, the importance of genomic stability for aging and the maintenance of protein functions is noted. Genetic methods for studying translation have been used to study the functional role of mammalian proteins using yeast as a model system. Since genetic instability is considered as one of the causes of aging and carcinogenesis, attempts are being made to study the aging mechanism in yeast cells (McMurray, Gottschling, 2004; Laun et al., 2007; Fontana et al., 2010). It was found in the works that the aging of the yeast *Saccharomyces cerevisiae* is explained by their connection with genome instability. Telomeres are nucleoprotein structures located at the linear ends of eukaryotic chromosomes, the integrity of which is required for cell proliferation and survival. It has been shown that the recombination of telomeric DNA accelerates the aging of *Saccharomyces cerevisiae* cells (Chen et al., 2009). It has been demonstrated that, although it is a unicellular organism, yeast is characterized by the main mechanisms of various types of cell death that occur in higher eukaryotes – apoptosis, regulated necrosis, and autophagy (Falcone, Mazzoni, 2016). The authors (Liu et al., 2016), studying the effect of telomerase on yeast lifespan, showed that active telomerase increases genome instability and promotes yeast aging. The authors of the work (Kyryakov et al., 2016) obtained 3 long-lived mutant strains of *Saccharomyces cerevisiae* by prolonged exposure to lithocholic acid. Each mutant strain could maintain an increased lifespan after multiple passages in a medium without lithocholic acid. The authors used these long-lived yeast mutants to empirically test theories of aging.

Mitochondria play a crucial role in eukaryotic cells – the mitochondrial electron transport chain generates adenosine triphosphate (ATP), which serves as a source of energy for cell life. In parallel, reactive oxygen species are generated as a natural product of oxidative phosphorylation, which are considered as one of the causes of aging, because they damage proteins, lipids and DNA. The authors (Kwon et al., 2015) analyzed lifespan, growth phenotype, membrane potential of mitochondria, as well as intracellular levels of ATP and mitochondrial superoxide in 33 yeast strains with a deletion of one component of the electron transport chain during cell aging. This mutation has been shown to reduce the life span of the yeast.

Oxidative stress is the process of cell damage as a result of oxidation (Degtyareva et al., 2008). It is believed that oxidative damage to DNA may be related to the etiology of cancer and accelerates tumor onset by increasing the frequency of mutations. Most malignant cells acquire a specific type of genomic instability characterized by large-scale genomic rearrangements. The authors used the yeast *Saccharomyces cerevisiae* as a model system to determine the relationship between genotoxic stress and chromosomal instability. Increased levels of chronic oxidative DNA damage have been found to cause chromosomal aberrations at high frequencies in both selective and non-selective growth conditions. Despite the fact that the exact role of this process in tumorigenesis is still the subject of serious debate, there is no doubt that the induction of large-scale genome rearrangements can contribute to the amplification of oncogenes and the deletion of tumor suppressor genes through the loss of heterozygosity, as well as through changes in the transcriptional regulation of genes. Chronic exposure to oxidative DNA damage from chronic inflammation or exposure to reactive oxygen species has been associated with an increased risk of cancer. The results of the work (Degtyareva et al., 2008) show that such an increased risk can be caused by an increase in chromosomal instability, as demonstrated in a yeast cell model.

A delay in the formation of colonies by mammalian cells (Sinclair, 1964; Nias et al., 1965) and yeast cells (Korogodin, 1966) has been described a relatively long time ago. Later, this phenomenon was associated with genetic instability (Chang, Little, 1992a, b; Little et al., 1997). Delayed colony formation occurs in cultured cells of various origins (Klein et al., 1990; Morgan, Murnane, 1995; Little, 1998; Korogodin et al., 2007; Korogodin, 2010; Mulla et al., 2014). It has been shown that the delay in reproductive cell death correlates with chromosome instability (Limoli et al., 1997). There is evidence that double-stranded DNA breaks are important for delayed reproductive death and delayed colony formation by cells that survived exposure to ionizing radiation (Chang, Little, 1992a, b; Brosh, Bohr, 2007). Then it was to be expected that this effect should be more pronounced in diploid than in haploid yeast cells. Indeed, the delay in the formation of colonies by the surviving yeast cells after irradiation was more pronounced for diploid than for wild-type haploid strains (Korogodin, 1966; Kapultsevich et al., 1972; Korogodin, 1993; Korogodin et al., 2007; Petin et al., 2008). This implies that diploid yeast cells, characterized by sigmoid survival curves and able to recover from radiation damage, inherit a number of non-lethal subdamages responsible for long-term cell death and genetic instability. A similar effect was less noticeable for wild-type haploid yeast cells showing exponential survival curves. Based on these facts, it was concluded that the shape of the survival curve and the ability of cells to repair double-stranded DNA breaks mainly determine the delayed formation of clones. However, in our work (Evstratova et al., 2016), preliminary data were obtained showing that genetic instability determined by the delayed colony formation by cells surviving after irradiation was well expressed and reached 100% for diploid cells, regardless of the shape of the survival curve and the ability of cells to recover from radiation damage. This data needs additional confirmation. It has long been known that the shape of the survival curves after irradiation with UV light, in contrast to the action of ionizing radiation, does not depend on ploidy and has a sigmoid shape for both haploid and diploid yeast cells. Then a comparison of the genetic instability of haploid and diploid yeast cells that survived after exposure to UV light can help better understand whether this effect is really related to the shape of the survival curve and the ability of cell to recover radiation damage or it is determined mainly by cell ploidy. One of the advantages of determining the delay in the formation of colonies by cells that survived after irradiation is the possibility of a quantitative assessment of the phenotypic manifestation of genetic instability. Using the delay in the formation of colonies by cells that survived after irradiation as a test of genetic instability, it is of interest to compare quantitatively RBE of densely ionizing radiation for both lethal effects and genetic instability of wild-type haploid and homozygous diploid cells and their radiosensitive mutants.

Chapter 2

Genetic Instability of Yeast Cells Irradiated with Sparsely Ionizing Radiation

Abstract

The interaction of ionizing radiation with matter is briefly described with an emphasis on the fact that the action of sparsely ionizing radiation is accompanied by UV light arising from the movement of electrons at a speed exceeding the speed of light in a biological environment. It was shown that the total number of primary radiation damage increases in proportion to cell ploidy. No correlation was found between cell ploidy and the recovery constant, which characterizes the probability of recovery per unit of time while the irreversible component of radiation damage decreases with increasing cell ploidy. As a quantitative test of genetic instability, the delay in the colony formation by cells surviving after irradiation was used. It was demonstrated that he genetic instability of yeast cells is mainly determined by cell ploidy, but not by the sigmoid shape of the survival curve or cell ability to post-irradiation recovery, as was repeatedly postulated earlier by many authors.

2.1. Interaction of Ionizing Radiation with Matter

Before proceeding to the description of the genetic instability of yeast cells after irradiation with ionizing radiation, we will briefly consider the interaction of ionizing radiation with matter. It is emphasized that the action of ionizing radiation of high energies of γ-rays is accompanied by a flux of UV light. More detailed information on these issues can be found in the works (Dertinger, Jung, 1973; Kogl, 1986; Radiation, 1990; Sivintsev, 1991; Tsyb et al., 2005; Petin, Morozov, 2015).

There are two types of electromagnetic ionizing radiation – X-rays and γ-rays. X-rays are emitted during deceleration of electrons in the field of the nucleus (continuous spectrum) and during reverse transitions of excited electrons to the inner shells of atoms (line spectrum). The line spectrum is called characteristic emission that is specific for the anode material. Gamma radiation occurs during the transition of a nucleus from an excited state, arising, for example, during β-decay, to a more stable one. This means that X-ray radiation occurs outside the nucleus, and γ-rays – inside the nucleus. Depending on the energy of ionizing electromagnetic radiation, as well as on the composition and density of the irradiated substance, ionizing radiation penetrates it to different depths. The higher the penetrating power, the greater the energy of the incident photon ($h\nu$) spent on ionization and excitation of atoms and molecules of the irradiated substance. There are three types of interaction of X-ray and γ-radiation with matter: photoelectric effect, Compton scattering and electron-positron pair formation.

The photoelectric effect, the main type of low-energy photons absorption, occurs as a result of the collision of a photon with the internal electrons of the atom, mainly located on the K-shell. This process is similar to the process of characteristic X-ray formation. The likelihood of a photoelectric effect is proportional to Z^4 or Z^5. In biological objects, the most common element with a large Z is carbon, in which the K-shell has a low binding energy (0.3 keV). If the incident photon has an energy of 75 keV, then the knocked-out photoelectron has an energy of 74.7 keV, i.e., almost all the energy of the photon is transferred to the electron, which causes further ionization and excitation.

Compton scattering occurs happens as a result of the collision of a photon with the outer electrons of the atom, which have a low binding energy, so they can be considered as free. At this interaction, the incident photon undergoes inelastic scattering and knocks out the electron from the atom. There is no energy loss in this process – the energy of the incident photon is equal to the sum of the energies of the knocked-out electron and the scattered photon. A scattered photon is characterized by a lower frequency (longer wavelength) and lower energy. Unlike photoelectrons, the energy of a Compton electron varies widely from zero to a certain maximum value. The Compton effect is proportional to Z of the substance. American scientist Compton, who received the Nobel Prize for this research, discovered this effect.

When the photon energy exceeds 1.02 MeV, electron-positron pairs may be produced. This process is a wonderful illustration of Einstein's principle of equivalence of mass and energy ($E = mc^2$). In a strong electric field around the nucleus, the photon disappears, turning into an electron and a positron. The energy of the incident photon hv is distributed between the particles so that hv = 1.02 MeV + the energy of the electron + the energy of the positron. The probability of such a process is proportional to the Z^2 of the substance. A positron surrounded by a large number of electrons in tissue has a high probability of annihilation. Therefore, after its motion slows down because of interactions similar to those experienced by electrons, it collides with the absorber's electrons and annihilates. Annihilation is the reverse transformation of mass into energy, while two photons with an energy of 0.51 eV are emitted at an angle of 180° to each other, which are called annihilation radiation.

From the foregoing, it is clear that, depending on the photon energy of ionizing radiation, various types of its interaction with matter prevail. Table 2.1 shows the percentage ratios of various processes of interaction of photon radiation with matter. It can be seen that in the region of low energies (0.01–0.1 MeV) the photoelectric effect predominates, and in the region of high energies (more than 10 MeV) – the process of pair formation. In most cases, when biological objects are irradiated, a significant contribution conditioned by the absorption of photon energy through the Compton interaction.

The interaction of electrons with matter leads to ionizations and excitations of the atoms of the medium. Due to its low mass and a single negative charge, every time when a free electron approaches the orbiting electrons, it deviates from its path. Due to multiple deviations of electrons from their main direction, the true path length differs from the penetration depth. The primary electron is able to transfer part of its energy to electrons knocked out from atoms (delta electrons). Delta electrons with energies above 100 eV create about 50% of ionizations. Slow delta electrons (energy less than 100 eV) form clusters – local accumulations of ions that can be important for the manifestation of radiobiological reactions. As the energy of the electron decreases, its speed decreases and the probability of interaction with atoms of the substance through which it passes increases. At high energies of electrons (above 500 keV), their LET (Linear Energy Transfer) is small (about 0.2 keV/μm). The highest ionization density (LET ~ 33 keV/μm) is observed at the end of the electron trajectory (Bragg peak), when its energy drops to hundreds of eV.

Table 2.1. Percentage of different interaction processes of ionizing electromagnetic radiation with a matter

Energy	Photoeffect	Compton effect	Pair formation
10 keV	100	0	0
20 keV	75	25	0
40 keV	50	50	0
200 keV	1	99	0
400 keV	0	100	0
1 MeV	0	100	0
10 MeV	0	50	50
100 MeV	0	0	100

Densely ionizing radiations presented by the nuclei of various elements (including protons and α-particles). Charged particles ionize matter by pulling or repelling orbiting electrons. The high-LET of these emissions results in their high RBE (Relative Biological Effectiveness). Due to their mass (7,500 electron masses), α-particles move relatively slowly (with a velocity of 20,000 km/s), so they have sufficient time to ionize the atoms of the medium through which they pass. They spent all their energy at a short distance and their path is presented in the form of short, straight tracks with high-LET. As α-particles penetrate into matter, more and more interactions with matter occur. This decreases the velocity of the particle, which in turn increases the likelihood of further ionization. At the end of the path, the LET (Bragg peak) sharply increases due to a decrease in the particle velocity. A stopped alpha particle attracts two electrons and becomes a neutral helium atom. In living matter, due to its higher density, α-particle with an energy of 5 MeV is able to travel a path of several tens of microns. In addition to ionization, all types of radiation cause excitation of atoms and molecules of the irradiated substance by transferring to them energy that is insufficient for ionization. Ionization and excitation are the main processes due to which all the energy of ionizing radiation penetrating into the irradiated object is used up.

In conclusion of this section, we note that electrons arising as a result of the described interactions of photons with matter and moving in a medium at a speed exceeding light speed in this medium emit the Čerenkov emission, one of the components of which is UV light, that can cause excitation of atoms and molecules of the environment. It is known that the rest energy for electrons is $E = mc^2 = 0.512$ MeV, and the speed of light in water is $v = c/n = 0.74c$, where n is the refractive index, which is 1.34 for water. The expression $v = 0.74c$ characterizes the threshold speed of electrons, above which particles move with a speed exceeding the light speed in a given medium. Substituting this quantity into the relativistic expression for the energy of a charged particle, we have

$$E = \frac{mc^2}{\sqrt{1-v^2/c^2}} - mc^2 = \frac{mc^2}{\sqrt{1-(0,74c)^2/c^2}} - mc^2 = 0.49mc^2. \quad (2.1)$$

Then for electrons E (threshold) = 0.49·0.512 = 0.25 MeV. The range of such electrons in air is 75 cm, in water – 0.64 mm. Used in many radiobiological experiments, photons of ionizing radiation create electrons with an energy of more than 0.25 MeV. Consequently, in such experiments, UV light from Čerenkov emission accompanies the action of ionizing radiation. Naturally, the output of UV light depends on the radiation energy and the volume of the irradiated object. In monograph of Petin and Morozov (2015), a method is described for the quantitative assessment of the relative contribution of molecular excitations to the lethal effect of ionizing radiation. In the same work, a method was proposed for estimating the dose of equivalent UV light accompanying a dose of 100 Gy of ionizing radiation. Using this method, the relative contribution of excitations (*k*) to the inactivation of *Escherichia coli* bacteria (strain B_{s-1}) hypersensitive to UV light was calculated depending on the energy of ionizing radiation (Table 2.2).

Table 2.2. The dependence on the energy of ionizing radiation on the relative contribution of excitations (k) to the lethal effect of *Escherichia coli* bacteria (strain B_{s-1}), as well as the dose of equivalent UV light accompanying ionizing radiation dose of 100 Gy

Ionizing radiation	**Energy, MeV**	***k***	**UV light dose, J/m²**
X-rays	0.04	0.03	0.04
X-rays	0.15	0.04	0.05
X-rays	0.20	0.08	0.11
γ-rays of ^{137}Cs	0.67	0.19	0.24
γ-rays of ^{60}Co	1.25	0.29	0.40
Bremsstrahlung	25.0	0.43	0.56

This table shows the dependence of UV light dose accompanying a dose of 100 Gy of ionizing radiation on its energy. It is clear that the contribution of excitations to the lethal effect of ionizing radiation increases from 3% for X-rays with an energy of 0.04 MeV to 43% for 25 MeV bremsstrahlung radiation, while the dose of equivalent UV light accompanying a dose of 100 Gy of ionizing radiation increases from 0.04 to 0.56 J/m^2.

Table 2.3 shows the relative contribution of excitations (*k*) to the inactivation of *Escherichia coli* bacteria (strain B_{s-1}) hypersensitive to UV light, depending on the volume of the irradiated suspension.

Table 2.3. The dependence of the irradiated suspension volume on the relative contribution of excitations (k) in the lethal effect of Escherichia coli bacteria (strain B_{s-1}), as well as the dose of equivalent UV light accompanying the irradiation with γ-rays of ^{60}Co at a dose of 100 Gy, on the volume of the irradiated suspension

Suspension volume, ml	***k***	**UV light dose, J/m²**
0.5	0.26	0.33
1.0	0.33	0.43
5.0	0.38	0.50
10	0.41	0.53
100	0.46	0.60
500	0.49	0.64

This table shows the dependence of the UV light dose accompanying irradiation with ^{60}Co γ-rays at a dose of 100 Gy on the volume of the irradiated suspension. It can be seen that the contribution of excitations to the lethal effect of radiation increases from 26% for a 0.5 ml suspension to 40% for 500 ml of suspension, while the dose of equivalent UV-light accompanying 100 Gy of ionizing radiation increases from 0.33 to 0.64 J/m^2. Note that to simplify the calculations, we compared the entire spectrum of Čerenkov UV radiation with the equivalent flux of the UV light wavelength most effective for lethal and genetic effects (254 nm).

2.2. Role of Ploidy and Repair in Cell Radiosensitivity

As shown in Chapter 1, genetic instability of cells is closely related to radiation damage in DNA and the reliability of their recovery, which depend on the ploidy and radiosensitivity of cells. Therefore, this section will present some data demonstrating the role of ploidy and cell recovery from radiation damage in their radiosensitivity. One of the central problems of radiobiology is the study of factors that control the radiosensitivity of cells, as well as its modification using various endogenous and exogenous factors (Chadwick, Leenhouts, 1981; Dethlefsen, Dewey, 1982; Joiner, van der Kogel, 2009; Tsyb, Mardynsky, 2010; Kaprin, Mardynsky, 2018). The interest of researchers to this problem is due to the following issues. First, – to the practical needs of applied radiobiology, including the optimization of combined effects in radiation therapy and other fields of knowledge. Secondly, – the possibility of developing new theoretical concepts of the primary mechanisms of ionizing radiation action may be expected.

It is known that according to the organization of genetic material, living organisms are divided into two large groups – prokaryotes and eukaryotes. The nuclear apparatus of prokaryotes is not surrounded by a membrane that would be separate it from the cytoplasm; it divides without visible rearrangement of the structure and without spindle formation. Eukaryotes have a well-developed chromosomal apparatus that changes in the phases of the cell cycle. The nucleus of eukaryotic cells is surrounded by a membrane that disappears at some stages of cell division. In eukaryotic chromosomes, DNA is associated with histone proteins that form nucleoprotein complex. The nucleus contains a set of chromosomes, the number of which is constant for cells of a given species. The nucleus can contain several sets of homologous chromosomes, which are identical to each other. The number of such sets determines cell ploidy. When eukaryotic cells divide, all chromosomes are reproduced simultaneously and only once in the course of one division. A special apparatus of eukaryotic cells (spindle) ensures an equal distribution of chromosomes between daughter cells in the process of nuclear division during mitosis. Doubling of chromosomes occurs before nuclear division, and already in prophase, each chromosome consists of two chromatids. In anaphase, these chromosomes branch off to different poles. During reduction division (meiosis), the number of chromosomes is halved, since during meiosis the nucleus divides twice while the chromosomes only once. Meiosis is a prerequisite for the mechanism of genetic recombination. At the beginning of meiosis, there is a pairwise union of homologous chromosomes (conjugation), after which the conjugated chromosomes diverge to the poles of the cell during the first division. At the moment of conjugation, each chromosome is split into two chromatids. In the second division, the chromatids of one chromosome diverge.

An example of the simplest unicellular eukaryotes is yeast, the structure of the cell and nucleus of which is in many ways similar to higher organisms (Zakharov et al., 1980, 1984; Frankenberg-Schwager et al., 1994; Game, 2000; Friedberg et al., 2006; Fontana et al., 2010; Feldman, 2012). Unlike the cells of these organisms, which are diploid throughout their entire life, with the exception of the unicellular stage of gametes, vegetatively propagating yeasts can be either haploid or diploid. The genome of yeast cells, such as *Saccharomyces cerevisiae*, is very small compared to other eukaryotes. Thus, haploid cells contain about $1.2 \cdot 10^{10}$ Da of DNA, 80–90% of which is in the nucleus, and the rest in mitochondria. For comparison, the only chromosome of bacterial cells of *Escherichia coli* contains $0.25 \cdot 10^{10}$ Da. However, yeast DNA is distributed among 16 chromosomes, so that a single yeast cell chromosome is only 1/5 the size of an *E. coli* chromosome and 120 times smaller than a human chromosome (Haynes, 1975). To date, about 250 separate genetic loci that control various metabolic processes have been mapped in yeast chromosomes (Game, 2000; Feldman, 2012).

Eukaryotes, in turn, can be subdivided into haplonts and diplonts. In haplonts, the vegetative phase of the life cycle is represented by haploid cells; diploid strains were obtained artificially in laboratory conditions. On the contrary, in diplonts the vegetative phase of the life cycle is represented by diploid cells, while haploid strains were obtained artificially by laboratory methods. Initially, in the course of the evolution, apparently, unicellular haplonts arose. Diplonts could have come from haplonts later (Korogodin, 2010). All higher plants and multicellular animals are diplonts. It was suggested that it was the diploid nature of the vegetative phase that served as the foundation on which the progressive evolution of eukaryotes is based (Korogodin, 1982; 2010; Korogodin et al., 1996). Most of the experimental results were obtained on the most common and studied diplont yeast, in which the vegetative phase of development in natural conditions is represented by diploid cells, and the haploid strains were obtained artificially. Yeast is one of the classical cellular systems studied in radiation biology, and the first hypotheses concerning the existence and nature of DNA repair mechanisms were based precisely on the analysis of the kinetics of radiation death and post-radiation recovery of yeast (Korogodin, 1958a, b; Patrick et al., 1964; Korogodin et al., 1968; Korogodin, 1993).

The exponential shape of the survival curves of haploid and the sigmoid shape of diploid yeast cells, as well as changes in the radiosensitivity of cells depending on their ploidy, were associated at first with recessive and dominant mutations. The idea that recessive lethal mutations play a leading role in radiation inactivation of haploid cells, and dominant ones for diploids, was based on the results of assessing the contribution of recessive and dominant lethal events to radiation inactivation of yeast cells (Mortimer, 1955). According to this estimation, the contribution of recessive lethal events to radiation inactivation of haploid cells was 95%, and that of dominant cells was only 5%. However, new quantitative estimates showed that the contribution of recessive mutations to the inactivation of haploid strains was initially significantly overestimated. For example, the determination of the yield of recessive lethal events after irradiation of diploid cells, performed by the tetrad analysis method (James, Werner, 1966), makes it possible to calculate that no more than 30% of irradiated haploid cells can die due to such mutations. Approximately the same result (35%) was obtained by the assessment of recessive lethal events in the inactivation of haploid cells, performed in experiments on crossing irradiated and unirradiated *S. cerevisiae* haploid cells carrying the *rad*52 mutation (Ho, Mortimer, 1973). This mutation weakly increases the radiosensitivity of haploid yeasts in the G1 phase of the cell cycle, but greatly increases the radiosensitivity of diploid yeasts. In recent years, data have appeared confirming the idea that the inactivation of cells induced by ionizing radiation is based on lesions of the same type (but not recessive and dominant mutations in the classical sense of these terms). Differences in the radiosensitivity of polyploid cells may be due to recovery processes. In addition, the shown independence of cell radiosensitivity from ploidy for some *rad* mutants and haplont yeasts (Korogodin et al., 1977;

Petin, 1987; Korogodin, 1993) may also indicate the uniformity of lethal radiation damage (for example, DNA damage) in yeast of different ploidy.

Many data show that a double set of chromosomes is required for the effective functioning of cell repair systems. Due to the ability of diploid and polyploid wild-type cells to recover from radiation-induced double-stranded DNA breaks and almost absence of such recovery in haploid cells, the survival curves of diploid cells are characterized by a sigmoid shape, while those of haploid cells are exponential. The presence of diploid-specific recovery is confirmed by experimental data that the differences in radiosensitivity between haploid and isogenic diploid strains are significantly reduced for radiosensitive mutants defective in certain repair systems (Saeki et al., 1980). For some mutants, in which repair systems are suppressed or weakly expressed, haploid cells are even more resistant to the action of ionizing radiation than diploid ones (Petin, 1987; Petin, Komarov, 1989). The need for a diploid state for more effective recovery from radiation damage of yeast cells irradiated in the G1 phase of the cell cycle may mean that the interaction of homologous chromosomes, for example, their conjugation, plays an important role in yeast cell recovery processes. The possibility of conjugation of homologous chromosomes in diploid yeasts during vegetative reproduction follows from the firmly established fact that they have mitotic crossingover (Resnik. 1976; Zakharov et al., 1984; Friedberg et al., 2006; Storchova, 2014). It can be assumed, therefore, that the recovery processes are associated with the mechanisms involved in the implementation of mitotic recombination of chromosomes. The need for a diploid state for the implementation of recovery processes is in good agreement with the fact that an increase in the cell survival of irradiated cells when they are kept after irradiation in a non-nutritive environment is pronounced in diploid and other polyploid strains, but it is absent in haploid strains (Korogodin, 1966, 2010), This point of view is supported by the molecular data indicating that repair of DNA double breaks requires a double set of chromosomes (Luchnik et al., 1977; Frankenberg-Schwager et al., 1987).

At present, it can be considered firmly established that everything that has been said above regarding the shape of the survival curves is applicable not only to different strains of *Saccharomyces cerevisiae*. It was shown that most yeast vegetatively propagating in the haplophase give exponential survival curves and sigmoid curves in the diploid state (Sokurova, 1975). This situation remained until the discovery of radiosensitive mutants, which even in the diploid state could be characterized by exponential survival curves.

Phenomenologically, several types of recovery after exposure to ionizing radiation have been identified in yeast. These include recovery upon exposure to a non-nutritive medium (recovery by the delayed mitotic division); recovery responsible for increased resistance of budding cells; recovery answerable for the increased resistance of wild-type diploid strains compared to isogenic haploid strains; recovery accountable for the increased resistance of both haploid and diploid wild-type strains compared to the radioresistance of sensitive mutants. At present, it is impossible to answer which of the known repair pathways are responsible for this or that type of recovery, which manifests itself phenomenologically at the cellular level. Moreover, until the molecular mechanisms of these phenomena are deciphered, the question of their difference or identity cannot be unambiguously resolved.

The process of DNA damage repair can be subdivided (Korolev, 2009) into six separate, often overlapping systems: (i) direct repair, which restores the original DNA structure by a chemical reaction; (ii) excisional base repair is dominant in the repair of modified DNA bases and is carried out by deletion of the damaged nucleotide from DNA and filling the gap with

DNA polymerase; (iii) excisional repair of nucleotides by cleavage of DNA fragments together with damaged nucleotides and building up the gap with DNA polymerases; (iv) the system for correcting mismatched bases removes wrong bases arising from various types of DNA synthesis; (v) recombination repair removes the most cytotoxic damage, such as double-stranded DNA breaks and various kinds of cross-linking of DNA strands; (vi) post-replicative repair solves the problems of DNA synthesis when replicative forks meet with matrix damage. In the repair of DNA damage induced by ionizing radiation, the latter two systems play the main role. All living orgasms have sophisticated DNA damage repair systems. Recovery by repairing damaged nucleotides is also important. This type of recovery, which occurs in many organisms from yeast to mammals, is discussed in detail (Hoeijmakers, 1993).

The repair of double-strand breaks by the mechanism of homologous recombination can occur in the presence of two or more homologous chromosomes or chromatids in a cell (Frankenberg-Schwager et al., 1987; Blöcher, 1988; Frankenberg, 1994; Friedberg et al., 2006). Consequently, this process takes place in the late S-phase of haploid cells (there was a doubling of molecules) and in diploid (polyploid) cells. In the late S-phase of a haploid cell, homologous recombination between sister chromatids is predominant. When a double-strand break occurs in one of the sister chromatids, the other homologous chromosome can be used to repair damage. In eukaryotes (yeast, mammalian cells), as in prokaryotes (bacteria), double-strand DNA breaks lead to the most dangerous consequences of exposure to ionizing radiation. According to many authors, the double-strand breaks cause the formation of chromosomal aberrations, which lead to cell death. The radioresistance of cells also reflects their ability to repair (Zhestyanikov, 1968, 1979). When accounting one set of chromosomes, diploid cells of the yeast *Saccharomyces cerevisiae* are 10–15 times more resistent than haploid ones. It was found that haploid and diploid cells die when exposed to ionizing radiation due to the same type of radiation damage – most likely, chromosomal lethal mutations (Korogodin, 2010). Differences in the radioresistance of haploid and diploid yeasts are due to the ability of diploid cells to recover from double-stranded DNA breaks, which are to a much lesser extent or not at all repaired by haploid cells (Lewis, Resnick, 2000; Mao et al., 2008).

The performed analysis shows that the doubling of the chromosomes set can have different effects on the radiosensitivity of various yeast species after the action of ionizing radiation. In wild-type diploid yeast, the transition from the haploid state to the diploid state is accompanied by a significant increase in radioresistance. In diploid yeast with defects in the repair system, diploid cells are only slightly more resistant than their isogenic haploid strains. In haplont yeast, the transition from a haploid state to a diploid one may even be accompanied by a decrease in radioresistance. The enumerated set of data can be explained from the standpoint of the role of repair in the radiosensitivity of yeasts of different ploidy. This concept would be supplemented with the assumption (Korogodin et al., 1977a, b) that the haplont yeast is defective in repair systems. If the leading role in radiation damage to haploid and diploid cells belongs to the same type of radiation damage, for example, chromosomal aberrations, then such damage, naturally, in diploid cells should occur more often than in haploid cells due to the larger size of the target (volume of chromosomes). In the absence of repair processes, diploid cells may be more radiosensitive, which was observed in experiments with haplont yeasts (Korogodin et al., 1996).

Due to the activity of repair enzymes, structural disorders in the DNA molecule caused by ionizing radiation can be corrected, or defects in DNA synthesized on a damaged DNA matrix are eliminated (Zhestyanikov, 1968, 1979; Ryabchenko, 1979; Savina et al., 2015). For eukaryotic cells, double-stranded DNA breaks are the most severe damage, which serve as the basis for the formation of chromosomal aberrations (Luchnik, 1968; Bochkov, 1971; Sevankayev, 1987). Cells of diploid eukaryotes, unlike haploid ones, efficiently repair double-stranded DNA breaks due to the presence of diploid-specific repair (Resnick, 1976; Luchnik et al., 1977; Frankenberg et al., 1981; Frankenberg-Schwager et al., 1987; Karpenshif, Bernstein, 2012). Note that in recent years, information has appeared on the possibility of repairing DNA double strand breaks in haploid yeasts through nonhomologous reunification of broken DNA ends (Lewis, Resnick, 2000; Mao et al., 2008), the performance of which is significantly inferior to efficiency of repair of double stranded DNA breaks including the recombination of homologous chromosomes. On the basis of a large set of studies, the idea was developed that the radiation inactivation of haploid and diploid yeasts is based mainly on the same type of damage, for example, chromosomal aberrations, rather than recessive and dominant mutations in the classical sense of these terms. The differences in the radiosensitivity of haploid and diploid cells of the wild-type (with respect to radiosensitivity) are mainly due to recovery processes (Korogodin, 2010).

For a quantitative interpretation of experiments related to cell recovery from radiation damage, we will briefly describe the mathematical model of post-radiation recovery and a method for finding its main parameters. In the process of post-radiation recovery, the survival rate of cells irradiated with ionizing radiation (or after exposure to some other agents) increases. It can be drawn as a decrease in the effectiveness of the primary dose D_1 in which the cells were irradiated. The curve of change in cell survival, depending on the dose of the acting agent D, we will call, for brevity's sake, the survival curve and, depending on the duration t of keeping the cells in the post-radiation recovery mode, we will further call the recovery curve. In the process of such cell recovery, the number of potentially lethal damage decreases (Korogodin, 1966; Wang, Iliakis, 1992). It is known (Davidson, 1956; Korogodin, 1966; Kapultsevich, 1978) that post-radiation cell recovery is carried out by the type of effective dose reduction. For each value of the survival in the recovery curve $S(t)$, one can find the corresponding value of the effective dose $D_{\text{eff}}(t)$, i.e., the dose, at which the cells should be irradiated initially to ensure the survival $S(t)$ achieved after recovery during t time (Figure 2.1).

The equation

$$K(t) = D_{eff}(t)/D_1 \tag{2.2}$$

shows the relative part of the dose, or in other words, the relative part of primary damage, from which the cells did not have time (or could not) to recover during time t. If t is large enough (for example, for yeast cells it is 2–3 days), then the recovery curve reaches a plateau. Let this moment of time denote by the sign of infinity (∞). For this moment

$$K = K(\infty) = D_{eff}(\infty)/D_1. \tag{2.3}$$

In this expression, $D_{eff}(\infty)/D_1$ is the effective dose corresponding to the plateau of the recovery curve. $D_{eff}(t)$ tends to $D(\infty)$ as t tends to infinity. The ratio $D_{eff}(\infty)/D_1$ is called an irreversible component of radiation damage, which characterizes the proportion of primary damage from which the cell is unable to recover. The irreversible component is the reciprocal of the dose modifying factor traditionally used in radiobiology. The convenience of introducing the concept of irreversible component lies in the fact that its value directly indicates the fraction of irreversible radiation damage from which cells are unable to recover.

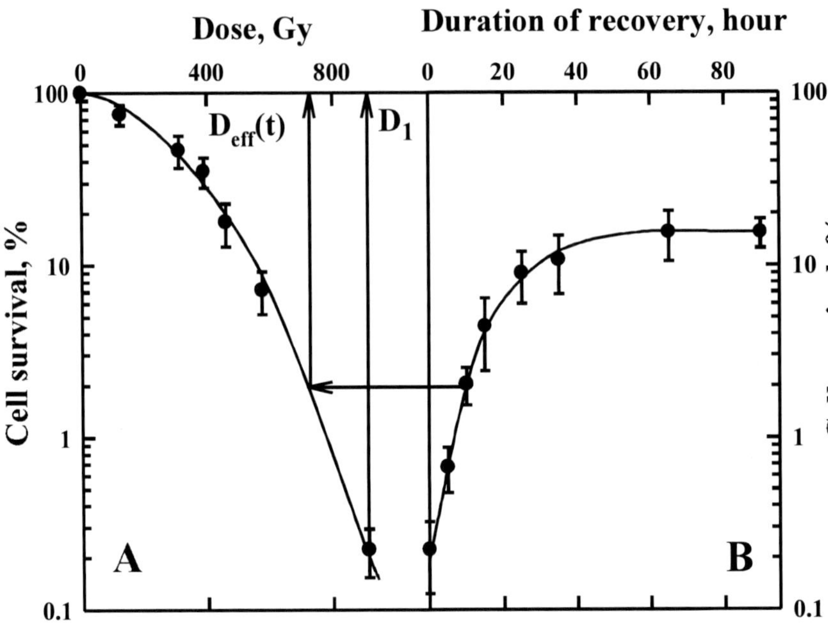

Figure 2.1. The dependence of the survival of the diploid yeast cells of S*accharomyces cerevisiae* (strain 5a3bα) on the dose of ionizing radiation (A) and the duration of recovery (B) of irradiated cells in a non-nutrient medium at 30ºC; examples of the evaluation of the effective dose $D_{eff}(t)$and the initial dose D_1 in which the cells were irradiated, are indicated by arrows.

To describe the kinetics of post-radiation recovery of yeast cells, it was proposed (Davidson, 1956; Korogodin, 1966; Kapultcevich, 1978) the equation

$$D_{eff}(t) = D_1[K + (1 - K) \cdot e^{-\beta \cdot t}], \quad (2.4)$$

where K is the irreversible component of radiation injury determined by equation (2.2), e – is the base of natural logarithms, and β is the recovery constant characterizing the probability of recovery from radiation damage per unit time. The applicability of this equation was previously demonstrated for recovery of various objects after the action of ionizing radiation (Korogodin, 1966, 2010; Komarova, Petin, 2007; Petin et al., 2012, 2020; Petin, Kim, 2016). This equation shows that for a quantitative description of the post-radiation recovery, it is necessary to know two parameters, K and β.

Taking into account equations (2.2–2.4), it can be shown that

$$e^{-\beta t} = \frac{D_{eff}(t) - D_{eff}(\infty)}{D_1 - D_{eff}(\infty)}. \quad (2.5)$$

We denote the right-hand side of equation (2.5) by $A(t)$

$$A(t) = \frac{D_{eff}(t) - D_{eff}(\infty)}{D_1 - D_{eff}(\infty)}. \quad (2.6)$$

The $A(t)$ function characterizes the change in time the proportion of radiation damage, which the cell is able to recover. From equations (2.5) and (2.6), we have

$$\beta = -\frac{lnA(t)}{t}. \quad (2.7)$$

Thus, the experimental survival curves obtained immediately after exposure to ionizing radiation or after combined exposure to ionizing radiation and another factor, as well as the recovery curves $S(t)$ obtained experimentally after different modes of exposure, make it possible to find D_1 and $D_{eff}(t)$ as shown by arrows in Figure 2.1. This allows using equation (2.2) to plot the curves of the dependence of the irreversible component $K(t)$ on the recovery time, using equation (2.3) to calculate the values of the irreversible component K and using equation (2.6) and the dependence of $lnA(t)$ on the recovery time, to calculate the recovery constant β for different irradiation regimes (equation (2.7)).

Next, we present the results of experimental studies of the characteristics of post-radiation recovery of isogenic polyploid yeast strains obtained by Laskowski (1960), who kindly provided us with these strains. Cells in the stationary stage of growth were irradiated with ^{60}Co γ-radiation on a Gammasell-220 device at a dose rate of 20 Gy/min. Figure 2.2 shows the survival curves (Figure 2.2, A) and curves of survival changes in the process of post-radiation recovery in a non-nutritive environment at 30°C (Figure 2.2, B).

It can be seen that the survival curve of the haploid strain is exponential. The rest of the strains are characterized by a sigmoid shape of the survival curve. The diploid strain is the most resistant, and with a further increase in ploidy, radiosensitivity increases with an increase in the number of chromosomes sets in the cell.

Table 2.4 shows the designation and ploidy of the strains, as well as the following radiobiological parameters: the extrapolation number n, the mean lethal dose D_0, calculated on the exponential part of the survival curves, recovery constant β, characterizing the probability of cell recovery from radiation damage per unit time (equation (2.7)), irreversible component of radiation damage K (equation (2.3)). The table shows a sharp increase in cell resistance during the transition from a haploid strain to a diploid one with a further gradual increase in radiosensitivity during the transition to higher yeast ploidy. The shape of the survival curves for all polyploid strains was sigmoid. An increase in the extrapolation number from one for a haploid strain to two for a diploid strain and three for a triploid strain is noted. A further increase in the number of chromosome sets led to a decrease in the degree of the survival curves curvature.

It can be noted that no correlation was found between the recovery constant β and ploidy. At the same time, the irreversible component of radiation damage K decreases with increasing ploidy, i.e., the proportion of non-repairable radiation damage decreases and, accordingly, the proportion of cells that can be recovered increases. If post-radiation recovery in a non-nutritive environment can to some extent serve as an indicator of the activity of other repair systems, these results mean that with an increase in the number of chromosomes sets in a cell, the total volume of repaired damage increases.

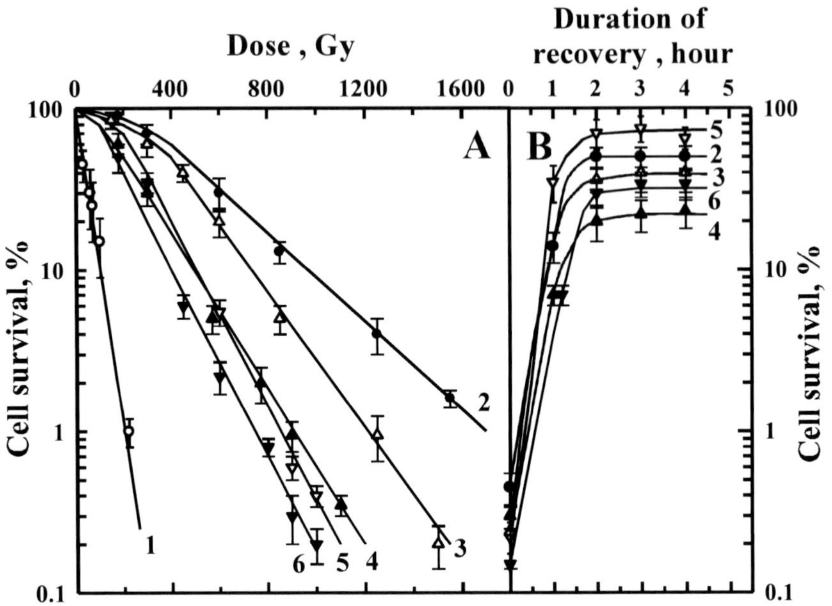

Figure 2.2. The dependence of *Saccharomyces cerevisiae* yeast cells survival on the dose of ionizing radiation (A) and the duration of post-radiation recovery of irradiated yeast cells in a non-nutrient medium at 30°C (B): 211-1a (haploid, curve 1), 211-1a/2 (diploid, curves 2), 312 (triploid, curves 3), 415 (tetraploid, curves 4), 500 (pentaploid, curves 5) and 600 (hexaploid, curves 6)

Table 2.4. Radiobiological characteristics of polyploid yeast strains *Saccharomyces cerevisiae*

Strain, ploidy	n	D_0, Gy	β, h^{-1}	K	Relative number of radiation damages	
					irreversible	repairable
211-1a haploid	1.0	40	–	1.00	1.00	0.00
211-1a/2 diploid	2.3	310	0.06	0.31	0.62	1.38
312 triploid	2.9	210	0.08	0.28	0.84	2.16
415 tetraploid	1.5	200	0.07	0.25	1.00	3.00
500 pentaploid	2.3	170	0.08	0.15	0.75	4.25
600 hexaploid	1.8	150	0.05	0.12	0.72	5.28

The data presented in Table 2.4 allow us to quantitatively evaluate the role of post-radiation recovery processes in the radiosensitivity of polyploid strains from the standpoint of the same type of primary radiation damage leading to inactivation of yeast cells of different ploidy. We will assume that, in a first approximation, the total number of primary radiation damage increases in proportion to the growth of ploidy due to an increase in the size of the target (DNA, chromosome) responsible for cell inactivation. Let us take the number of such lesions per unit in a haploid strain. Then, at the same radiation dose, the total number of formed lesions in polyploid strains in the first approximation will be equal to 2, 3, 4, 5, and 6 for diploid, triploid, tetraploid, pentaploid, and hexaploid strains, respectively. Multiplying these values by the corresponding magnitude of the irreversible component K, which characterizes the fraction of irreversible damage, we obtain the relative number of irreversible damages. The values of the recovered damage will be determined by the difference between the total and irreversible number of radiation damage. These characteristics for all strains are also summarized in Table 2.4. One can see a constant increase in the number of repairable damages with an increase in ploidy. These data show that the dependence of cell radiosensitivity on ploidy is largely determined by cell ability to recover from radiation damage, and with an increase in ploidy, the relative number of recovered damage increases, although diploid cells remain the most resistant to the action of ionizing radiation.

2.3. Radio- and UV-Sensitive Yeast Mutants

Yeast has a nuclear and mitochondrial genetic system; both mitotic and meiotic recombination is inherent to yeast, including gene conversion, which can be quantitatively measured in yeast cells. In recent years, genetic engineering methods have been developed for yeast cells: genetic transformation, DNA cloning, etc. A number of radiosensitive mutants defective in various repair systems have been discovered, and radioactive preparations have been obtained that make it possible to label yeast DNA and thereby to study the molecular processes of DNA damage and repair. With the discovery of radiosensitive mutants defective in repair systems, it was shown that there are cases when the sensitivity of haploid and diploid strains is approximately the same or even diploid cells are more sensitive than haploid ones (Petin, 1987). The effect of the xrs1-5 mutation, which increases the radiosensitivity of cells, on the radioresistance of a polyploid series of strains (from haploid to tetraploid) to UV and X-ray radiation was studied on the yeast *S. cerevisiae* (Zakharov et al., 1980, 1984). It was found that diploidization did not always lead to a sharp increase in resistance to both types of radiation. On the contrary, a mutant diploid homozygous for the xrs1-5 gene has a slightly reduced resistance to ionizing radiation. A further increase in ploidy with the xrs1-5 mutation led to an even greater sensitization of cells to the action of ionizing radiation in comparison with the mutant haploid strain. It is assumed that the xrs1-5 mutation abolished the protective effect of ploidy by blocking recovery from radiation damage. Comparison of the radiosensitivity of some haploid and diploid wild type and *rad*-mutant strains by Japanese researchers (Saeki et al., 1980) showed that in some cases diploid strains are comparable in radiosensitivity to haploid ones or even more sensitive.

Using genetic tests of complementation and recombination, it was shown that radiosensitive mutants of yeast cells contain mutations corresponding to at least 30 separate genetic loci (Game, Cox, 1973; Game, Mortimer, 1974; Haynes, 1975). The term *rad* was introduced to denote such mutants. Genetic loci that are responsible for sensitivity to UV radiation are numbered from *rad*1 to *rad*49, *rad*50 and higher numbers denote mutant genes that are responsible for the increased sensitivity of cells to the action of ionizing radiation. Most of these loci have been mapped, and it has been shown that they are all randomly distributed throughout the genome and do not form clusters. These mutants can be subdivided into three phenotypic categories (Haynes, 1975; Game, 2000; Friedberg et al., 2006): the first, – 17 strains sensitive to UV radiation; the second, – eight strains that are sensitive to ionizing radiation, and the third, – five strains that are sensitive to both agents. Mutants *rad*50, *rad*51, *rad*52, *rad*53, *rad*54, *rad*55, *rad*56, and *rad*57 are combined into one group (RAD52 group), which is characterized by increased sensitivity to the action of ionizing radiation (Saeki et al., 1980; Haynes, Kunz, 1981). Some of them, although they were more sensitive to the action of ionizing radiation in comparison with the parent wild-type strain, were characterized by a sigmoid shape of the survival curve and were slightly able to recover when irradiated cells hold in a non-nutrient medium (Petin, 1987; Petin, Komarov, 1989). These data served as the basis for identifying two subgroups that differ in the efficiency of modifying their radiosensitivity by various endo- and exogenous factors. It was shown that the relative radiosensitivity of mutants, quantitatively assessed by the ratio of the mean lethal doses for wild and radiosensitive strains homozygous for the *rad*50, *rad*51, *rad*52, and *rad*54 loci, was more pronounced than the relative radiosensitivity of the other strains (*rad*53, *rad*55, *rad*56, and *rad*57). The strains of the first subgroup, in contrast to the strains of the second one, were not capable of recovery in a non-nutritive environment, their radiosensitivity was less dependent on ploidy, the stage of growth, the radiation quality. They could be modified under the influence of radioprotectors. oxygen, hypoxic sensitizers and physical environmental factors (ultrasound, hyperthermia) in a lesser extent in comparison with those agents effecting to strains of the second subgroup. The modification of the radiosensitivity of strains homozygous at the *rad*2 (RAD3 group) and *rad*18 (RAD6 group) loci, UV-sensitive and typical representatives of the other two groups of mutants, resembles in its ability to modify the radiosensitivity of strains with *ra*d53, *rad*55, *rad*56, and *rad*57 mutations. Therefore, it can be assumed that the post-replicative (minor) pathway of repair is to a greater extent responsible for the mechanism of manifestation of yeast cell radiosensitivity modification. On the contrary, excisional (blocked by *rad*2 mutation) and mutagenic (blocked by *rad*18 mutation) repair pathways play a lesser role in modifying the radiosensitivity of these strains. This could be because these strains are more sensitive to UV light.

Based on the summation of all the data related with the sensitivity of double mutants to various mutagenic factors, a repair scheme in the yeast of *S. cerevisiae* was proposed. It includes four pathways (Haynes, 1975; Haynes, Kunz, 1981): (i) excisional; (ii) mutagenic; (iii) repairing of single DNA breaks and blocked by the *rad*51 mutation; (iv) repairing of double-stranded DNA breaks and blocked by *rad*52 and *rad*54 mutations. However, the *rad*51, *rad*52, and *rad*54 mutations are characterized by epistatic interactions (Game, Mortimer, 1974) and, therefore, block different stages of the same repair pathway. Thus, it has been firmly established that there are only three repair pathways in yeast cells: excisional, mutagenic, and minor (postreplicative). They have common stages with recombination processes and the removal of single- and double-stranded DNA breaks (Hunnable, Cox, 1971; Brendel, Haynes, 1973; Game, Cox, 1973; Game, Mortimer, 1974; Haynes, 1975; Saeki et al., 1980; Zakharov et al., 1980; Game, 2000; Friedberg et al., 2006).

The excisional repair pathway functions in the dark. With its help, various types of DNA damage can be repaired – damaged bases, nucleotides, reunification of single-strand breaks also may occurred. More than 16 different types of proteins are involved in the mechanism of this type of repair, ensuring the recognition of damage in the DNA structure and their subsequent removal, filling the gap with intact bases, followed by reparation of the integrity of the polynucleotide DNA strand. Excisional DNA repair is under the genetic control of a number of genes (Prakash, Prakash, 2000). For instance, in *E. coli* bacteria, the genes *uvr*A, *uvr*B, and *uvr*C are involved in the stage of excisional repair. In human cells, 9 genes were identified, in rodents – 11 genes, mutations in which reduce the efficiency of DNA excision repair and increase the sensitivity of mutant cells to radiation and various chemical factors. Excisional repair is not associated with DNA replication and can occur both before and after the passage of replicative synthesis. This type of repair is a universal system for repairing the original DNA structure, with the help of which both violations in the structure of nitrogenous bases and breaks in DNA strands are repaired. Excisional repair works practically without errors and is capable, for example, in UV-irradiated DNA, to remove about 100 thousand pyrimidine dimers before one lethal damage appears in the cell (Zhestyanikov, 1979; Ryabchenko, 1979; Zakharov et al., 1980; Pelevina et al., 1985).

There are numerous publications indicating that post-radiation incubation of bacteria, yeast, and mammalian cells in certain situations leads to a decrease in the number of radiation-induced DNA single-strand breaks. The repair of single DNA breaks in yeast cells is blocked by the *rad*51 mutation. It has been shown that in mammalian cells about 50–70% of single-strand breaks are removed during rapid repair within 3–5 min of incubation at 37°C, and the remaining breaks are reconnected within 90 min. Repair of single-strand breaks depends on metabolic processes and is inhibited at low temperatures (0–5°C). There is also evidence that the efficiency of repair of single-strand breaks decreases when cells are irradiated with densely ionizing radiation (Krasavin, 1989).

Along with the repair of single strand breaks, the repair of double DNA breaks is well known (Luchnik et al., 1977; Frankenberg-Schwager et al., 1987; Blöcher, 1988; Frankenberg, 1994). In diploid yeast cells, this repair pathway is blocked by the *rad*52 and *rad*54 mutations. The lack of repair of double-strand breaks in these mutants indicates the participation of recombination processes in such repair. Therefore, the repair of double breaks is more efficient in diploid rather than haploid yeast strains. This mechanism is called homologous recombination repair, since it uses the recombination exchange of damaged regions with homologous DNA regions in sister chromatids. It has been shown that radiosensitive *rad*51 mutants of the yeast *Saccharomyces cerevisiae* are defective in repair due to exchanges in sister chromatids, but they demonstrate an increased yield of translocations, in which a chromosome region is transferred to a nonhomologous chromosome (Fasullo et al., 2001). During the repair of double-strand breaks, the DNA region containing the double break is degraded by endonuclease. In the repair of double-strand breaks, the main role is played the recombination mechanism of exchange with homologous regions of intact DNA.

The post-replicative (minor) repair pathway is carried out by means of bypassing DNA damage. Using this pathway for repairing daughter DNA, cells bypass the remaining unrepaired damage in the parental DNA. The process of post-replicative DNA repair can be carried out by recombination of the newly synthesized DNA with the parental intact strand. Along with the elimination of gaps in the newly synthesized DNA using recombination mechanisms, some of the gaps can be filled in a non-recombination way. While the DNA excision repair system works practically without errors, the post-replicative DNA repair system often makes mistakes. Post-replicative DNA repair increases cell survival; however, in the distant offspring of these cells, the level of mutagenesis increases (Mattern, Painter, 1979; McEachern, Haber, 2006).

Note that bypassing damage to the parental DNA by eliminating damage in the newly synthesized daughter DNA can also be carried out at the time of replicative synthesis. During replication, opposite the damage, DNA synthesis stops and a double-stranded structure is formed between the daughters DNA strands, because the longer strand serves as a template for the subsequent synthesis of the shorter daughter DNA strand. The carried-out analysis of the damaged DNA repair pathways shows that the process of enzymatic DNA repair can proceed through two main mechanisms – by removing damage from the DNA structure, as well as by bypassing them. In the latter case, the damage remains in the DNA structure, but the cell as though eliminates the functional consequences of their existence.

Therefore, for further research, along with wild-type haploid strains of *Saccharomyces cerevisiae* (211-1a and S288C), their isogenic diploid strains (211 and XS800), as well as the following group of haploid and isogenic diploid mutants sensitive to the action of ionizing radiation (*rad*51, *rad*51/*rad*51, *rad*52, *rad*52/*rad*52, *rad*54, *rad*54/*rad*54) were used in experiments. Loci of the RAD3 group are sensitive to UV radiation and are characterized by increased mutagenesis under the influence of UV radiation. It is known that at least nine of these loci control the excision of pyrimidine dimers. Loci of the RAD52 group affect the sensitivity of cells to ionizing radiation, but do not significantly affect mutagenesis. The loci of this group also control the minor pathway for the reparation of UV damage. Some loci of this group control the repair of double strand breaks in DNA. Loci of the RAD6 group impair sensitivity to both UV and ionizing radiation and control trial-error recovery; some of them also control error-independent recovery. It should be noted that loci of this group are pleiotropic in the sense that some of them are required for mitotic recombination, sporulation, successful completion of meiosis, etc.

2.4. Participation of Recovery Processes in the Modification of Cell Radiosensitivity

All living organisms have complex systems for repairing DNA damage. One of the most important of these is restoration through the repair of damaged nucleotides. This type of repair, which is realized in many organisms from yeast to mammals, is discussed in detail (Hoeijmakers, 1993; Friedberg et al., 2006). Molecular mechanisms of DNA excision repair in yeast cells were discussed (Resnick, 1976; Prakash, Prakash, 2000: Frankenberg-Schwager et al., 1987). Many authors have repeatedly emphasized and studied the importance of recovery processes in modifying the cell radiosensitivity. For example, the role of recovery processes in the mechanism of radioprotectors action has been prove for bacterial cells (Noskin, 1984; Bresler et al., 1978). In these papers, when studying the molecular processes of the radioprotective effect of various compounds on the survival of *Escherichia coli* bacteria, it was shown that radioprotectors do not change the total amount of γ-damage to DNA, but decrease the rate of enzymatic incision. On this basis, it was concluded that radioprotectors interact with damaged DNA regions so that they become partially shielded from the action of endonuclease. Then a decrease in the incision rate leads to a decrease in the enzymatic induction of double DNA breaks, which may be important only for wild-type cells and those mutants, who were able to recover from radiation damage.

The participation of recovery processes in the mechanism of modifying the effectiveness of radioprotective action and radiosensitizers has been widely discussed in many works (Kiefer, 1974, 1976; Burlakova et al., 1975; Eidus, 1984; Eidus, Korystov, 1984; Rozhdestvensky, 1984, 1985; Hanson, Komar, 1985; Petin, 1987; Boothman et al., 1989; Petin, Komarov, 1989; Petin, Kim, 2016). In these works, schemes of radiation damage development at the molecular and cellular levels are analyzed, taking into account the recovery processes both at the chemical stage of damage formation and in the post-radiation period of biochemical reparation and fixation of radiation damage. When studying the dependence of the radioprotective effect of cysteamine on the state of repair enzymes in wild-type cells and three mutant diploid strains of the yeast *Saccharomyces cerevisiae* (homozygous for loci *rad2*, *xrs4*, *rad54*), no significant differences were found in the radioprotective effect of cysteamine on cells of wild-type and radiosensitive strains defective on excisional and recombination repair systems (Guiliano et al., 1981). However, the yeast *S. cerevisiae* contains at least eight distinct loci that control the sensitivity of cells to ionizing radiation. Therefore, below are the results of our study of the radioprotective effect of cysteamine on wild-type *S. cerevisiae* yeast cells and on various mutant cells that control the sensitivity of cells to ionizing radiation. In addition, to demonstrate the participation of recovery processes in the mechanism of the radioprotective action of the T1023 compound, this section of the work will present the results on the survival of clonogenic bone marrow cells in mice irradiated with ionizing radiation (Filimonova et al., 2012, 2014, 2015, 2019).

The experiments described below were performed on haploid and diploid yeast strains of wild-type and radiosensitive mutants of various genotypes. A 0.01 M aqueous solution of cysteamine, which has no toxic effect, was used as a radioprotector. The cells were placed in an aqueous solution of this compound 30 min before irradiation. A suspension of yeast cells with a concentration of 10^6 cells/ml was irradiated with ^{60}Co gamma rays on a "Researcher" device (Russia) at a dose rate of 20 Gy/min. The cells were irradiated at the stationary stage of growth. The length of pre-radiation cultivation was determined by the cessation of cell budding. The proportion of budding cells was about two percent. Each experiment was repeated 2–5 times. Results are presented as a mean and standard error.

The effect of synthesized in the laboratory of radiation pharmacology of A.F. Tsyb Institute of Medical Radiology compound T1023 (75 mg/kg), which is a modifier of endogenous nitric oxide synthesis, on the survival of clonogenic bone marrow cells of mice (CBA-C57BL6j) irradiated with ^{60}Co γ-rays at the “Luch-1” installation (Russia) at a dose rate of 0.4–0.8 Gy/min. The animals were irradiated in groups of 15 individuals. The radioprotector was administered to the animals intraperitoneally 15–20 min before irradiation. The survival rate of hematopoietic clonogenic bone marrow cells of mice was assessed by the test of exogenous spleen colonies (Filimonova et al., 2015). The radioprotective effect of the compound under study was judged by the relative number of spleen colonies formed in the control and irradiated groups (Filimonova et al., 2012, 2014, 2019).

The efficacy of cysteamine for haploid yeast was quantified by the dose modifying factor (DMF) determined by the ratio of the mean lethal dose D_0 (the dose that reduces survival by a factor of e, where e is the base of natural logarithms) for the survival curves obtained with cysteamine DMF = D_0(with cysteamine)/D_0(without cysteamine). The same ratio was used for diploid yeast cells, with D_0 calculated on the exponential part of the survival curves. The DMF parameters for clonogenic bone marrow cells were assessed by the ratio of isoeffective doses on the survival curves obtained with and without a protector (Filimonova et al., 2019). Figure

2.3 shows the survival curves of haploid *rad*-mutants and their parent wild-type strain, irradiated without a radioprotector and in a 0.01 M solution of cysteamine.

The effectiveness of the protection of haploid yeast is summarized in Table 2.6 for some other strains, the results with which are not shown in Figure 2.3. It can be seen that the efficiency of cysteamine protection was almost the same for wild-type haploid cells and the most mutant cells. The effectiveness of the protective action of cysteamine varied randomly around the mean DMF = 1.75 ± 0.16.

Table 2.5. Efficiency of protection of haploid yeast *S. cerevisiae* of various genotypes with cysteamine (0.01 M)

Strain	Mutant locus	D_0, Gy without cysteamine	D_0, Gy with cysteamine	DMF
S288C	RAD	85.0	154.0	1.81 ± 0.2
g218/6b	*rad*50	37.0	72.0	1.95 ± 0.3
g218/7a	*rad*51	60.0	101.0	1.69 ± 0.1
g160/2b	*rad*52	57.0	91.0	1.59 ± 0.1
g151/12c	*rad*53	79.0	137.0	1.73 ± 0.2
g218/7c	*rad*54	58.0	83.0	1.42 ± 0.3
X1687/101b	*rad*55	57.0	126.0	2.20 ± 0.4
30-57-rs31	*rad*56	62.0	98.0	1.58 ±0.1
g246/6b	*rad*57	70.0	123.0	1.78 ± 0.2

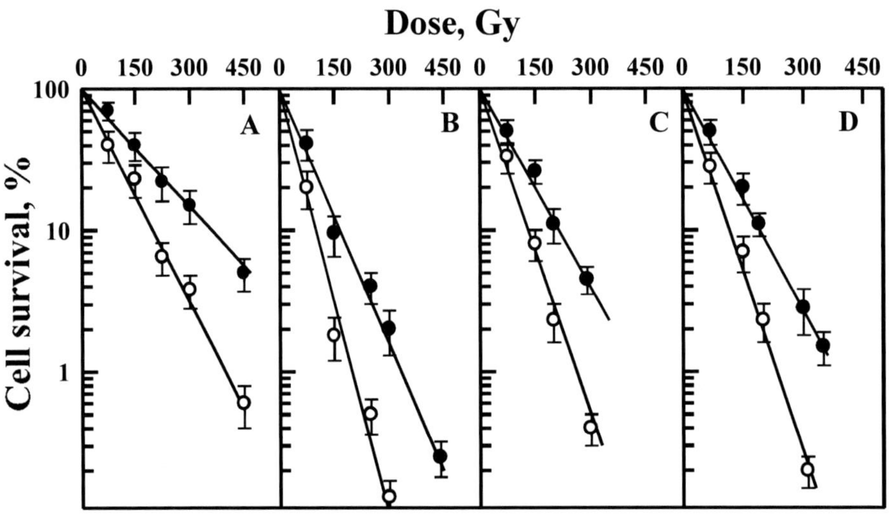

Figure 2.3. Survival curves of *S. cerevisiae* haploid yeast cells of various genotypes: A – S288C (RAD), B – g218/4d (rad50), C – g218/7a (rad51), D – gl60/2b (rad52); cells were irradiated with ^{60}Co gamma rays in the absence (open symbols) and presence (closed symbols) of 0.01 M cysteamine.

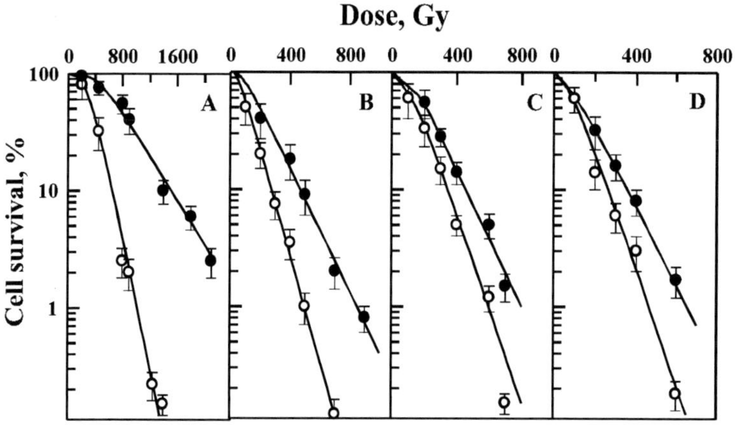

Figure 2.4. Survival of *S. cerevisiae* diploid yeast cells irradiated with ^{60}Co γ-rays without cysteamine (open symbols) and in the presence of 0.01 M cysteamine (closed symbols): A – XS800 (RAD/RAD); B – XS1806 (*rad*50/*rad*50); B – XS806 (*rad*51/*rad*51); D – XS1898 (*rad*52/*rad*52).

Similar experiments were performed with diploid yeast cells. Figure 2.4 shows the survival curves of various wild-type *S. cerevisiae* strains and radiosensitive strains homozygous for the mutant locus. Table 2.7 summarizes the radiobiological parameters of the survival curves of these and some other strains, the results of experiments with which are not shown in Figure 2.4.

Table 2.6. Efficiency of protection of haploid yeast *S. cerevisiae* of various genotypes with cysteamine (0.01 M)

Strain	Mutant locus	D_0, Gy without cysteamine	D_0, Gy with cysteamine	DMF
S288C	RAD	85.0	154.0	1.81 ± 0.2
g218/6b	*rad*50	37.0	72.0	1.95 ± 0.3
g218/7a	*rad*51	60.0	101.0	1.69 ± 0.1
g160/2b	*rad*52	57.0	91.0	1.59 ± 0.1
g151/12c	*rad*53	79.0	137.0	1.73 ± 0.2
g218/7c	*rad*54	58.0	83.0	1.42 ± 0.3
X1687/101b	*rad*55	57.0	126.0	2.20 ± 0.4
30-57-rs31	*rad*56	62.0	98.0	1.58 ±0.1
g246/6b	*rad*57	70.0	123.0	1.78 ± 0.2

It is known that wild-type diploid cells, as well as some *rad* mutants (*rad*53, *rad*55, *rad*56, *rad*57), although they were more sensitive to the action of ionizing radiation in comparison with that for cells of the original parental strain, retained the ability to recover during the post-radiation period in a non-nutritive environment at 30°C. Other strains homozygous for *rad* mutations (*rad*50, *rad*51, *rad*52, *rad*54) did not have such ability (Petin, Kim, 2016).

From the presented data, it can be seen that the effectiveness of radiation protection with cysteamine was higher for wild-type cells and some mutants that retained the ability to recover. Calculations for a large set of diploid yeast cells of various genotypes, including the results shown in Table 2.7, show that the parameters D_0 with cysteamine and D_0 without this protector correlate with each other (correlation coefficient r = 0.98). This fact is an additional confirmation of the point of view that the effectiveness of the action of radioprotectors is associated with cell radiosensitivity and cell ability to recover from radiation damage, and thereby confirms the hypothesis under discussion about the participation of post-radiation recovery systems in the mechanism of action of radioprotectors.

Table 2.7. Efficiency of radioprotection action of cysteamine (0.01 M) for diploid yeast *S. cerevisiae* of various genotypes

Strain	Mutant locus	D_0, Gy without cysteamine	D_0, Gy with cysteamine	DMF
XS800	RAD/RAD	178.0	495.0	2.78 ± 0.4
XS1806	*rad50/rad50*	89.0	178.0	2.00 ± 0.3
XS806	*rad51/rad51*	34.0	47.0	1.38 ± 0.1
XS1898	*rad52/rad52*	32.0	48.0	1.40 ± 0.2
XS1889	*rad53/rad53*	72.0	137.0	1.90 ± 0.2
T3	*rad54/rad54*	23.9	30.4	1.27 ± 0.1
XS1935	*rad55/rad55*	80.0	169.0	2.11 ± 0.2
XS1943	*rad56/rad56*	137.0	300.0	2.19 ± 0.3
XS1988	*rad57/rad57*	134.0	268.0	2.00 ± 0.2

Figure 2.5 demonstrates the dependence of the survival of clonogenic bone marrow cells in mice (Figure 2.5, A) and the effectiveness of the radioprotective action (DMF) of the T1023 compound (Figure 2.5, B) on the dose of ionizing radiation. It can be seen that the exponential curve of the survival of bone marrow cells after irradiation of mice without radioprotector transforms into a sigmoid one after irradiation of mice in the presence of a radioprotector. Due to this fact, the DMF is about 1.4 in the region of high doses and more than 4.0 in the region of the sigmoid shoulder of the dose-effect curves recorded at low doses of ionizing radiation. In accordance with the classical concepts of radiobiology (Dertinger, Jung, 1973; Yarmonenko et al., 1992; Joiner, van der Kogel, 2009; Chapman et al., 2012; Hall, Giaccia, 2018), these data indicate the recovery of cells from sublethal damage in the area of low doses of ionizing radiation only in the presence of a radioprotector. This demonstrates the role of cell recovery from radiation damage in the mechanism of the radioprotective action of the T1023 compound on the survival of clonogenic cells in irradiated animals.

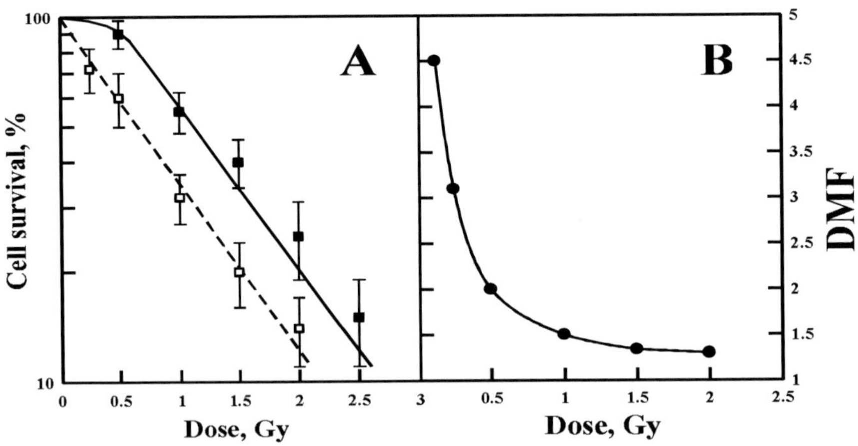

Figure 2.5. The dependence on the radiation dose of the survival of clonogenic bone marrow cells in mice (A) and the effectiveness of the radioprotective action (DMF) of the T1023 compound (B); mice were irradiated in the absence (open symbols) and presence (closed symbols) of the T1023 compound.

Data upon the dependence of the radioprotective effectiveness on cell genotype that controls the recovery processes in bacterial and yeast cells, served as the basis for the development of kinetic schemes for the development of radiation damage, in which considerable attention is paid to the participation of cell recovery from radiation damage in the radiosensitizing and radioprotective action of chemical agents (Kiefer, 1974, 1976; Eidus, 1984; Eidus, Korystov, 1984; Rozhdestvensky, 1984, 1985; Petin, Kim, 2016). In a simplified form, this scheme is shown in Figure 2.6. Because of the direct (*hv* – quantum energy) or indirect ($OH^{\cdot}$ radical) action of ionizing radiation on the target M, primary radiation damage $M^{\cdot}$ produce, which can interact with the protector (SH) or with oxygen (O_2). In the first case, MH states are formed, which are lethal in themselves, but amenable to enzymatic repair (recovery constant β). In the case of interaction of the primary damage $M^{\cdot}$ with oxygen, it becomes irreversibly lethal. At the physicochemical stage of radiation damage formation, the proportion of potentially lethal damage (α) can be determined as $\alpha = MH/M^{\cdot}$ that can be restored by enzymes at the biochemical stage of repair and fixation of radiation damage. So, according to this mechanism, the probability of manifestation of the radiation effect is determined by the product of $\alpha \cdot \beta$ probabilities of physicochemical and enzymatic reparation. The presented results on the effect of cysteamine on yeast cells of various genotypes are qualitatively explained by the described mechanism. The presence of radioprotector at the time of irradiation increases the probability α of damage recovery at the physicochemical stage of damage formation to the same extent for cells of different genotypes. The difference in the effectiveness of the protection of cells capable of recovery and cells without this ability to a lesser extent is explained by the different ability of cells to recover from potentially lethal damage (probability β) at the biochemical stage of damage formation.

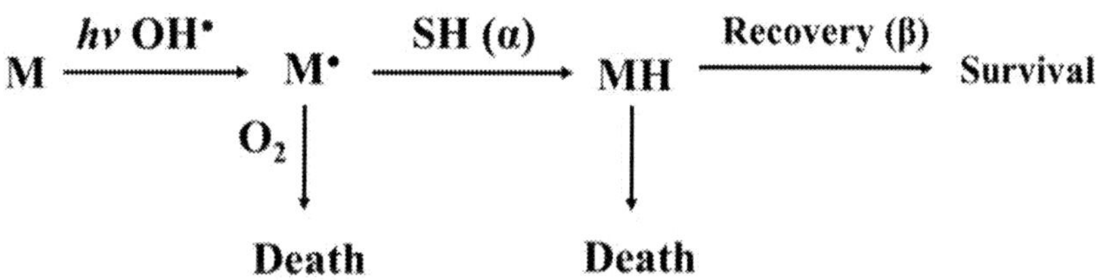

Figure 2.6. Kinetic scheme of the radiation damage development with the participation of the processes of enzymatic recovery of cells.

It is logical to assume that in order to increase the efficiency of radioprotector action, the possibilities of cell to recover from radiation damage should be taken into account. In a recent publication, it was shown that, after X-ray irradiation, the recovery of *S. cerevisiae* cells enriched with the magnetic isotope of magnesium (^{25}Mg) proceeds twice as fast as the recovery of cells enriched with the nonmagnetic isotope (^{24}Mg) (Avdeeva et al., 2018). At the same time, the proportion of irreversible damage in cells enriched with ^{25}Mg is 60% less than in cells enriched with ^{24}Mg. Since the recovery processes take part in the mechanism of the implementation of radioprotectors, the authors admit in principle the possibility of creating new radioprotectors based on stable magnetic isotopes.

Note that the action of many radioprotectors at the organismic level is explained by the formation of hypoxia in tissues. In such a case, the so-called oxygen effect is observed, the mechanism of which is also associated with the ability of cells to recover from radiation damage

(Kiefer, 1974, 1976; Burlakova et al., 1975; Eidus, 1984; Eidus, Korystov, 1984; Rozhdestvensky, 1984, 1985; Petin, Komarov, 1989; Petin, Kim, 2016). That is why the value of the coefficient of oxygen enhancement ratio, which characterizes the efficiency of radiation sensitization by oxygen, has higher values for diploid yeast cells with a full-fledged repair system and lower values for wild-type haploid cells and radiosensitive mutants, both haploid and diploid types (Petin, Ryabchenko, 1982; Petin, 1983; Petin, Komarov, 1989; Petin et al., 2020). In these studies, for yeast cells of various species and genotypes, the relationship between the oxygen effect and hypoxic sensitizers with the ability of cells to post-radiation recovery was demonstrated. It has been shown that a differential increase in radiosensitivity for cells with different β is observed under the influence of both oxygen and hypoxic sensitizers.

The analysis shows that the effectiveness of the radioprotective action, as well as the efficiency of enhancing of ionizing radiation action by oxygen and electron-acceptor compounds are determined not only by a change in the probability of the lethal and potentially lethal radiation damage production due to interaction at the physicochemical level, as was traditionally assumed, but also by a decrease (protectors) or an increase (oxygen, hypoxic sensitizers) of the likelihood of manifestation of already formed potentially lethal damage due to the functioning of post-radiation recovery systems. In other words, these data indicate the existence of a mechanism for modifying the radiosensitivity of cells, which is common for the studied compounds, and this mechanism is associated with the ability of cells to recover from radiation damage and is more pronounced for cells capable to recover from radiation damage.

2.5. Radiobiological Examples of Yeast Cell Genetic Instability

The dependence of yeast cells survival on the dose of ionizing radiation is only one of many cell responses to radiation. In accordance with the probability model (Kapultsevich, Petin, 1977; Kapultsevich, 1978), which combines the hit and target principles (Timofeeff-Ressovsky, Zimmer, 1947; Lea, 1963) with biological stochastics (Hug, Kellerer, 1969), both the dying and surviving parts of irradiated cells are characterized by the heterogeneity of the observed cell responses to radiation. Not very successful terms "probability model" and "biological stochastics" were introduced by the authors of these mathematical models and were fixed in radiobiology. The variety of effects associated with the action of ionizing radiation on the reproductive capacity of cells is manifested in the fact that cells can die both immediately after irradiation (interphase death) and after a different number of divisions (proliferative death). Diploid yeast cells surviving after exposure to ionizing radiation produce on a solid nutrient medium visible macrocolonies, which differ from each other and from control by their size, morphology, the presence of dead cells in them, the timing of macrocolony formation, the ability of cells to post-radiation recovery (Korogodin, 1966; Kapultsevich, 1978; Korogodin et al., 2007; Petin et al., 2008; Zhurakovskaya et al., 2009). In somatic cells of mammals that survived after irradiation, a high yield of lethal mutations was also observed (Seymour et al., 1986). Consequently, the cells that survived after irradiation and form colonies later than non-irradiated cells inherit some sub-damage, the consequences of which are manifested long time after irradiation.

It is known that the effect of the delayed colony formation by yeast cells surviving after irradiation, which is one of the manifestations of genetic instability, is well pronounced for diploid cells capable of repair double-strand DNA breaks induced by ionizing radiation. On the contrary, this effect is expressed to a much lesser extent for haploid cells incapable of post-radiation recovery when exposed to irradiated cells in a non-nutritive medium (Kapultsevich et al., 1972; Kapultsevich, 1978; Korogodin et al., 2007; Korogodin, 2010).

The most of damage induced by ionizing radiation, including damage leading to mutations, occurs during cell division. Then it can be expected that in the distant descendants of cells that survived after irradiation, the delayed formation of colonies on a nutrient medium and an increased yield of various mutations should be observed. It is reasonable to assume that such offspring contains some damage ineffective for cell inactivating. By analogy with ineffective damages, which are responsible for the synergistic interaction of various environmental factors (Petin et al., 2012, 2020; Evstratova et al., 2018 b; Petin, Kim, 2016), we will also call them sub-damages. Moreover, the later the macrocolonies are formed in comparison with the control, the more such sub-damages they should contain. Accordingly, the yield of mutant cells should also be increased. It is of interest to check the fulfillment of this conclusion.

Data on the content of respiratory mutants in colonies formed at different times after the control cells are given in Table 2.8. These experiments were performed with the diploid yeast *Saccharomyces ellipsoideus*, strain Megri 139-B (Korogodin et al., 2007; Petin et al., 2008).

Table 2.8. The content of respiratory mutants in clones produced by diploid yeast cells surviving after exposure to ionizing radiation (γ-rays of ^{60}Co, dose rate 10 Gy/min)

Dose, Gy	Time of colony formation later than control, days	Colonies containing different numbers of respiratory mutants, %			
		0–2%	3–9%	10–49%	50–100%
0	0	100	–	–	–
200	1	100	–	–	–
200	2	83	12	5	–
200	3	20	61	19	–
600	1	80	16	2	2
600	2	53	12	20	15
600	3	32	16	28	14

To identify respiratory mutants, the technique of colonies flooded with tetrazole (2,3,5-triphenyltetrazolium chloride) was used. Colonies with normal respiration colored in red, while colonies consisting of respiratory mutants remain white (Ogur et al., 1957). It can be seen that as a result of irradiation with ionizing radiation (γ-rays of ^{60}Co, 10 Gy/min), the content of respiratory mutants increases in all clones formed after the control. This is especially noticeable for clones, the formation of which is delayed for several days in comparison with the control. The content of respiratory mutants is several times higher than that in clones formed one day later than control.

To study mitotic recombination, we used the diploid yeast *Saccharomyces cerevisiae*, strain 5a3bα, heterozygous for adenine loci (Korogodin et al., 2007; Petin et al., 2008). This strain was synthesized by us by cultivating haploid strains with different types of crossing in a liquid nutrient medium. The yield of diploid strains was 30–40%. The cells of this strain produced smooth white colonies on a solid nutrient medium, while recombinants for the *ade*1 and *ade*2 mutations formed smooth, wholly or partially red colonies. The cells were irradiated with ^{60}Co gamma rays at a dose of 300 Gy. Table 2.9 shows the data on the content of recombinant cells in the control and in the offspring of yeast cells that survived after irradiation and that appear at different times later than in control. It can be seen that in this case, the same regularity is fulfilled as in experiments with the yield of respiratory mutants – the later the colonies are produced in comparison with the control, the greater number of recombinant colonies can be observed.

Table 2.9. The content of recombinant cells produced by yeast cells surviving after irradiation

Time of colony formation later than in control, days	Total number of clones used	Summary number of colonies produced	Content of colored recombinants			
			Full	Sectoral	Sum	% of all colonies
0	13	768	1	2	3	0,4
1	59	6966	181	33	214	3,1
2	34	3695	267	29	296	8,0
3	16	927	158	32	190	20,0

The appearance among macrocolonies produced by irradiated cells various morphologically modified colonies containing altered regions (sectors) or completely altered colonies as compared to the control may also indicate chromosomal (genetic) cell instability. In some works, such modified colonies were called mutants. To designate them, Nadson often used the term "saltants" (Korogodin, 2010). Radiation damage inherited by distant descendants of cells that survived after irradiation can be attributed to the manifestation and persistence of some non-lethal sub-damages inherited by irradiated cells. Table 2.10 shows the content of morphologically altered colonies among clones formed by irradiated (γ-rays of ^{60}Co, 600 Gy) diploid yeast cells *Saccharomyces ellipsoideus*, strain Megri 139-B.

Table 2.10. The content of morphologically altered colonies among colonies, formed by irradiated diploid yeast cells *Saccharomyces ellipsoideus*, strain Megri 139-B

Time of colony formation later than in control, days	Total number of clones used	Summary number of colonies produced	The number of morphologically changed colonies	
			Total number	Percent
0	28	4763	53	1.1
1	50	2978	496	16.6
2	22	2399	2265	94.4
3	28	3717	3077	82.8

In general, the presence in distant offspring of cells that survived after irradiation respiratory mutants (Table 2.8), recombinant cells that arise after irradiation with γ-rays (Table 2.9), or morphologically altered colonies (Table 2.10) can be considered as radiobiological evidence (illustrations) of the genetic instability of yeast cells surviving after irradiation.

Figure 2.7 shows an example of colonies formed by diploid yeast cells *Saccharomyces cerevisiae*, strain XS800, on a solid nutrient medium by control unirradiated cells (A) and cells grown after irradiation at a dose that reduces the survival rate to 10% (B).

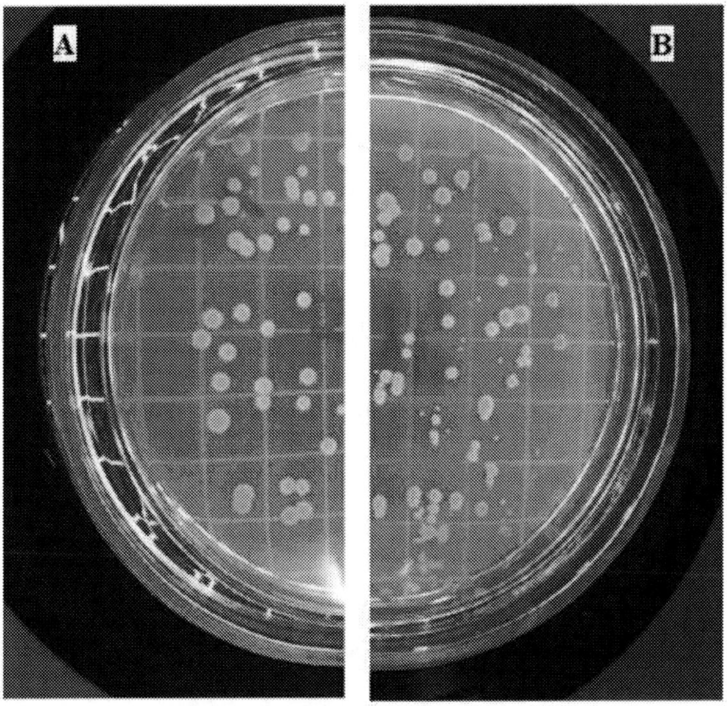

Figure 2.7. Colonies produced on a solid nutrient medium by cells from a control suspension of *S. cerevisiae* diploid yeast cells (A) and cells that survived after irradiation at a dose that reduces the survival to 10% (B).

Unirradiated yeast cells appear approximately synchronously in 2–3 days and form large colonies of approximately the same size, round, smooth, white or light cream in color. After irradiation, the formation of colonies by surviving cells is delayed, and they become visible at different times after irradiation (3–5 days); therefore, the produced colonies are heterogeneous in size. Some of these colonies in color and shape remain similar to the control ones, while many others differ from the control cells either by the presence of altered sectors in the colony, or in size, original shape and color of the entire colony. It is of interest that this heterogeneity is retained during subsequent long-term cell growth (up to 20–30 days). The resulting colonies V.I. Korogodin calls saltants and refers as radioraces (Korogodin, 2010). Some radioraces retained their properties for tens of passages (replating on a fresh nutrient medium). Other radioraces were unstable – during subsequent replating, they continuously produce new forms of colonies. This process was called cascade mutagenesis (Korogodin et al., 2007; Korogodin, 2010), which, in principle, is a phenotypic manifestation of the genetic instability of yeast.

It is clear that the occurrence of heterogeneity of colonies produced by cells surviving after irradiation can be based on the events of a very different nature. It has long been noted that saltants in different organisms can be formed due to one or several gene mutations, various chromosomal rearrangements, cleavage or recombination of genes. These considerations are quite acceptable for diploid yeasts, in which, in addition to the indicated factors, a significant role in race formation may also belong to heritable cytoplasmic variability (Korogodin, 2010).

It should be emphasized that the subsequent re-plating of colonies, produce new radioraces of yeast, originated from colonies grown for long periods after irradiation, and these experiments were performed with yeast cells *Saccharomyces ellipsoideus*, strain Megri 139-B

(Korogodin 2010), characterized by high heterozygosity, when in homologous chromosomes some genes are represented by different alleles. Therefore, we carried out experiments on re-plating of colonies formed by the surviving diploid yeast cells of *Saccharomyces cerevisiae*, strain XS800, characterized by high homozygosity, which was confirmed by tetrad analysis (Saeki et al., 1980). The results are shown in Figure 2.8. In this figure, one can see on the Petri dish the colonies formed by the surviving diploid yeast cells of *Saccharomyces cerevisiae*, strain XS800. In the second row of this figure, the results of repeated re-plating of cells from the colonies designated as 0, 1, 2, and 3. It can be seen that after repeated seeding cells from the colony formed simultaneously with the control (colony 0 in Figure 2.8), most of clones did not differ in their shape and time appearance from the control ones. On the contrary, seeding cells from colonies that grew later than in control (colonies 2, 3, and 4 in Figure 2.8) did not retain uniformity in their appearance and become visible at different times after irradiation (3–5 days), therefore the colonies are very heterogeneous in size.

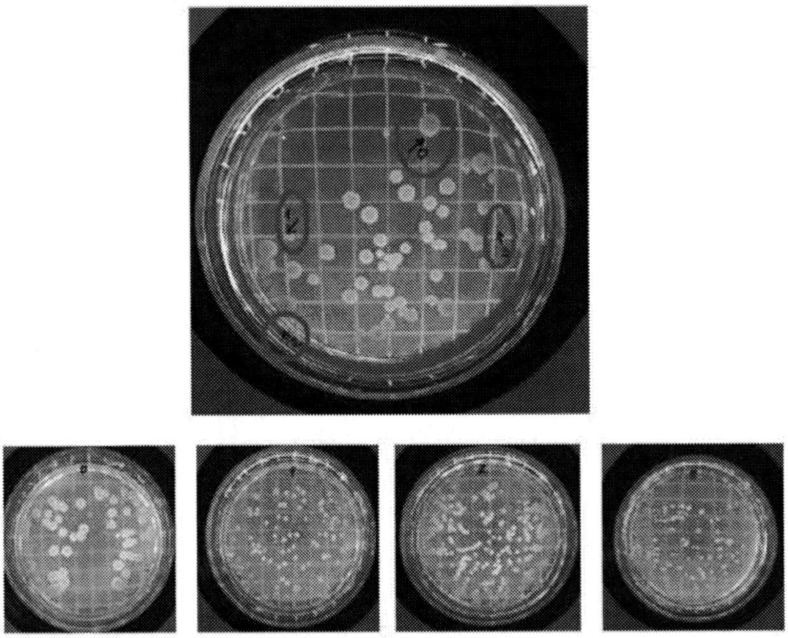

Figure 2.8. Repeated re-place on nutrient media of *Saccharomyces cerevisiae* diploid yeast cells from colonies produced by cells, surviving after irradiation; explanations in the text.

The data on re-plating of the colonies formed by the yeast cells surviving after irradiation indicate the genetic instability of cells from the macrocolonies produced later than in control and are an addition to the above radiobiological illustrations of the genetic instability of yeast.

2.6. Radiation-Induced Genetic Instability and Ploidy of Yeast Cells

Currently, the role of DNA damage in genetic instability, aging, and cancer formation is being actively studied (Cheng, Loeb, 1993; McMurray, Gottschling, 2004; Korogodin, 2010; Shen, 2011; Mulla et al., 2014; Evstratova, Petin, 2018 a, b, c). There are evidence that double-stranded DNA breaks are important for the induction of the delayed reproductive cell death and, as a consequence, for the delayed colony formation by cells surviving exposure to ionizing radiation (Chang, Little, 1992 a, b; Brosh, Bohr, 2007; Korogodin et al., 2007). The slow appearance of colonies, along with other delayed effects induced in individual cells that survive after exposure to ionizing radiation, is a further manifestation of genetic instability of the offspring. This effect occurs in cultured cells of various origins (Klein et al., 1990; Morgan, Murnane, 1995; Little, 1998). Most of the results on genetic instability were obtained with cultured mammalian cells in a diploid state. Yeast cells, as the simplest example of eukaryotic cells, are a convenient object for comparing the lethal effect and genetic instability of haploid and diploid cells, which makes it possible to obtain new fundamental data.

It was shown (Korogodin, 1966; Kapultsevich et al., 1972; Korogodin et al., 2007) that the effect of the delayed colony formation by yeast cells surviving after irradiation is more pronounced for diploid than for haploid cells. This means that radioresistant diploid yeast cells exhibiting sigmoid dose-response curves inherit a number of non-lethal sub-damages responsible for genetic instability. A similar effect is observed to a lesser extent for wild-type haploid yeast cells, characterized by exponential dose-dependent survival curves. On this basis, it was concluded that the shape of the survival curve and the ability of cells to repair DNA damage determines the delayed appearance of colonies. However, in our recent work with wild-type haploid and diploid yeast cells and their radiosensitive mutants, it was noted (Evstratova et al., 2016) that genetic instability, determined by the delay in colony formation, is well pronounced and reaches up to 100% with an increase in the radiation dose for diploid cells regardless on the shape of the survival curve and the ability of cells to recover from radiation damage. This effect was poorly expressed (20–30%) for all haploid strains characterized by exponential survival curves. Based on this, a preliminary assumption was made that the genetic instability caused by ionizing radiation is largely determined by cell ploidy, and not by the shape of the survival curve and the ability of cells to recover from radiation damage. This hypothesis requires additional confirmations. Figure 2.9 presents data on the survival of the haploid yeast *Zygosaccharomyces bailii* (strain K7, curves 1) and the diploid yeast *Saccharomyces ellipsoideus* (strain Megri 139-B, curves 2).

Hereinafter, in the described experiments of this section, the cells were exposed to γ-rays of ^{60}Co at a dose rate of 10 Gy/min in the stationary stage of growth and were plated on a nutrient medium immediately after irradiation. Figure 2.9A shows the dependence of cell survival on the dose of irradiation with sparsely ionizing radiation. It can be seen that the survival curve of haploid yeast is exponential, while for diploid strain is sigmoid. It is known that double-stranded DNA breaks are the main cause of cell inactivation. For yeast cells, it has been shown that even one unrepaired or mistaken repaired double-strand DNA break is lethal (Frankenberg et al., 1981).

Figure 2.9B shows the dependence of the delayed formation of colonies produced by survivors after irradiation of the same haploid (curve 1) and diploid yeast cells (curve 2) on the radiation dose. As a measure of this delay, we used the ratio of the number of colonies formed later than in the control to the total number of colonies. Colonies produced by unirradiated diploid cells appeared on Petri dishes after 24–28 h of growth at 30°C and in haploid cells – after 46–54 h. It can be seen that with an increase in the irradiation dose, the proportion of surviving cells that form colonies later than control increases for haploid yeast strain up to 20-25% and almost reaches 100% for diploid cells.

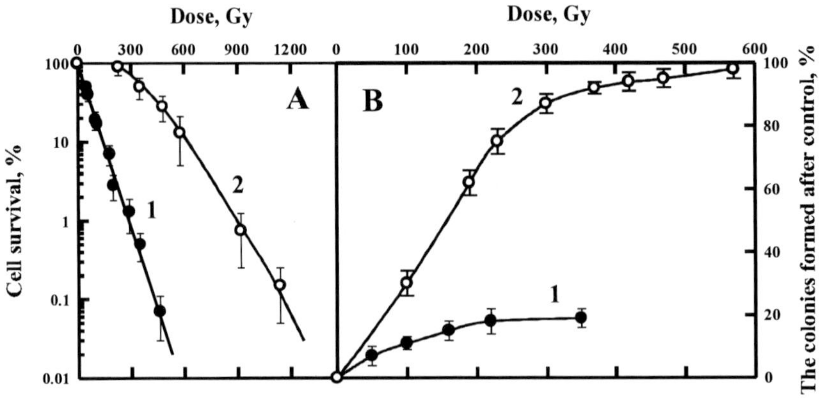

Figure 2.9. A: the dependence of cell survival on the dose of ionizing radiation for haploid yeast *Zygosaccharomyces bailii*, strain K7 (RAD, curve 1), and diploid yeast *Saccharomyces ellipsoideus*, strain Megri 139-B (RAD/RAD, curve 2); B: the dependence of the delayed formation of colonies by cells surviving after cell irradiation on the dose of ionizing radiation for haploid *Z. bailii* yeast cells (curve 1) and diploid yeast cells of *S. ellipsoideus* (curve 2).

It was found that the greater the shape of survival curves differs from exponential, the more pronounced the delayed colony forming ability (Kapultsevich et al., 1972). On this basis, it was suggested that this effect is due to the same damage that leads to inactivation. If the number of such damage is greater than or equal to the critical one, the cell dies. If there are fewer such damage, the cell forms a macrocolony, but more slowly than in the control. Obviously, the more damage is required for cell death, i.e., the greater the shoulder of the survival curve, the more cells with some residual damage should be preserved among the viable individuals, which lead to a delay in the formation of macrocolonies by the cells that survived after irradiation. We call such cell dsmsge, ineffective for cell inactivation by sub-damage. The more such sub-damage persist in the distant descendants of the cell that survived after irradiation, the later the visible colonies are formed and the greater the effect of the delayed colony formation should be observed. The similarity of quantitative patterns of recovery of diploid yeast from lethal damage and sub-damage is also evidence in favor of the assumption about identity of the lesions causing cell inactivation and the delayed growth of colonies (Korogodin et al., 1977a, b). These papers provide evidence of the existence of a recovery effect of irradiated diploid during their incubation after irradiation in a non-nutritive medium. It has been shown that postradiation mitotic dormancy is accompanied by a decrease in the biological consequences of irradiation associated with both inhibition of the first mitosis and a delay in the formation of macrocolonies by cells that survived the irradiation. It is concluded that primary radiation damage can have a potential form and reversible character when the irradiated diploid cells are kept in a non-nutritive environment.

The above reasoning allows us to imagine the following hypothetical scheme of radiation damage to cells. Cells that did not receive effective (lethal) damage or sub-damage during irradiation form macrocolonies at the same time and the same size as in the control. Cells that have not received lethal damage but contain a different number of sub-damages form colonies more slowly than non-irradiated cells. The greater such sub-damage in cell, the slower the formation of a colony occurs. Note that in the experiments just described, genetically non-isogenic haploid and diploid strains of different species were used. Therefore, the following figures (Figure 2.10 and 2.11) show the results of similar experiments with isogenic haploid and diploid yeast cells carrying identical alleles in homologous chromosomes.

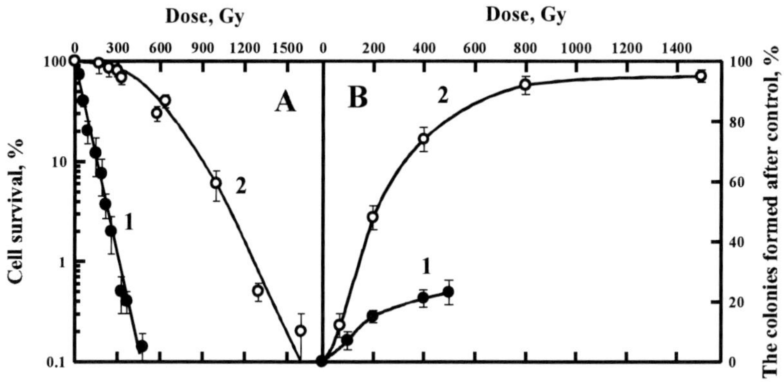

Figure 2.10. A: the dependence of cell survival on the dose of sparsely ionizing radiation for homozygous wild-type yeast cells of *S. cerevisiae*: haploid strain S288C (RAD, curve 1), diploid strain XS800 (RAD/RAD, curve 2); B: the dependence of the delayed colony formation by cells surviving after irradiation on the dose of ionizing radiation for wild-type *S. cerevisiae* yeast cells: haploid strain S288C (curve 1), diploid strain XS800 (curve 2).

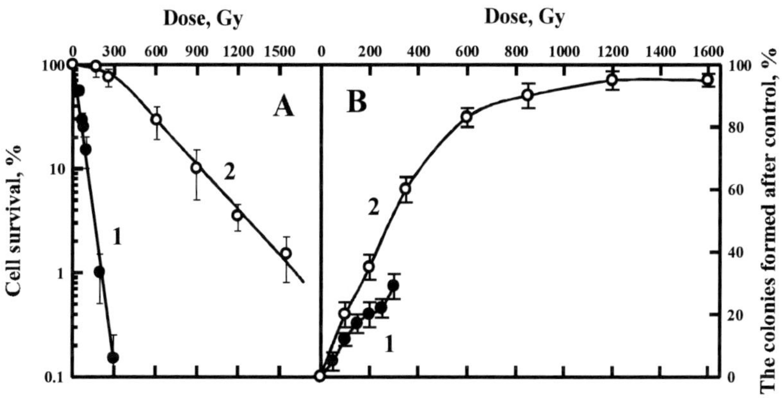

Figure 2.11. A: the dependence of cell survival on the dose of sparsely ionizing radiation for haploid yeast cells of *Saccharomyces cerevisiae*, strain 211-1a (RAD, curve 1) and diploid *Saccharomyces cerevisiae* yeast cells, strain 211 (RAD/RAD, curve 2); B: the dependence of the delayed colony formation by cells surviving after irradiation on the dose of ionizing radiation for wild-type *S. cerevisiae* yeast cells: haploid strain 211-1a (curve 1) and diploid strain 211 (curve 2).

The presented data show that the effect of the delayed colony formation by yeast cells that survived after irradiation is well pronounced for diploid cells of both homozygous strains capable of recovery double-strand DNA breaks induced by ionizing radiation. On the contrary, this effect is expressed to a much lesser extent for haploid cells, which are less capable of restoring radiation damage. These data are consistent with the previously described results (Kapultsevich et al., 1972; Korogodin et al., 1974; Kapultsevich, 1978; Korogodin, 1993; Korogodin et al., 2007; Petin et al., 2008) showing that the delay in formation colonies by cells surviving after irradiation is due to the ability of cells to recover from damage induced by ionizing radiation.

The results shown in Figure. 2.9, 2.10, 2.11 and in Table 2.11 show that wild-type diploid cells are 4.1–6.6 times more resistant than homozygous haploid ones that is due to the greater efficiency of diploid cells in recovering from radiation damage.

Table 2.11. Quantitative parameters characterizing the radiosensitivity of haploid and diploid yeast cells of various types and strains

Strain	The greatest number of colonies produced later than in control, %	LD_{90}, Gy	$\frac{LD_{90}(\text{dipolid})}{LD_{90}(\text{hapolid})}$
K7	15	155 ± 10	4.1
Megri 139-B	100	640 ± 44	
S288C	20	160 ± 15	5.8
XS800	100	930 ± 28	
211-1*a*	17	140 ± 12	6.6
211	100	920 ± 30	

These results are logically interpreted by the scheme for the development of radiation damage in cells described above (Figure 2.6). Cells that did not receive effective (lethal) damage or sub-damage during irradiation produce macrocolonies at the same time and the same size as in the control. Cells that have not received lethal damage but contain a different number of sub-damages, produce colonies more slowly than non-irradiated cells. The more such sub-damages a cell contains, the slower the macrocolony formed after irradiation. The delay in the formation of colonies can be consistently explained from the standpoint of classical concepts based on the principle of hit and target.

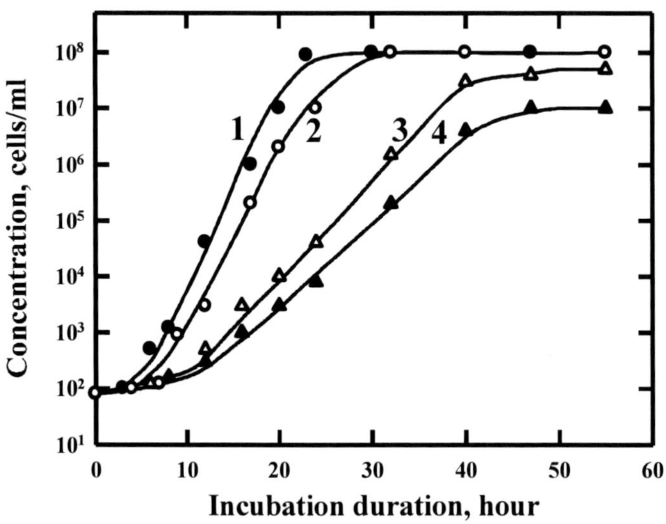

Figure 2.12. Reproduction dynamics of control cells (curves 1, 2) and distant offspring of the diploid yeast *Saccharomyces cerevisiae* surviving after irradiation (curves 3, 4). The cells were cultured in a liquid nutrient medium at 30°C (curves 1, 3) and 37°C (curves 2, 4).

The presence of sub-damages in cells surviving after irradiation is manifested not only in a delay in the formation of colonies. An increased radio- and thermal sensitivity of cells from colonies formed by cells that survived after irradiation was demonstrated, and the effect was more pronounced for colonies formed at a later date as compared to the control (Korogodin et al., 2007; Petin et al., 2008). Note that the probability model of cell inactivation (Kapultcevich, Petin, 1977; Kapultcevich, 1978) predicts that an increase in the cultivation temperature above the optimal one can reduce the rate of reproduction of distant descendants of irradiated cells containing non-lethal sub-damage. Figure 2.12 shows the data on the dynamics of reproduction of control cells and offspring of cells surviving after irradiation in a liquid nutrient medium at various temperatures.

These data show that the rate of control cells reproduction at 37°C is even slightly higher than at 30°C. In the offspring of irradiated cells, taken from their colonies formed 2 days later than the control, the reproduction rate, on the contrary, is higher at a cultivation temperature of 30°C compared to the reproduction rate at 37°C, which corresponds to the prediction of the probability model and indicates the presence of some sub-damages in distant offspring, which are manifested to a greater extent at a non-optimal cultivation temperature. This effect may be due to the appearance of non-viable offspring in the cells surviving after irradiation during their reproduction. These data may also indicate the manifestation of genetic instability – some cells continue to multiply, as in the control, others undergo proliferative death – death after several divisions, and others are characterized by interphase death – die without any division (Petin et al., 2008).

It was shown (Korogodin et al., 2007; Petin et al., 2008) that the content of non-viable cells in clones formed by cells that survived after irradiation depends on the degree of delay in the formation of macrocolonies by irradiated cells and on the temperature of cultivation of irradiated cells. Colonies formed 1 day later than the control contained 20% of non-viable cells under optimal cultivation conditions (30°C, standard nutrient medium) and 43% under non-optimal cultivation conditions (37°C, nutrient medium contained 7% sodium chloride). For colonies formed 2 days later than control, these values were 27 and 54%, and on the 3rd day – 36 and 87%. It is clear that the greater the delay in the formation of colonies, the more the relative content of non-viable cells in the colony, formed by surviving cells. These observations are in good agreement with the assumption that ionizing radiation induces in the cell some sub-damage ineffective for cell inactivation, disrupting the genetic stability and reliability of cells surviving after irradiation.

The consequences of the presence of sub-damage in the cells are also manifested in other deviations of the cultivation conditions from the optimal ones, for example, the composition of the nutrient medium. Figure 2.13 shows the dependence of cell survival on the radiation dose for haploid yeast *Z. bailii*, strain K7 (A) and diploid yeast *S. ellipsoideus*, strain Megri 139-B (B), grown at different temperatures of post-radiation cultivation. It can be seen that an increase in the temperature of cultivation increases the sensitivity of diploid, but not haploid cells. These data also indicate an increase of radiation damage in diploid cells due to the interaction of damage induced by elevated temperature with pre-existing sub-damage in diploid yeast, which are manifested to a much lesser extent in haploid strains.

Note that in some publications the term "lethal sectors" was introduced for inactivated cells within the former colony, which are more common in diploid rather than haploid cells (James, Werner, 1966; Korogodin, 2010). It is clear that such "partial" inactivation, especially if it spreads to many cell generations, should lead to a slowdown in the formation of macrocolonies. A certain role in the effect of the delayed colony formation should be played by the formation of slowly multiplying clones by irradiated cells, which in some cases may be associated with a change in their heredity. Consequently, the effect of the delayed colony formation can be complex in nature.

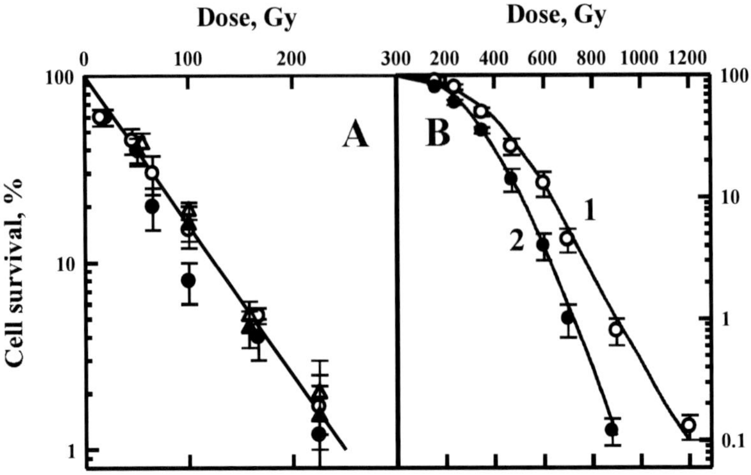

Figure 2.13. The dependence of cell survival on the dose of sparsely ionizing radiation (γ-rays of ^{60}Co, 10.8 Gy/min) of the haploid yeast *Z. bailii* (A) and diploid yeast *S. ellipsoideus* (B), grown at different temperatures of post-irradiation cultivation – 20 (open triangles), 30 (open circles), 34 (closed triangles) and 37°C (closed circles).

Starting to analyze the shapes of the survival curves of haploid and diploid yeast cells, their ability to post-irradiation recovery and the delayed colony formation by surviving cells after exposure to sparsely ionizing radiation, we proceeded from the simplest assumption based on hit and target theory, according to which different radiobiological reactions can be caused by a different number of damages. From the standpoint of this theory, we were able to explain the relationship between the forms of the survival curves of yeasts of different species and strains and the peculiarities of the manifestation of the delayed colony formation by cells that survived after irradiation. Because of irradiation, a different number of qualitatively different lesions (damage to different targets) can arise in a cell. The critical number of damages resulting to cell inactivation can be depend on the cultivation conditions, that lead to inactivation of the cell after one or several divisions (reproductive death). If the number of damages is not lethal for cell under the given conditions, it can produce a colony that grows more slowly than in control. Irradiated cells that have not received any damage produce colonies of the same size and at the same time as non-irradiated ones.

2.7. Genetic Instability of Wild-Type Yeast Strains and Their Radiosensitive Mutants

The role of genetic instability in cell radioresistance is widely discussed in various publications (Morgan, Murnane, 1995; Kaplan, Morgan, 1998; Petin et al., 2008; Lee et al., 2015). A number of radiobiological effects recorded with some delay after irradiation, for example, chromosomal aberrations, reproductive cell death, malignant transformation, decreased cloning efficiency, and heterogeneity of the offspring of irradiated cells illustrate the genetic instability of irradiated cells (Stirling et al., 2011; Kumaran et al., 2013; Skoneczna et al., 2015). However, there are some contradictions on this issue. Thus, the observation that double-strand DNA breaks are not restored in the CHO mutant (xrs-5) and genetic instability (delayed appearance of colonies) is not recorded confirms the idea that DNA repair should play an important role in genetic instability (Chang, Little, 1992 a, b). On the other hand, radiosensitive mutant human cell lines (xeroderma pigmentosa, Fanconi anemia, and ataxia-telangiectasia), which are defective in repairing radiation damage in DNA, demonstrate genomic instability (Gatti et al., 1991; Kalb et al., 2006; Sugasawa, 2008).

The results described above for the survival and delayed appearance of colonies by isogenic haploid and diploid wild-type *Saccharomyces cerevisiae* yeast cells, illustrating genetic instability, partially confirm the previously published data (Korogodin, 1966; Kapultsevich et al., 1972; Korogodin et al., 2007; Petin et al., 2008). The results of these studies show that this effect is more pronounced for diploid than for haploid yeast strains (100% versus 20%). In our publications (Evstratova, Petin, 2018 a, b, c) it was noted that these data require clarification and it is not excluded that the time has come to change this paradigm. We assumed that the effect of the delayed colony formation is not associated with the dynamics of survival, characterized by the sigmoid shape of the survival curve and the ability of cells to recover from radiation damage, but is completely determined by the ploidy of the cells. To test this assumption, one should use haploid and isogenic diploid mutant strains characterized by exponential survival curves regardless of their ploidy. Such studies are not described in the literature. Therefore, it is of interest to compare the delay in the formation of colonies (genetic instability) of wild-type haploid and isogenic diploid yeast strains surviving after irradiation with their radiosensitive mutants, the ability of which to post-irradiation recovery is reduced or completely absent.

Figure 2.14 shows the results of our experiments for strains with *rad*51 mutation. Figure 2.15 for strains with *rad*52 mutation and Figure 2.16 – for strains with *rad*54 mutation. These strains were selected as the most sensitive among the known mutants of the *Saccharomyces cerevisiae* yeast cells.

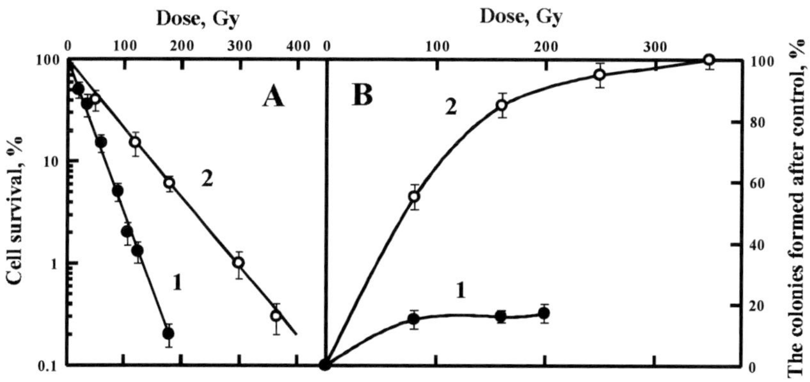

Figure 2.14. A: the dependence of cell survival on the dose of ionizing radiation for radiosensitive mutants of yeast cells *S. cerevisiae:* haploid strain g218/7a (*rad*51, curve 1) and isogenic diploid strain XS806 (*rad*51/*rad*51, curve 2). B: the dependence of the delayed colony formation by cells surviving after irradiation on the dose of ionizing radiation for radiosensitive mutant of *S. cerevisiae* yeast cells: haploid strain g218/7a (*rad*51, curve 1) and isogenic diploid strain XS806 (*rad*51/*rad*51, curve 2).

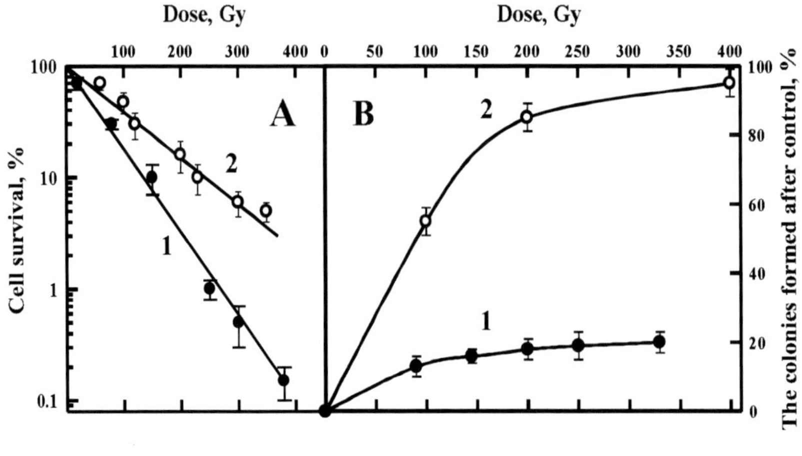

Figure 2.15. A: the dependence of cell survival on the dose of ionizing radiation for radiosensitive mutant yeast cells *S. cerevisiae*: haploid strain g160/2d (*rad*52, curve 1) and isogenic diploid strain XS1898 (*rad*52/*rad*52, curve 2). B: the dependence of the delayed colony formation by cells surviving after irradiation on the dose of ionizing radiation for radiosensitive mutant yeast cells *S. cerevisiae*: haploid strain g160/2d (*rad*52, curve 1) and isogenic diploid strain XS1898 (*rad*52/*rad*52, curve 2).

These figures show the survival curves of isogenic haploid and diploid radiosensitive yeast cells that have lost the ability to recover due to the absence of diploid-specific recovery (Saeki et al., 1980). Table 2.12 shows the quantitative parameters characterizing the radiosensitivity of haploid and diploid yeast cells *S. cerevisiae* of different genotypes. It can be seen that in this case the difference in radiosensitivity between the diploid and haploid strains was 2.3, 1.7, and 2.1 for the strains with the *rad*51, *rad*52, and *rad*54 mutations, respectively. The same difference for wild-type cells was 5.8. These data show that the increase in the radiosensitivity of mutant strains is largely due to the increase in the sensitivity of the diploid rather than the haploid strains. This can be explained by the absence of the diploid-specific repair in them, which is characteristic of wild-type diploid cells irradiated at the stationary stage of growth and kept in a non-nutrient medium during the post-irradiation period (Korogodin, 1966, 2010).

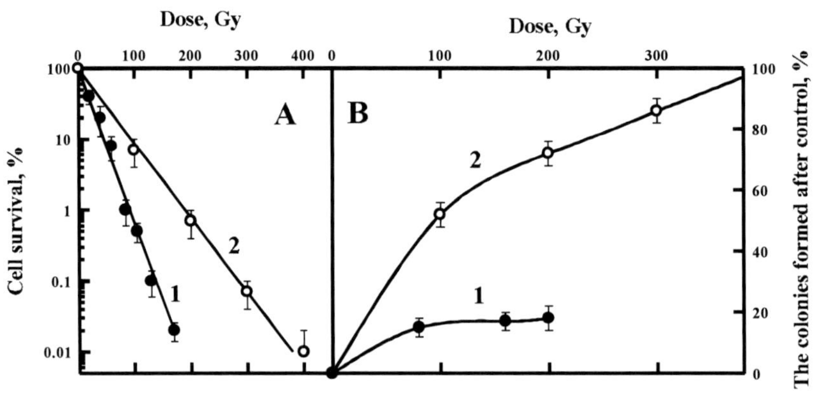

Figure 2.16. A: the dependence of cell survival on the dose of ionizing radiation for radiosensitive mutant yeast cells *S. cerevisiae*: haploid strain g218/7c (*rad*54, curve 1) and isogenic diploid strain L9 (*rad*54/*rad*54, curve 2). B: dependence of the delayed formation of colonies that survived after irradiation on the dose of ionizing radiation for radiosensitive mutant *S. cerevisiae* yeast cells: haploid strain g218/7c (*rad*54, curve 1) and isogenic diploid strain L9 (*rad*54/*rad*54, curve 2).

Table 2.12 shows the quantitative parameters characterizing the radiosensitivity of haploid and diploid yeast cells *S. cerevisiae* of different genotypes. It can be seen that in this case the difference in radiosensitivity between the diploid and haploid strains was 2.3, 1.7, and 2.1 for the strains with the *rad*51, *rad*52, and *rad*54 mutations, respectively. The same difference for wild-type cells was 5.8. These data show that the increase in the radiosensitivity of mutant strains is largely due to the increase in the sensitivity of the diploid rather than the haploid strains. This can be explained by the absence of the diploid-specific repair in them, which is characteristic of wild-type diploid cells irradiated at the stationary stage of growth and kept in a non-nutrient medium during the post-irradiation period (Korogodin, 1966, 2010).

Table 2.12. Quantitative parameters characterizing the radiosensitivity of haploid and diploid yeast cells *S. cerevisiae*

Strain	Genotype	*n*	LD_{90}, Gy	$\frac{LD_{90}(\text{dipl.})}{LD_{90}(\text{hapl.})}$	$\frac{LD_{90}(wild)}{LD_{90}(\text{mutant})}$
S288C	RAD	1	160 ± 15	5.8	–
XS800	RAD/RAD	30	930 ± 28		–
g218/7a	*rad*51	1	65 ± 5	2.3	2.5
XS806	*rad*51/*rad*51	1	150 ± 12		6.2
g160/2d	*rad*52	1	132 ± 14	1.7	1.2
XS1898	*rad*52/*rad*52	1	220 ± 19		4.2
g218/7c	*rad*54	1	45 ± 6	2.1	3.5
L9	*rad*54/*rad*54	1	93 ± 11		10.0

The results are shown in Table 2.12 and in Figures 2.14, 2.15, and 2.16 show that the mutant strains were more sensitive than their parent wild-type strains by a factor of 6.2 and 2.5 for diploid and haploid strains with the *rad*51 mutation, 4.2 and 1.2 for strains with the mutation *rad*52, and 10.0 and 3.5 for *rad*54 strains. Figures 2.14, B, 2.15, B, and 2.16, B show the dependence of the delayed formation of colonies formed by survivors after irradiation with the same haploid (curves 1) and diploid (curves 2) yeast cells on the radiation dose. It can be seen that with an increase in the irradiation dose, the proportion of surviving cells that form colonies later than the control increases for mutant haploid strains with an increase in the irradiation dose up to 20%, as was the case for the wild-type haploid strain. If the main postulate about the relationship between the effect of the delayed colony formation and the dynamics of survival, characterized by a certain shape of the curve were valid, then one would expect a weak manifestation of this effect for radiosensitive diploid mutants, also characterized by exponential curves of survival. However, this turned out not to be the case (Figures 2.14, B, 2.15, B, and 2.16, B) – the delay in the formation of colonies by cells that survived after irradiation is well pronounced and reaches 100% for radiosensitive diploid cells characterized by exponential survival curves.

The set of our new results presented in this section for radiosensitive mutants show that irradiated cells of radiosensitive diploid strains, characterized by an exponential shape of the survival curve, are also characterized by a significant delay in the formation of colonies by cell surviving after irradiation. This effect reaches 100% with increasing dose, as well as for cells capable of recovering from radiation damage. In other words, the observed effect of the delayed colony formation, reflecting the phenotypic manifestation of genetic instability, is not unambiguously associated with a certain shape of the survival curve and cell ability to recover from double-stranded DNA breaks. These data directly show that the genetic instability of yeast cells is mainly determined by cell ploidy, but not by the sigmoid shape of the survival curve and the cell ability to post-irradiation recovery, as was repeatedly postulated earlier (Korogodin, 1966; Kapultsevich et al., 1972; Kapultsevich, 1978; Korogodin, 2010; Korogodin et al., 2007; Petin et al., 2008). Indeed, despite the defects that block the recovery of cells from radiation damage, diploid cells, in contrast to haploid ones, demonstrate significant genetic instability – 100 and 20%, respectively. These data definitely indicate that the genetic instability caused by sparsely ionizing radiation is largely determined by the cell ploidy, but not by the shape of the survival curve and cell ability to recover from radiation damage, as traditionally postulated.

Before proceeding to the analysis of possible mechanisms of genetic instability, it is useful to consider whether the revealed new regularities persist under the action of densely ionizing radiation and UV light.

Chapter 3

Genetic Instability of Yeast Cells after Exposure to Densely Ionizing Radiation

Abstract

The main regularities of biological effects after exposure to ionizing radiation with high LET are presented. The mechanisms of the different shapes of the survival curves after exposure to radiation with various LET as well as the dependence of RBE on the LET of radiation are described and discussed. The differences in the change of the survival curves shape for yeast cells and cultured mammalian cells with an increase in LET are qualitatively explained. A distinctive feature of the effect of charged particles on cell genetic structures is the formation of cluster DNA damage, from which it is difficult to cell recover due to problems with the repair of severe DNA damage. New experimental data are presented that confirm the participation of recovery processes in the manifestation of the RBE of densely ionizing radiation both for cell inactivation and for radiation-induced genetic instability. The identity of the RBE of α-particles for both of these tests means that the formation of lethal damage is accompanied by the proportional formation of sub-damage responsible for the genetic cell instability.

3.1. Patterns of Biological Effects after Ionizing Radiation with High LET

The study of the RBE of densely ionizing radiation for various biological tests is an urgent problem of modern radiobiology. Indeed, half of the natural human exposure to ionizing radiation is due to high LET radiation resulting from the decay of the widespread radionuclide radon and its daughter products. It is well-known the advantages of application of densely ionizing radiation in the oncological disease treatment – increased RBE values, a significant decrease in the ability of cells to recover from potentially lethal and especially from sublethal radiation damage. The circumstance is extremely important at low doses of ionizing radiation used in dose fractionation in radiation therapy, which can significantly reduce the radiation load on patients (Yarmonenko et al., 1992; Lindenbraten, Korolyuk, 2000; Martin, 2001; Saha, 2006; Dale et al., 2009; Gulidov, Mardynskiy, 2010; Tsyb, Mardynskiy, 2010; Kaprin et al., 2017; Hall, Giaccia, 2018; Kaprin, Mardynskiy, 2018). Of particular interest are heavy ions, which create very high LET values at the end of their way (Bragg peak) (Raju, 1980; Choi, Kang, 2012). Another advantage is the low scattering of charged particles along the track, which reduces the likelihood of normal tissue damage during radiation therapy. RBE values for low doses of various types of ionizing radiation are of considerable relevance, both from the point of view of their use in radiation therapy (Yarmonenko et al., 1992; Lindenbraten, Korolyuk, 2000; Dasu, Toma-Dasu, 2008; Joiner, van der Kogel, 2009; Franken et al., 2011; Hall, Giaccia, 2018), for radiation safety (Kogl, 1986) and radioecology (Budarkov et al., 2017). The RBE problem is currently important for ensuring the safety of space flights (Grigoriev, 1982; Krasavin, 1989; Kudryashov, 2004). Assessment of biological effects induced by densely ionizing radiation has acquired particular relevance after the accident at the Chernobyl nuclear power plant and in Fukushima (Glushchenko, 2010; Ivanov, Kaprin, 2015; Anisimov et al., 2018).

Many works are devoted to the study to the biological action of ionizing radiation with various LETs (Frankenberg et al., 1981; Petin, Kabakova, 1981; Ward, 1981; Frankenberg-Schwager et al., 1984; Petin, 1987; Krasavin, 1989; Petin, Petin, 1995; Martin, 2001; Urushibara et al., 2008; Jones et al., 2011; Fujii et al., 2013; Mairani et al., 2013; Petin, Kapultcevich, 2014; Krasavin, 2016; Petin, Kim, 2016; Hall, Giaccia, 2018; Petin et al., 2020). RBE is quantitatively determined by the ratio of doses of standard radiation with the lowest possible values of LET (~0.2 keV/μm) and the studied radiation for equal values of biological effects. As standard radiation, γ-rays of ^{60}Co or any other electromagnetic ionizing radiation of higher energy is often used. For cell survival, RBE is quantitatively determined by the ratio D_0 of the standard radiation to the same parameter for the studied radiation, where D_0 is the mean lethal dose, which reduces cell survival by a factor of e on the straight part of the survival curves depicted on a semi-logarithmic scale. Here e is the base of natural logarithms. It is believed that for exponential survival curves at a dose of D_0 in a hypothetical target of a cell, on average, one damage is formed, and the survival of cells S is related to the radiation dose by the expression

$$S = \exp(-D/D_0). \tag{3.1}$$

The D_0 value can serve as a measure of the radioresistance of cells, and the inverse value $1/D_0$ is a measure of their radiosensitivity.

By analogy with the concept of the cross section of interaction of neutrons with matter (Saha, 2006), well-known in nuclear physics, a similar approach can be introduced for ionizing radiation in radiobiology. In the case of non-ionizing electromagnetic radiation, this concept was introduced because the radio waves might go around the object and slightly interact with it (Zhavoronkov, Petin, 2018). Why did it become necessary to introduce such a concept for ionizing radiation? If, in accordance with classical concepts, there really is a radiosensitive target inside the cell that is responsible for radiation damage, then its cross section should be revealed in various experiments. The reasoning concerning the cross section of the interaction relates primarily to corpuscular radiation. For this case, the hitting of ionizing radiation tracks into the target volume V can be associated with the passage of the flow of charged particles through the formally determined cross section. Let us denote this section by σ. Then the exponential (one-hit) survival curve can be described by the expression

$$S = N/N_0 = e^{-VD}, \tag{3.2}$$

where S is the probability of survival, N and N_0

and in the control, respectively, and $V = 1/D_0$ and VD is the target volume multiplied by dose. In fact, this value is equal to the number of hits, i.e., the number of particles crossing the target. This number can be written as the product of the particle flux density (fluence) n by the effective cross section of the target σ, i.e.,

$$VD = n\sigma. \tag{3.3}$$

Then $S = N/N_0 = e^{-VD} = e^{-n\sigma}$. Let $S = e^{-1} = e^{-VD} = e^{-n\sigma}$. From here

$$V = 1/D_0 \text{ and } \sigma = 1/n_0. \quad (3.4)$$

Therein n_0 is the particle flux density required to form one damage. It can be seen from the last expression that the smaller the particle flux required for the formation of one damage, the larger the cross section of the target σ. The dose can be represented in terms of particle flux and their LET (L)

$$D = nL/\rho, \quad (3.5)$$

where ρ is the density of the substance. Taking into account expressions (3.3) and (3.5), we have

$$VD = n\sigma = VnL/\rho. \quad (3.6)$$

From here

$$V = 1/D_0 = \sigma\rho/L \quad (3.7)$$

or for biological tissue $\rho = 1$

$$\sigma = VL. \quad (3.8)$$

Equation (3.8) shows that the cross section of the target depends on the target volume and the quality of radiation (LET). The dependence of the target cross section and consequently the cross section of cell inactivation on the LET is qualitatively explained by the fact that the amount of energy transferred during the passage of a particle through the target increases with an increase in the LET of radiation to a certain maximum corresponding to the critical energy that must be released in the target to produce an effective damage.

In accordance with these ideas, one should expect that with an increase in LET, the target cross section would approach the geometric cross section of the target. Such patterns were obtained for bacteria *E. coli* (Krasavin, 1989), yeast cells (Korogodin, Krasavin, 1982), and Chinese hamster cells (Todd, 1967). These data show that the cross section for the action of ionizing radiation behaves like the geometric cross section of a hypothetical target.

For the simplest variant of the independence of the survival curve shape on radiation quality, one can write for α-particle

$$RBE(L) = D_0(\gamma)/D_0(\alpha) = 1/D_0(L), \quad (3.9)$$

where $1/D_0$ is radiosensitivity. If we take it as a unit for sparsely ionizing radiation, then the dependence of $1/D_0$ on LET will characterize the dependence of RBE on LET. In this case, the ratio $1/D_0$ is expressed in relative units with respect to $D_0(\gamma)$. Taking into account equation (3.7), we obtain

$$V = 1/D_0 = \sigma\rho/L = RBE \tag{3.10}$$

or

$$RBE = \sigma(L)\rho/L \tag{3.11}$$

Since $\sigma(L)$ increases to a certain maximum that equal to the cross section of the target, the RBE value reaches its greatest value when $\sigma(L) = \max$. It follows that RBE should increase with an increase in LET, reach the greatest value, and then decrease. The presence of the greatest value means that at the corresponding LET value (120–130 keV/μm) the energy released in the sensitive target corresponds exactly to that which is necessary to inactivate the cell. A further increase in LET leads to excessive energy release, due to which the RBE begins to decrease. Naturally, with an increase in LET, this probability increases. However, after some LET values, saturation occurs, and each subsequent particle loses energy already in the killed cell; consequently, the efficiency of radiation with such LET decreases.

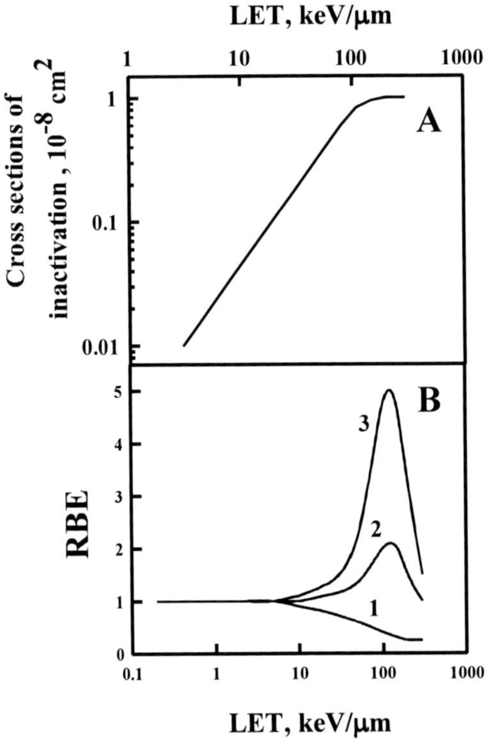

Figure 3.1. The dependence on LET radiation: A – cross sections for inactivation of haploid yeast (Dertinger, Jung, 1970) and B – RBE radiation for bacteriophage (curve 1) (Brustad, 1962), haploid and diploid yeast cells (curves 2 and 3, respectively) (Sayeg et al., 1959; Petin, Kim, 2016; Petin et al., 2020).

The radiobiological reactions of yeast cells, the simplest model of eukaryotes, are qualitatively identical to the biological effects of cultured mammalian cells; therefore, yeast cells are considered a suitable model system for many radiobiological studies (Petin, Kabakova, 1981; Critchlow, Jackson, 1978; Resnisk, Cox, 2000; Petin, Kapultcevich, 2014; Petin, Kim,

2016). For yeast cells, as well as for mammalian cells, RBE increases with increasing LET radiation until it reaches the greatest value at LET about 120–130 keV/μm, after which RBE decreases (Figure 3.1). This figure shows an example of a change in the cross section for inactivation of haploid yeast cells (Figure 3.1 A) and the dependence of RBE of charged particles on the LET of corpuscular radiation (Figure 3.1 B). It can be seen that the cross section for cell inactivation increases with increasing LET, then it approaches a constant value corresponding to the geometric cross section of cell nucleus.

To explain the increase in the RBE with increasing LET of ionizing radiation, numerous mathematical models have been proposed (Kappos, Pohlit, 1972; Kapultsevich, Petin, 1974; Kellerer, Rossi, 1972, 1978; Kapultsevich, 1978; Chadwick, Leenhouts, 1981; Obaturov, 1987), in which groups of ionization of molecules inside a sensitive target can lead to the formation of radiation damage. The RBE maximum indicates that at this LET value, each passage of a particle through the nucleus is effective for transforming the energy corresponding to cell inactivation (Dertinger, Jung, 1970). A decrease in RBE with an increase in LET means that a particle passing through a critical structure releases more energy in this structure than it is required to produce damage. Figure 3.2 shows the dependence of RBE on LET for cultured human cells. It is seen again that the RBE first increases, but then, starting from the LET ~ 120–130 keV/μm decreases.

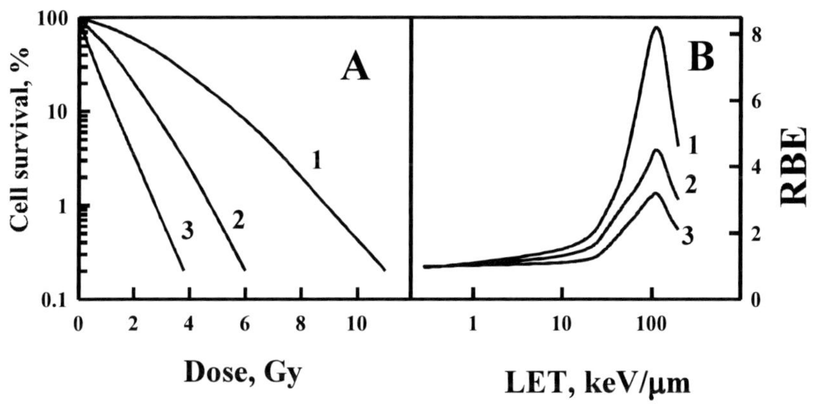

Figure 3.2. A – the dependence of the survival of human kidney cells in culture exposed to 250 kV X-ray irradiation, LET = 2 keV/μm (curve 1), deuterons, LET = 61 keV/μm (curve 2) and α-particles, LET = 120 keV/μm (curve 3); B – the dependence of RBE on LET radiation and the level of survival, for which it is calculated for human kidney cells: curves: 1 – 80%, 2 – 10%, 3 – 1% (Barendsen et al., 1966; Barendsen, 1968).

In eukaryotes (yeast, mammalian cells), as in prokaryotes, double-strand DNA breaks are the most dangerous for cell viability. The radiosensitivity of these cells, therefore, should also reflect their ability to repair double-strand breaks. For bacterial cells, according to works (Bresler, Noskin, 1978; Bresler et al., 1978, 1981; Noskin, 1984), radioprotectors, slowing down the stage of endonuclease clearing of gaps in DNA strands or the rate of incision of γ-sites, create conditions for better balancing between the processes of degradation of damaged DNA regions and their re-synthesis, which leads to a decrease in the likelihood of the formation of enzymatic double-strand DNA breaks. Since the contribution of direct double-strand breaks to the total yield of double-strand breaks increases with an increase in LET of radiation, the effectiveness of radioprotectors decreases with an increase in LET (Krasavin, 1989). In this work, it was shown that when *E. coli* bacteria are irradiated with heavy ions together with radioprotector β-mercaptoethylamine, a transformation of the dependence of RBE on LET is observed.

For diploid yeast cells, which have a double set of chromosomes, like mammalian cells, the survival curve remains sigmoid regardless of the quality of radiation (Laskowski, 1960; Petin, 1969, 1979; 1987; Kapultsevich, Petin, 1974; Pohlit, Schäfer, 1974; Schneider, Kiefer, 1976; Petin, Kabakova, 1981; Frankenberg et al., 1981; Hannan et al., 1986; Korogodin, 1993; Petin, Petin, 1995; Kiefer et al., 2002; Petin, Kapultcevich, 2014; Petin, Kim, 2016). For this case, RBE for diploid yeast cells capable of post-radiation recovery varies in the cited works in the range of 3.5–4.5 and does not depend on the level of survival. As an example, Figure 3.3 shows the survival curves for the diploid yeast cells of *Saccharomyces ellipsoideus*, strain Megri 139-B, irradiated with γ-rays and α-particles.

Analysis of numerous published data shows some contradictions about the effect of radiation quality on the shape of the survival curves of yeast cells and cultured mammalian cells. For mammalian cells, sigmoid survival curves at low values of LET radiation are transformed into exponential curves after exposure to radiation with high LET (Figure 3.2 A). Such data were obtained by many authors (Todd, 1967; Barendsen, 1968, 1974; Neary, 1970; Alper, 1971, 1979; Raju, 1980; Fox, McNally, 1988; Krasavin, 1989; Franken et al., 2011; Tracy et al., 2015; Hall, Giaccia, 2018). This transformation means that the RBE of densely ionizing radiation for such cases strongly depends on the level of survival selected for calculation. In particular, the RBE value are very large (10–15) in the region of low doses of ionizing radiation, approaching the ratio of the slopes of the exponential parts of the survival curves (3–5) with an increase in the radiation dose (Figure 3.2 B). Many works with yeast and cultured mammalian cells have been cited here to emphasize that the observed differences in the responses of these cells to high LET radiation at the low radiation dose range are not accidental, since they were obtained by different scientists in different research laboratories. On this basis, it can be concluded that the theoretical RBE models, based mainly on the change in the shape of the survival curves with a change in the radiation quality, are not correct enough to be applied to various biological objects.

An increase in RBE with increasing radiation LET for exponential survival curves allows different interpretations, the general requirement for which is the formation of several pairs of ions by one ionizing particle in one or several targets or one pair of ions in several targets. The multi-hit model (many hits in one target) qualitatively explains the dependence of the shape of the survival curve on the radiation LET. In the case of sparsely ionizing radiation, many particle tracks are needed to produce the damage (sigmoid survival curve), while one densely ionizing particle forms many hits (damage) at once, and the survival curve becomes exponential.

The question arises, why RBE of densely ionizing particles increases with decreasing dose of ionizing radiation for cultured mammalian cells and does not depend on the dose for yeast cells? In other words, why is the radiation response to irradiation with high LET radiation at low doses qualitatively different for yeast and mammalian cells? A thorough mathematical analysis of this problem and the corresponding experimental studies (Kapultsevich, Petin, 1974; Petin, Petin, 1995; Petin, Kapultcevich, 2014) led to the conclusion that the radiobiological response of mammalian cells is consistent with theoretical concept, according to which the event hit is the formation of several ionizations by the same particle in a sensitive site, and the total number of sensitive sites is large enough, and they are distributed relatively evenly throughout the entire volume of the nucleus. In this case, each ionizing particle entering the sensitive volume crosses several sensitive sites. Similar conclusions were made in works (Kellerer, Rossi, 1971, 1978) based on more complex microdosimetric calculations. They hypothesized the existence of a large sensitive volume, which contains homogeneously dispersed sensory loci, and that at least two inactivated loci are required for the effect.

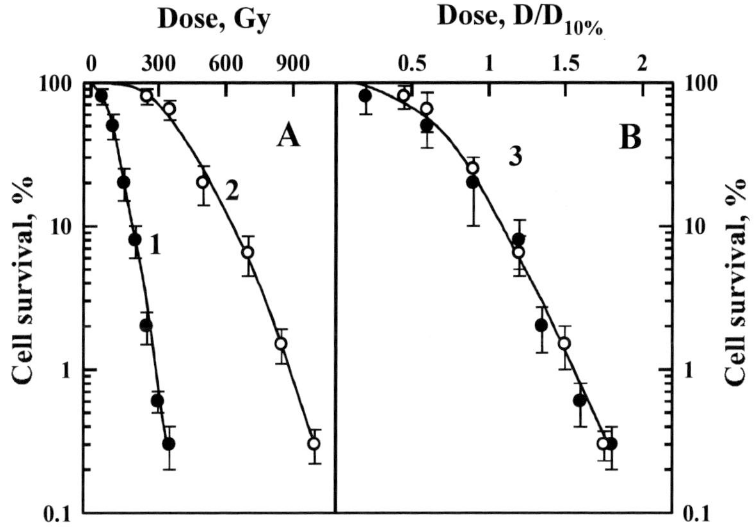

Figure 3.3. The dependence of the survival curves of *Saccharomyces ellipsoideus* diploid yeast cells, strain Megri 139-B, on the dose of irradiation with ^{60}Co γ-rays (LET = 0.2 keV/μm, open circles) and ^{210}Po α-particles (LET = 120 keV/μm, closed circles) in absolute (curves 2 and 1, respectively) and relative units (curve 3).

For objects, in which the sigmoid survival curve does not depend on LET radiation, each particle entering the effective volume must cross a limited number of sensitive sites. In this case, with an increase in LET, the likelihood of damage of one sensitive site will increase (increase in RBE). To inactivate a cell, several hits of ionizing particles are required even at very high LETs, as a result of which the survival curves retain a sigmoid shape with an increase in LET of the particles. Note that yeast cells are irradiated in the stationary stage of growth, when the survival curve remains sigmoid even at very high LET values. For this case, the shape of the target should be such that even a densely ionizing particle would cross a limited number of sensitive sites inside the nucleus, which should not be uniformly distributed over the entire nucleus, but concentrated in some places. As a working hypothesis, it was assumed that in a diploid yeast cell, the target has the shape of a spherical layer, the diameter of which is equal to the diameter of the cell nucleus, and the thickness is several tens of angstroms. This issue was discussed in more detail in (Petin, Kapultcevich, 2014; Petin, Kim, 2016; Petin et al., 2020).

The question remains, why in diploid yeast cells the shape of the survival curve does not depend on the LET radiation, while for cultured mammalian cells the sigmoid shape of the survival curve after exposure to sparsely ionizing radiation is transformed into exponential one after exposure to radiation with high LET? Note that yeast cells, as a rule, in radiobiological experiments are irradiated at the stationary stage of growth, while mammalian cells are irradiated at the logarithmic stage. Yeast cells in the stationary stage have 100% growth efficiency, are very homogeneous in size, sensitivity to the action of ionizing radiation and other damaging factors. A more complicated situation is observed for yeast cells irradiated at the logarithmic stage of growth. For this case, the survival curves have a resistant "tail" associated with the heterogeneity of the irradiated population. This heterogeneity is explained by the fact that some of the haploid cells have a double set of chromosomes, and the diploid cells have a tetraploid set, when the chromosomes have doubled, but did not pass into the daughter bud. Thus, when the population of haploid yeast is irradiated in the logarithmic stage of growth, some of these cells are actually diploid, have a double set of chromosomes and are characterized by increased resistance to the action of ionizing radiation. For this reason, yeast cells in radiobiological experiments are usually irradiated in a stationary rather than logarithmic stage of growth.

In this case, the question arises, why are cultured mammalian cells irradiated in the logarithmic and not in the stationary growth stage? This is due to a completely different reason. The point is that mammalian cells in the stationary growth stage are characterized by a reduced growth efficiency, which in some cases can reach 30–70%. Consequently, mammalian cells that have reached the stationary growth stage already bear some kind of damage, and their survival curves are often exponential even under irradiation with low LET radiation. Therefore, in radiobiological experiments, cultured mammalian cells are irradiated, as a rule, in the logarithmic stage of growth. Then, the transformation of the sigmoid survival curve into an exponential one with an increase in LET radiation can be explained by the uniform distribution of sensitive sites throughout the entire target volume (cell nucleus). In this case, under the action of sparsely ionizing radiation, several particles should be passed through a target to form a critical number of lesions, and therefore the survival curve will be sigmoid. Each passage of a densely ionizing particle will be accompanied by the defeat of a sufficient number of sensitive sites, and the survival curve will be exponential, which is observed for mammalian cells. For diploid yeast cells irradiated in the logarithmic stage of growth, a similar transformation of the sigmoid shape of the survival curve into an exponential one can be also observed only at the initial part of the dose-effect curve as it is observed for mammalian cells. Further, with an increase in the dose, the resistant tail on survival curve emerges, the origin of which was discussed above. This resistant tail remains after exposure to both sparsely and densely ionizing radiation. Therefore, the survival curve, for diploid yeast cells irradiated in the logarithmic stage of growth endures a similar transformation of the sigmoid shape of the survival curve into an exponential one if we take into account only the initial part of the dose-effect curve and do not take into account the resistant tail, the origin of which was discussed above.

Thus, the analysis of the survival curves of diploid yeasts and mammalian cells from the point of view of the hit principle showed that for the interpretation of the results obtained with these objects, a multihit model is acceptable, in which the hit event is the local formation of several ions by one ionizing particle in one of the sensitive structures, presumably a chromosome. Depending on the stage of cell growth, sensitive structures can be located either at the periphery of the cell nucleus, forming a thin spherical layer, or evenly occupy the inner regions of the cell nucleus.

The RBE of α-particles was compared for the formation of double DNA breaks, chromosomal aberrations, and inactivation of mammalian cells (SW-1573) in order to predict the radiosensitivity of tumors and normal tissues (Franken et al., 2011). The cells were irradiated with ^{137}Cs γ-rays and ^{241}Am α-particles. The obtained RBE values for the induction of two types of chromosomal aberrations (fragments and translocations) are close to those for cell inactivation. It is concluded that the analysis of these aberrations can provide relevant information on the biological effectiveness of radiotherapy with high LET radiation. Data on thermal radiosensitization of wild-type Chinese hamster CHO cells, as well as radiosensitive mutants for repair by homologous recombination and reunification of non-homologous ends after exposure to protons and carbon ions, were obtained (Maeda et al., 2015). The cells were exposed to ionizing radiation followed by hyperthermia treatment (42.5°C for 1 hour). Thermal sensitization was less after irradiation with densely ionizing than after sparsely ionizing radiation.

A distinctive feature of the effect of charged particles on cell genetic structures is the formation of cluster DNA damage (Krasavin, 1989; Sage, Shikazono, 2016; Zadnepryanets et al., 2018 a, b), including closely spaced double-strand DNA breaks, which is not typical for sparsely ionizing radiation. The results of work (Burch et al., 2011) dedicated to the study of genetic instability caused by a high yield of cluster mutations show that this effect is due to a combination of three factors: (i) a high ability to repair DNA damage, (ii) tolerance of large areas of DNA damage and (iii) high efficiency prone to errors of the translesion synthesis during the restoration of damaged DNA. The high multiplicity and density of mutations means that this new form of genetic instability may play an important role in the incidence of cancer and genetic diseases. It is known that with an increase in LET radiation, the number of double-stranded DNA breaks formed along the particle track and their complexity increase (Jenner et al., 1993; Goodhead, 1994; Nikjoo et al., 1999, 2001; Pastwa et al., 2003; Moore et al., 2014). Therefore, the efficiency of their repair is much lower than for single damage induced by low LET radiation. This is largely due to the high density of ionization in cell nuclei during the passage of charged particles through genetic structures (Prise et al., 2001; Saha et al., 2014). Differences in the induction and repair of double-strand DNA breaks are clearly reflected in the nature of radiation-induced effects in mammalian and human cells. Using various methods to study the formation of double-stranded DNA breaks, it was found that high LET radiation induces double-strand DNA breaks with a higher frequency compared to the action of sparsely ionizing radiation (Chausov et al., 2009; Lorat et al., 2015). It was shown that with an increase in LET radiation, the number of double-strand breaks per single track of a particle increases and their clustering is also increased (Leatherbarrow et al., 2006; Zadnepryanets et al., 2018 a, b). Complex DNA damage is a special class of damage, which is two or more individual DNA damage within one or two turns of DNA. Clustering events of DNA double-strand breaks can be manifested to a large extent under the action of high LET radiation and it can be influenced by the factor of DNA supercoiling (Ponomarev et al., 2008).

Under the action of sparsely ionizing radiation, modified bases may be located close to DNA breaks, as well as other damage can occur near the forming double-stranded DNA breaks, but several double-stranded DNA breaks in one site in the form of a cluster are rarely formed. For densely ionizing radiation, when a charged particle passes through 1–2 turns of DNA (10–20 nm), an increase in the frequency of formation of highly complex double-stranded DNA breaks is characteristic, up to 25 lesions in a cluster (Asaithamby et al., 2011). There is ample evidence that the intricacy and complexity of DNA damage increases with increasing LET radiation (Moore et al., 2014), which is reflected in the formation of clusters of radiation-induced foci, the nature of which is currently being intensively studied. A number of works indicate that the analysis of foci formed along the track of a charged particle does not allow differentiating individual foci (Costes et al., 2006; Antonelli et al., 2015) due to their overlap and limited optical resolution of experimental methods (Okumura et al., 2013). However, studies using the latest high-resolution microscopy methods have demonstrated the complex structure of radiation-induced foci in the tracks of accelerated ions and noted the slowed-down kinetics of their elimination (Nakajima et al., 2013; Reindl et al., 2017; Zadnepryanets et al., 2018 a, b).

While the repair of single DNA damage is carried out efficiently, complex DNA damage is repaired more slowly, and may even remain unrepaired, which leads to serious biological consequences (Lobrich, 1998). Numerous data indicate the complex nature of double-stranded DNA breaks under the action of accelerated heavy ions, containing, in addition to DNA breaks, damage of another type, and, accordingly, more difficult to repair compared to γ-induced damage (Asaithamby et al., 2011; Antonelli et al., 2015). A number of studies have shown that the frequency and complexity of DNA damage increases with an increase in LET radiation (Hada, Sutherland, 2006). This is due to the nature of the transfer of energy from charged particles to matter. For example, α-particles with an energy of 5 MeV, transferring ~ 90% of their energy within a radius of 10 nm, induce double-stranded DNA breaks located at a distance of only 10–20 nucleotide pairs (Eccles et al., 2011). In this limited area, direct ionization events along the DNA strand will lead to the formation of closely spaced double-strand breaks and highly reactive free radicals, causing the formation of DNA-DNA cross-links. Low LET radiation can induce a similar spectrum of DNA damage, but the damage will be located at a greater distance from each other, and only a small part of the cross-links or damaged bases can be so close as to create an obstacle for the repair of DNA double-strand breaks (Goodhead, 1999).

The idea of the formation of complex DNA lesions – several closely spaced lesions within a short DNA fragment – was first put forward in (Ward, 1981). Currently, complex DNA damage is defined as two or more damage formed in one or two turns of DNA when a single track crosses a particle (Hada, Georgakilas, 2008). Complex DNA lesions contain double-strand breaks, single-strand breaks, and damaged bases within the range of 20–30 nucleotide pairs in a DNA region with a length of about 10–20 nm (Watanabe et al., 2011). According to theoretical calculations, high LET radiation can induce up to 25 lesions per 1–2 DNA turns (Semenenko, Srewart, 2004), and clustering can occur both locally, within a region of 40 pairs of bases, and cover regions in several thousand base pairs (Asaithamby et al., 2011). It was shown (Brenner, Ward, 1992) that complex DNA damage containing double-strand breaks occurs as a result of energy release from 2–5 ionization events localized within 1–4 nm. When exposed to high LET radiation, at least 70% of DNA double-strand breaks are complex due to the formation of additional breaks, while for low LET radiation, the number of complex breaks decreases (Nikjoo et al., 1999). In turn, such unrepaired complex lesions can lead to the formation of enzymatic double-strand breaks and increase genetic instability. It is possible that the high density of DNA lesions within the cluster makes simple lesions inaccessible for repair enzymes (Paap et al., 2008). It is also possible that the functional activity of repair enzymes is reduced within the complex DNA damage. It was suggested that the formation of multiple closely spaced lesions leads to a change in DNA conformation (Lomax et al., 2005), which can prevent the free binding of repair enzymes to the site of damage, which ultimately leads to a decrease in the efficiency of repair.

At present, the regularities of the formation and elimination of radiation-induced foci are being intensively studied – areas formed after irradiation at sites of DNA damage or near them and consisting of proteins involved in the repair of double-stranded DNA breaks (Jakob et al., 2003; Ibañez et al., 2009; Bracalente et al., 2013; Zadnepryanets et al., 2018 a, b). The authors used the method of immunocytochemical analysis, which allowed them to establish quantitative patterns of the implementation of processes reflecting the formation and repair of double-strand DNA breaks under the action of ionizing radiation of different quality. Using human fibroblasts as an example, these regularities were studied after exposure to ^{60}Co γ-rays (standard radiation, LET = 0.2 keV/μm), accelerated boron (^{11}B) and neon (^{20}Ne) ions with energies of 8.1 and 46.6 MeV/nucleon and LET equal to 138 and 132 keV/μm, respectively. It was found that under the action of heavy ions, a greater number of such foci are formed than under the action of γ-radiation. The slowed down kinetics of their elimination in the tracks of accelerated ions in comparison with the action of γ-rays was established. It was shown that this effect was associated with the nature of the microdistribution of the energy of heavy charged particles in the genetic structures of cells and the formation of complex double-stranded DNA breaks, which difficult to repair. It is concluded that the formation of complex cluster damage is a characteristic feature of the action of heavy charged particles on genetic structures.

3.2. RBE of Densely Ionizing Radiation Dependence on Cell Radiosensitivity

Beginning from the classical works that laid the foundation for theoretical radiobiology, it was believed for many years that an increase in RBE of ionizing radiation with an increase in LET is due to an increased probability of the formation of primary radiation damage or due to the interaction between physical events of energy transfer by charged particles at the microgeometric level (Timofeeff-Ressovsky, Zimmer, 1947; Fritz-Neagley, 1961; Zimmer, 1962; Lea, 1963). Some indirect data were presented in favor of the point of view that an increase in RBE with an increase in LET radiation may be associated not only with the interaction of ionizing radiation at the physical level of the formation of ionizations and excitations of biologically important molecules, but also at the biochemical stage of the production, restoration and fixation of radiation damage (Neary, 1970; Eidus, 1977). More direct results on this problem were obtained in experiments with bacteria (Haynes, 1966; Munson et al., 1967; Alper, 1979; Krasavin, 1989). In these works, a negative correlation between radiosensitivity and RBE has been demonstrated. Extensive similar data were obtained on yeast cells of various species, strains and ploidy, differing in their ability to recover from damage induced by ionizing radiation (Petin, Kabakova, 1981; Petin, 1987; Glazunov, 1989; Korogodin et al., 1996; Korogodin, 2010; Petin, Kim, 2016; Petin et al., 2020). It has been noted that the nature of the recovery and RBE of densely ionizing radiation is determined by the energy "reservoir", which may depend on the pre-radiation conditions of cultivation (Pohlit, Schäfer, 1974). RBE values can be determined not only by the different energy required for the formation of primary radiation damage, but also by the ability of cells to recover from potentially lethal and sublethal damage (Petin, 1977, 1979, 1987; Korogodin, Krasavin, 1982; Krasavin, 1989; Petin, Petin, 1995; Weyrather et al., 1999; Petin, Kim, 2016). In these studies, it was shown that the RBE of densely ionizing radiation are largely determined by the ability of cells to recover from radiation damage.

There are some specific data obtained on cells of different genotypes and confirming this conclusion. A comparison was made of the radiosensitivity of cultured human cells, which are normally radioresistant and radiosensitive ataxia telangiectasia for high LET of heavy ions (Tobias et al., 1984). It was shown that the greatest RBE (LET = 120 keV/μm) were 3.2 for radioresistant cells and only 1.8 for radiosensitive ones. The RBE of fast neutrons was 2.1 for the radioresistant and only 1.3 for the radiosensitive mouse lymphoma cell line (Hesslewood, 1978). The relationship between LET radiation and the lethality of chicken lymphocyte DT40B cell line was investigated in the DNA defective for the repair of double-strand breaks cells with genes Ku70$^{-/-}$, Rad54$^{-/-}$ and Ku70$^{-/-}$Rad54$^{-/-}$, as well as in the wild-type of cells (Aoki-Nakano, Furusawa, 2013). The cells were irradiated with various heavy ions, providing a change in LET from 13 to 5800 keV/μm. The expected RBE–LET relationship was observed in cells capable of DNA repair, but the greatest RBE value for the survival of mutant cells was not found. These data indicate that the maximum RBE in the dependence of RBE on LET is due to the repair of DNA double-strand breaks. These results indicate that the erroneous repair of DNA double-strand breaks after exposure to heavy ion beams is responsible for the maximum in RBE versus LET relationship and cell death. In other words, these data also indicate the participation of recovery processes in the formation of the increased RBE values. It was shown above that with an increase in the number of chromosomes sets in a cell, both the total amount of radiation damage and the relative contribution of repaired radiation damage increase (Table 2.4). If the assertion about the relationship of RBE with the ability of cells to recover from radiation damage is correct, then one would expect a certain increase in RBE with an increase in the number of chromosomes sets in the cell. Using the data published by Laskowski (1960), we calculated that the RBE of ^{210}Po alpha particles was 2.1 for the haploid strain of yeast *S. cerevisiae*; 4.9 for an isogenic triploid strain and 6.1 for a tetraploid one (Petin et al., 2020). Consequently, these data also confirm the hypothesis under discussion that RBE of densely ionizing radiation, in addition to the increased probability of the formation of primary radiation damage, is also determined by the ability of cells to recover from radiation damage.

We studied the dependence of RBE of α-particle upon irradiation of yeast cells under conditions when their ability to recover from radiation damage significantly changes – cell ploidy, stages of growth, and cell genotype that blocks some stages of recovery (Petin, Kim, 2016; Petin et al., 2020). The most significant of these results are presented below. The following haploid and diploid yeast cells of *Saccharomyces cerevisiae* were used: wild type strains S288C (RAD), 211-1a (RAD), XS800 (RAD/RAD), 211 (RAD/RAD) and radiosensitive mutants XS774-4d (*rad*51) and XS806 (*rad*51/*rad*51), as well as the strain Megri 139-B (RAD/RAD). The radiobiological responses of these strains have been well studied (Korogodin, 1993; Petin, Petin, 1995; Petin, Kim, 2005 a, 2016; Korogodin, 2010). When cells were irradiated at the stationary stage of growth, we used 4–6-day cultures of diploid cells and 12–14-day cultures of haploid cells grown on a solid nutrient medium at 30°C. The duration of pre-irradiation cultivation was determined by the time of termination of cell budding. To irradiate cells in the logarithmic stage of growth, the cells were cultured in a liquid nutrient medium at 30°C. Under these conditions, the cells stopped multiplying after 48 hours when the cell concentration was about $5 \cdot 10^7$ cells/ml. In the experiments, we used cells that were in a liquid nutrient medium for 18–24 hours, when the cell concentration was about 10^6 cells/ml at an initial concentration of 10^2 cells/ml. Then the cells were washed from the nutrient medium by centrifugation and resuspended in water for subsequent irradiation of the cell suspension with ^{60}Co γ-rays (20 Gy/min) and the cell monolayer for exposure to ^{239}Pu α-particles (25 Gy/min). The average LET of these α-particles is 120 keV/μm. It is just at this LET value the greatest RBE is observed in the dependence of RBE on LET for most eukaryotic cells. Due to the low penetrating ability of α-particles, a monolayer of cells was irradiated: 0.02 ml of a suspension with a concentration of $5 \cdot 10^7$ cells/ml was applied to the surface of "hungry" agar and placed in a thermostat (30°C) for 20–30 min to evaporate water from drop. The dosimetry of the sources used was described earlier (Klyuch, Dubrovin, 1967; Petin, 1977, 1987). Cell survival was determined by counting the colonies formed after 3-5-day incubation on solid nutrient medium at 30°C. Each experiment was repeated 3–5 times. Results are presented as mean and standard error.

Figure 3.4 shows the survival curves for homozygous haploid (A) and diploid (B) *S. cerevisiae* yeast cells that were irradiated at the stationary stage of growth. RBE was assessed by the ratio of isoeffective doses of sparsely and densely ionizing radiation. For the haploid strain, RBE was 1.7, while for the isogenic diploid strain, this value was 3.8, regardless of the dose or survival for which it was calculated.

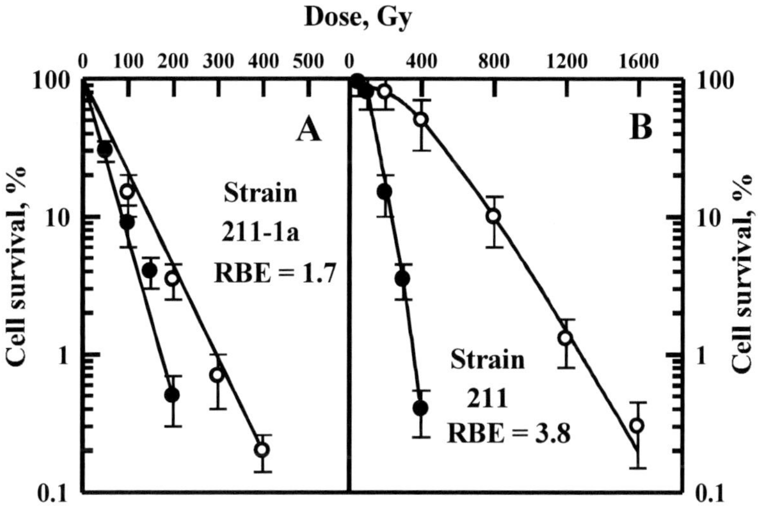

Figure 3.4. Survival curves of haploid (A, strain 211-1a) and homozygous diploid yeast cells of *S. cerevisiae* (B, strain 211) of wild-type irradiated at the stationary stage of growth with ^{60}Co γ-rays (open circles) and ^{239}Pu α-particles (closed circles).

Similar data were obtained for other pairs of homozygous haploid and diploid yeasts (Petin, 1977; Petin et al., 2020). These data indicate that for cells capable of recover from radiation damage, RBE values of high LET radiation were much higher than for haploid strains characterized by increased radiosensitivity due to the reduced ability of cells to repair DNA radiation damage (Luchnik et al., 1977; Frankenberg, 1984; Lewis, Resnick, 2000; Mao et al., 2008). Based on these data, it can be expected that RBE should be lower for radiosensitive mutants with a reduced ability to recover from radiation damage than for the original parental strains capable to repair radiation damage.

Figure 3.5 shows the survival curves of *rad*51 radiosensitive mutants of haploid (A) and diploid (B) yeast cells of *S. cerevisiae* irradiated at the stationary stage of growth with ^{60}Co γ-rays and ^{239}Pu α particles. It can be seen that the *rad*51 mutation practically did not change the radiosensitivity and RBE of densely ionizing radiation in comparison with the original wild-type strain.

The presented data for homozygous diploid cells exhibit that their radiosensitivity increased significantly and RBE values decreased from 3.8 to 2.3, not much different from RBE for the haploid strain. This is due to the decreased ability of the diploid mutant for recombination repair, which is involved in the recovery of cells from double-stranded DNA breaks (Luchnik et al., 1977; Saeki, 1980; Frankenberg, 1984). Note that similar results have been published for a number of other *rad* mutants (Petin, Kim, 2016). Considering these data as a whole one can conclude that the ability of cells to recover from radiation damage directly emphasize that RBE of densely ionizing radiation is largely determined by the ability of cells to recover from radiation damage.

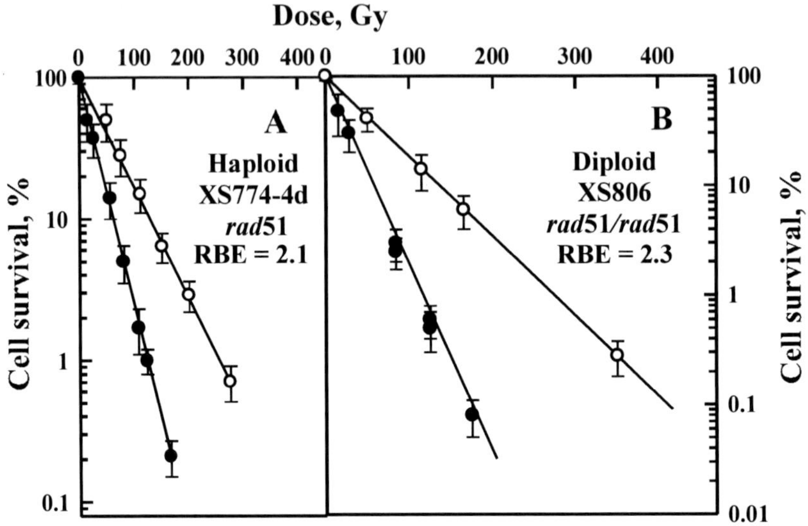

Figure 3.5. Survival curves of radiosensitive mutants of haploid (A) and diploid (B) yeast cells of *S. cerevisiae* irradiated at the stationary stage of growth with ^{60}Co γ-rays (open circles) and ^{239}Pu α-particles (closed circles).

For a clearer definition of the duration of cultivation of yeast cells to achieve the required stage of growth, Figure 3.6 shows examples of the multiplication of haploid and diploid yeast cells in a liquid nutrient medium. It can be seen that the delay in budding (lag-stage) of haploid cells (7–9 h) is greater than that of diploid ones (2–4 h), and haploid yeasts reproduce more slowly than diploid ones. In addition, it is obvious that, regardless of the initial cell concentration, after exponential growth, characterizing the logarithmic stage of growth, the cells reach a plateau that determines the stationary stage of growth, when cell multiplication stops. It is also seen that the maximum cell concentration in the plateau region for diploid cells is an order of magnitude higher than for haploid cells.

It is of interest that the cells in the logarithmic stage of growth did not recover when irradiated cells in the post-radiation period are kept in a non-nutritive medium. Extensive results concerning this problem have been published (Petin, 1987). To illustrate this pattern, one example is shown in Figure 3.7, which shows the survival curves obtained immediately after irradiation (curves 1) and after recovery in a non-nutrient medium at 30°C (curves 2). It can be seen that the cells irradiated in the stationary stage of growth (Figure 3.7 A) recovered well, while the cells irradiated in the logarithmic stage of growth (Figure 3.7 B), did not recover, moreover such cells even died out in the process of their keeping during the post-radiation period in a non-nutritive environment. Then one should expect a significant RBE variation for cells irradiated at the logarithmic stage of growth.

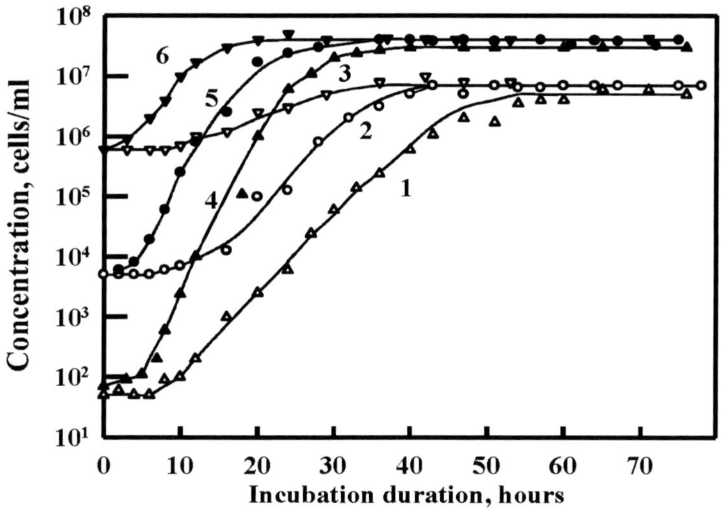

Figure 3.6. Examples of reproduction of the haploid yeast *Zygosaccharomyces bailii* (curves 1, 2, 3, open symbols) and the diploid yeast *Saccharomyces ellipsoideus* (curves 4, 5, 6, closed symbols) depending on the initial cell concentration and the duration of reproduction in a liquid nutrient medium at 30°C.

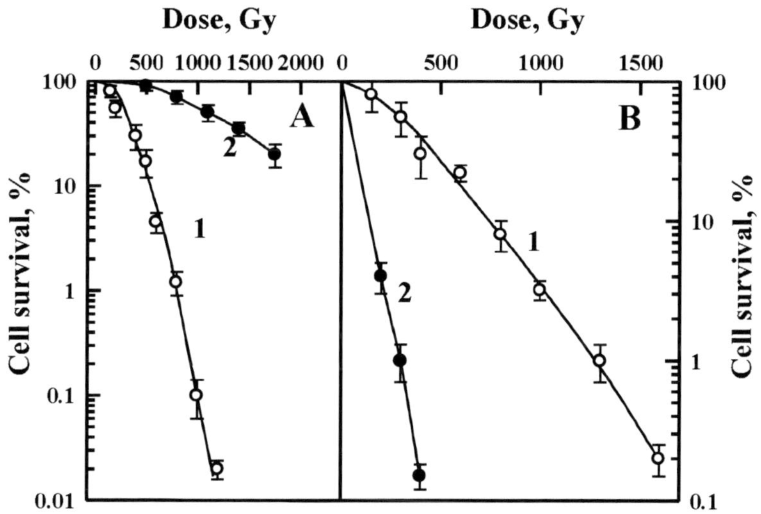

Figure 3.7. Survival of *Saccharomyces ellipsoideus* diploid yeast cells, strain Megri 139-B, irradiated at stationary (A) and logarithmic (B) stages of growth with ^{60}Co γ-rays (20 Gy/min) and placed on a solid nutrient medium immediately after irradiation (curves 1) and after a three-day recovery in a non-nutritive environment (curves 2).

It was previously demonstrated (Elkind, Beam, 1955; Petin, 1987) that the survival curves of haploid yeast strains irradiated in the logarithmic stage of growth are characterized by the presence of a resistant region ("tail") that is determined by the percentage of cells, which produced a bud of a certain size when chromosomes doubled, but did not move to the daughter cell. Such cells should respond to radiation in a similar way as diploid cells. Then one would expect that RBE should increase significantly in the region of the "tail" of the survival curve. The results of experiments with haploid and diploid strains are shown in Figure 3.8, which shows the dependence of the survival and RBE of α-particles on the radiation dose for yeast *S. cerevisiae*.

From the presented data, it can be seen that for haploid yeast cells in the region of the resistant tail of survival curves the RBE increases significantly, reaching a maximum value of 9.0. The main difference between the reactions of diploid yeasts irradiated in the stationary and logarithmic stages of growth is that in the first case, the shape of the survival curve did not depend on radiation quality (Figures 3.3 and 3.4), and in the second, on the contrary, in the region of low doses, the sigmoid shape of the survival curve becomes exponential (Figure 3.8 B), as was observed for mammalian cells. That is why, with a decrease in the dose, the RBE of densely ionizing radiation reaches large values equal to 12 (Figure 3.8 D). As noted above, a similar pattern is observed for sigmoid survival curves of cultured mammalian cells.

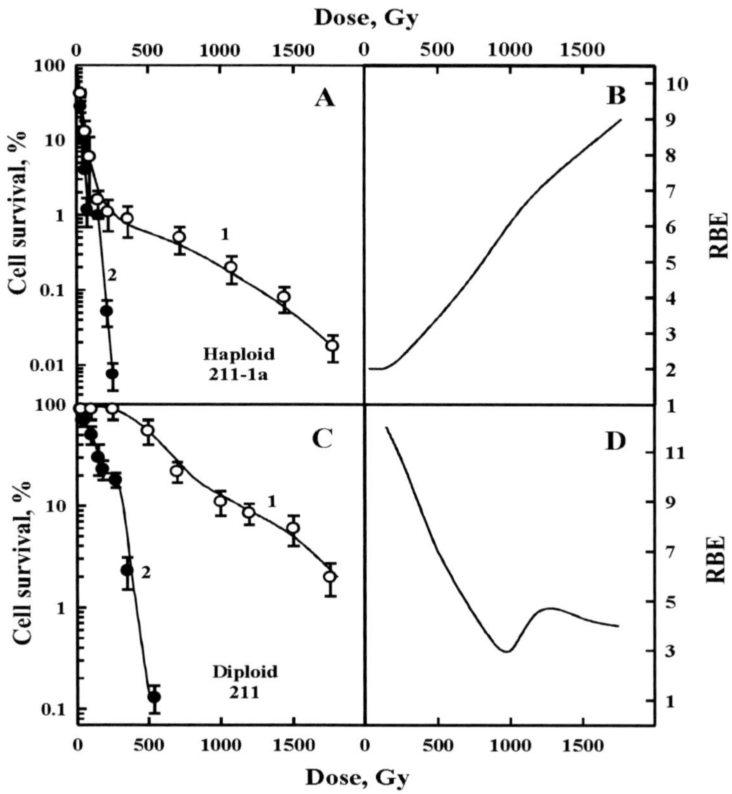

Figure 3.8. The dependence of cell survival (A, C) and RBE of α-particles (B, D) on the dose of ionizing radiation for haploid, strain 211-1a (A, B), and diploid (C, D), strain 211, *S. cerevisiae* yeast cells of wild-type irradiated at the logarithmic stage of growth with ^{60}Co γ-rays (open circles) and ^{239}Pu α-particles (closed circles).

Figure 3.9 shows the survival curves of radiosensitive mutants of haploid (Figure 3.9 A) and homozygous diploid (Figure 3.9 B) *S. cerevisiae* yeast cells irradiated at the logarithmic stage of growth. It can be seen that in this case the survival curves of both strains were exponential and did not have resistant "tails," since these strains could not recover from double-stranded DNA breaks.

It is obvious that the RBE of densely ionizing radiation for *rad*51 mutants irradiated in the stationary and logarithmic stages of growth slightly decreased – for the haploid strain from 2.1 to 1.6, for the diploid strain from 2.3 to 1.6. A certain increase in radiosensitivity observed in these experiments may be associated with an increase in the volume of sensitive targets in the logarithmic stage of growth, while a decrease in RBE may be due to a decrease in cell ability to recover from radiation damage.

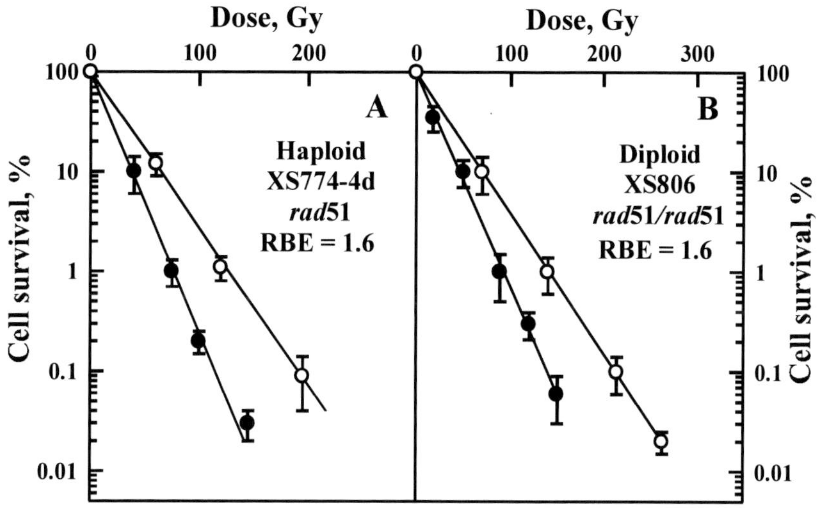

Figure 3.9. Survival curves of radiosensitive *rad*51 mutants of haploid (A) and diploid (B) yeast *S. cerevisiae* irradiated at the logarithmic stage of growth with ^{60}Co γ-rays (open circles) and ^{239}Pu α-particles (closed circles).

Figure 3.10 shows the relationship between the values of D_0 for all studied yeast strains after γ- and α-irradiation. Open circles indicate data for cells that are able to recover from radiation damage and closed circles denote the data for cells that do not have this ability. It was believed that cells are capable of post-radiation recovery if their survival increased after the cells were kept in a non-nutrient medium during the post-radiation period. For the convenience of quantitative understanding of the RBE, this figure contains straight lines of identical RBE values determined by the ratio $D_0(\gamma)/D_0(\alpha)$.

One can see from this data that the D_0 values for γ- and α-irradiation are related for all yeast strains investigated. RBE values increase with increasing in D_0, There is a correlation (r = 0.88) between RBE and cell radiosensitivity: those strains that are most sensitive to the action of ionizing radiation, i.e., the most deficient in their ability to recover from radiation damage, were characterized by the lower values of RBE (~ 2–3). The strains that were more resistant to the action of ionizing radiation due to their ability to recover from radiation damage had grater RBE values varying in the range from 3 to 5. These differences are even more pronounced when RBE is assessed after cell recovering in a non-nutrient medium (Figure 3.10 B). For such a case RBE varied in the range from 3 to 10 or even more.

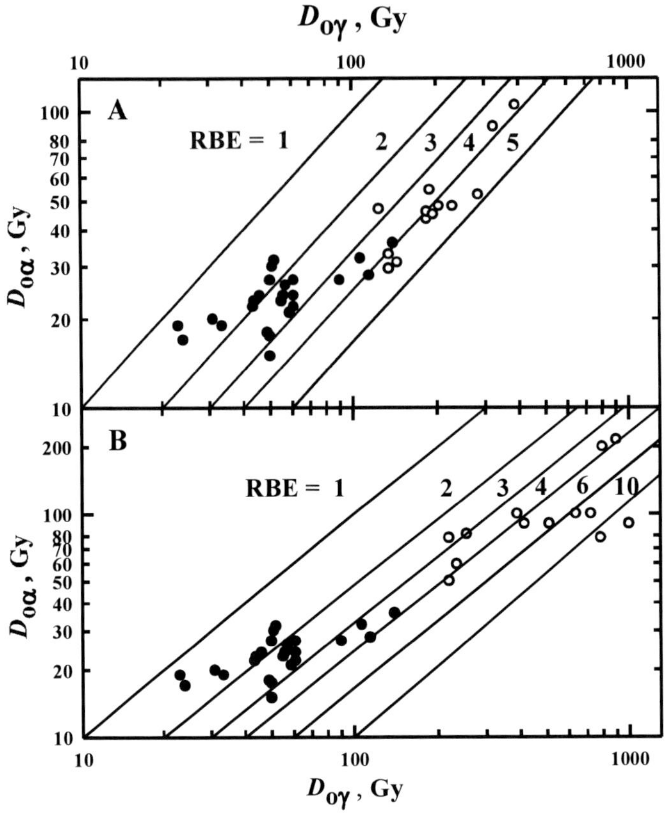

Figure 3.10. The relationship between D_0 values after irradiation with γ-rays and α-particles of various strains of *S. cerevisiae* yeast cells; open circles indicate results for cells that are able to recover from radiation damage, closed symbols – for cells that do not possess such ability; cells were plated on a nutrient medium immediately after irradiation (A) and after post-radiation recovery in a non-nutrient medium for 3 days (B).

In general, the presented data support the point of view that the RBE of densely ionizing radiation is lower for cells with a reduced efficiency to recover from radiation damage. This indicates that RBE values for radiation with high LET may be due not only to physical processes at the initial stage of the formation of primary radiation damage, but also due to the ability of cells to recover at the biochemical stage of radiation damage recovery and fixation.

It is important to discuss some possible molecular mechanisms explaining the observed patterns. It is known that post-radiation recovery in a non-nutritive environment was greater for γ-induced damage than for damage caused by α-particles (Petin, 1977, 1987; Korogodin, Krasavin, 1982). A comparison was made of the kinetics of changes in survival in the process of post-radiation recovery of yeast after exposure to isoeffective doses of γ-rays and α-particles (Petin, Kim, 2005 a). It is shown that the recovery constant characterizing the probability of cell recovery per unit time does not depend on the quality of radiation, while the irreversible component of radiation damage was significantly higher under the action of densely ionizing radiation. Therefore, the RBE value is expressed to a greater extent for cells that capable to recover from radiation damage than for radiosensitive mutants with a reduced ability to recover. That is why the RBE of α-particles is higher for situations where recovery processes are more pronounced.

It is known the attempts of some authors to find out what specific repair mechanisms are involved in the RBE manifestation of densely ionizing radiation. Analyzing the data on this problem, including some of our results (Alper, 1984), it was suggested that there are two classes of recovery – recovery from potentially lethal radiation damage (*P*) and recovery from sublethal damage (*Q*). It was shown that the reduction of *P* damage takes part in the display of an increase in the RBE with an increase in the LET of the radiation. However, subdividing the entire existing variety of repair processes into classes *P* and *Q* does little to clarify specific mechanisms. In this regard, the works of German researchers are of greater interest (Frankenberg, 1984; Frankenberg-Schwager et al., 1984). These authors attempted to elucidate the nature of primary radiation damage, which leads to an increase in RBE. It was shown that double breaks in DNA strands are critical damage for the death of yeast cells under the influence of ionizing radiation, since one double break per yeast cell, according to their data, corresponds to one lethal event. It was also shown that α-particles produce not only more double breaks per yeast cell in comparison with sparsely ionizing radiation, but also a greater number of DNA fragments after a three-day post-radiation recovery of irradiated yeast cells.

Thus, the formation of damages, which are less efficiently repaired after the action of densely ionizing radiation than those of sparsely ionizing radiation, are one of the mechanisms of the participation of recovery processes in the display of the RBE of high-LET radiation. Cell recovery from sublethal and potentially lethal radiation damage is also important in clinical radiology (Weichselbaum, 1986; Yarmonenko et al., 1992; Lindenbraten, Korolyuk, 2000; Hall, Giaccia, 2018; Kaprin, Mardynsky, 2019), reducing the effectiveness of tumor treatment with ionizing radiation. In these works, it is noted that the ability of cells to recover from radiation damage is the main reason for the resistance of some tumor cells to the action of ionizing radiation. Most of the modern methods for increasing the radiosensitivity of cancer cells are based on suppressing the cell ability to recover from radiation damage. One of such methods is the use of radiation with high LET in radiation therapy.

Since the processes of interaction of α-particles with living matter are qualitatively identical to the interaction of protons formed in tissues under neutron irradiation, the results obtained can be used for radiobiological analysis of the effectiveness of antitumor radiation therapy of hadron radiation – neutrons, protons and other heavy charged particles.

3.3. The Role of Recovery Processes in Genetic Instability of Yeast Cells Irradiated with Low- and High-LET Radiation

The relationship between the RBE of densely ionizing radiation and the ability of cells to recover from radiation damage, demonstrated in the previous section, can also be displaced when considering the cell genetic instability. However, this issue has practically not been studied. The role of DNA damage repair processes in genetic instability, cell aging, and carcinogenesis is currently being actively studied (Cheng, Loeb, 1993; Mazurik, Mikhailov, 2001; McMurray, Goltschling, 2004; Vorobtsova, 2008; Shen, 2011). There is evidence that double-stranded DNA breaks are important for the delayed colony formation by cells surviving exposure to ionizing radiation (Chang, Little, 1992 a, b; Brosh, Bohr, 2007; Korogodin et al., 2007). The mechanism of the delayed appearance of colonies, along with other remote effects of radiation, is realized in individual cells surviving the action of ionizing radiation, and may be revealed as genetic instability in the descendants of irradiated cells. This effect was found and described for cultured cells of various origins (Klein et al., 1990; Morgan, Murnane, 1995; Little, 1998). Evidence was presented to support the hypothesis that ionizing radiation can induce an inherited genomic process of instability, which leads to an increase in the frequency of genetic changes among the offspring of the original irradiated cell (Klein et al., 1990; Korogodin et al., 2007; Petin et al., 2008; Korogodin, 2010).

There are no publications in the literature on the comparison of genetic instability for wild-type haploid and diploid yeast cells and their radiosensitive mutants after exposure to low- and high-LET radiation. Although genetic instability is also known after irradiation with high LET radiation (Kadhim et al., 1994; Little et al., 1997; Dugan, Bedford, 2003; Ponnaiya et al., 2004), a little-studied aspect of this problem is the quantitative assessment of the RBE of radiation-induced genetic instability after the action of sparsely and densely ionizing radiation on haploid and diploid cells of various genotype. Moreover, there are no data in the literature on the comparison of the RBE of densely ionizing particles for both cell inactivation and genetic instability. Therefore, comparison of the RBE for lethal effects and for genetic instability of haploid and homozygous diploid cells, differing in radiosensitivity, remains completely unexplored. This section presents experimental data on the study of the RBE of α-particles for cell survival and the delayed colony formation by surviving wild-type haploid and diploid yeast cells and their radiosensitive mutants that survived irradiation with sparsely and densely ionizing radiation.

There is evidence that double-stranded DNA breaks initiate the mechanism of the delayed reproductive cell death (Chang, Little, 1992 a, b). It is well known that the repair of double-strand DNA breaks requires two homologous chromosomes (Luchnik et al., 1977; Frankenberg et al., 1981) and due to this ability diploid yeast cells are more resistant to ionizing radiation than haploid cells, the ability of which to recover from double-strand breaks is significantly reduced and can be realized due to less efficient reunification of breaks at non-homologous DNA ends (Lewis, Resnisk, 2000; Mao et al., 2008). This leads to the fact that radioresistant diploid yeast cells that survived after irradiation and exhibit sigmoid dose-effect curves retain a number of non-lethal sub-damages responsible for genetic instability in the distant offspring of cells that survived after irradiation. This effect was less pronounced for wild-type haploid yeast cells, characterized by exponential survival curves. Based on this fact, it was concluded in many works that the shape of the survival curve and the ability of cells to repair double-strand DNA breaks determine the delay in the appearance of colonies by irradiated cells (Kapultsevich et al., 1974; Korogodin et al., 1974, 1977 b; Kapultsevich, 1978; Tolstorukov et al., 1979, 1982; Tolstorukov, Effremov, 1984; Korogodin et al., 2007; Petin et al., 2008). However, it is not excluded that genetic instability caused by ionizing radiation is largely determined by cell ploidy, and not related with the shape of the survival curve and cell ability to recover from radiation damage. More research is needed to confirm this assumption.

That is why this section investigates the following aims. The first, to determine whether genetic instability is influenced by the ability of cells to recover from radiation damage and the shape of the survival curve, or whether this effect depends to a greater extent on cell ploidy. The second, to quantify and compare the RBE of α-particles for the survival and genetic instability of haploid and diploid yeast cells of various genotypes.

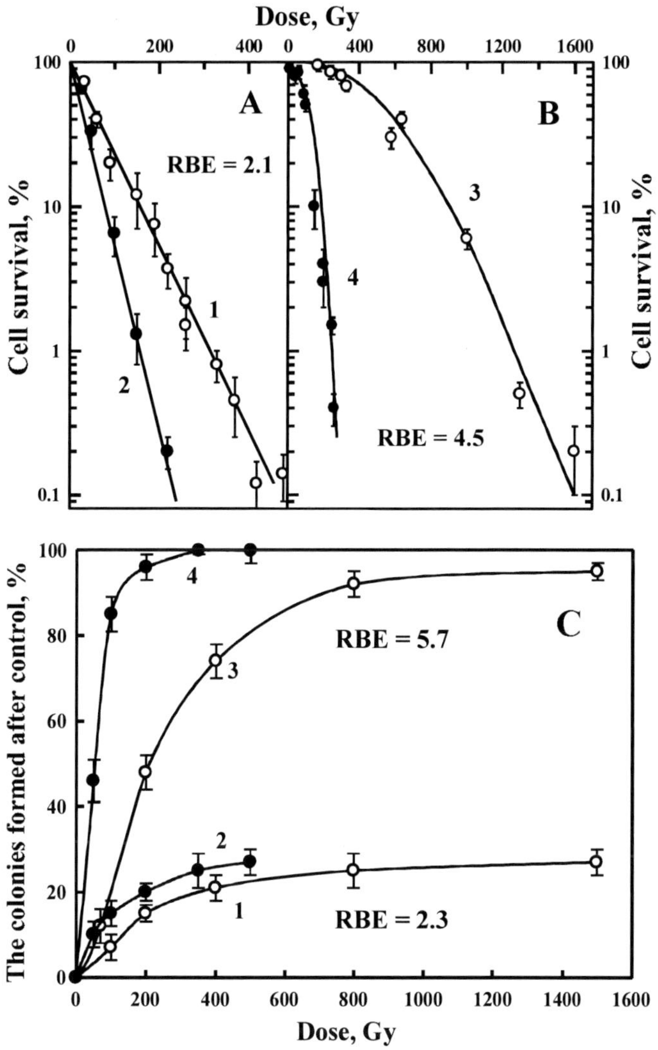

Figure 3.11. Radiobiological responses of haploid (strain S288C, RAD) and diploid (strain XS800, RAD/RAD) wild-type *S. cerevisiae* yeast cells: survival curves of haploid (A) and diploid (B) yeast cells and the delayed colony formation (C) by haploid (curves 1, 2) and diploid (curves 3, 4) yeast cells irradiated with γ-rays of ^{60}Co (open circles) and α-particles of ^{239}Pu (closed circles).

Figure 3.11 shows the dependence of the survival of *S. cerevisiae* yeast cells on the dose of ionizing radiation of haploid (Figure 3.11 A) and diploid (Figure 3.11 B) wild-type *S. cerevisiae* yeast cells irradiated with γ-rays (open circles) and α-particles (closed circles) at the stationary stage of growth.

Figure 3.11 C shows the dependence of colonies produced later than control by surviving haploid (curves 1, 2) and diploid (curves 3, 4) cells after irradiation with γ-rays of ^{60}Co (open circles) and α-particles of ^{239}Pu (closed circles). It can be seen that the survival curves, regardless of the radiation quality, remain exponential for the haploid strain and sigmoid for the diploid one. These data correspond to the above data for another pair of homozygous haploid and diploid cells (Figure 3.4). The exponential shape of the survival curve of haploid cells indicates a lower efficiency of the formation in these cells of some sub-lethal damage responsible for the sigmoid shape of the survival curves of diploid cells. Then one should expect a less pronounced effect of the late appearance of colonies for haploid cells that survived after irradiation, which is well illustrated by the data shown in Figure 3.11 B. This consequence was previously confirmed for the haploid yeast *S. cerevisiae*, strain 2GU-78 (Kapultsevich et al., 1972). In this work, the maximum effect of the delayed growth of colonies was 20% for haploid cells and 100% for diploid ones after exposure to sparsely ionizing radiation.

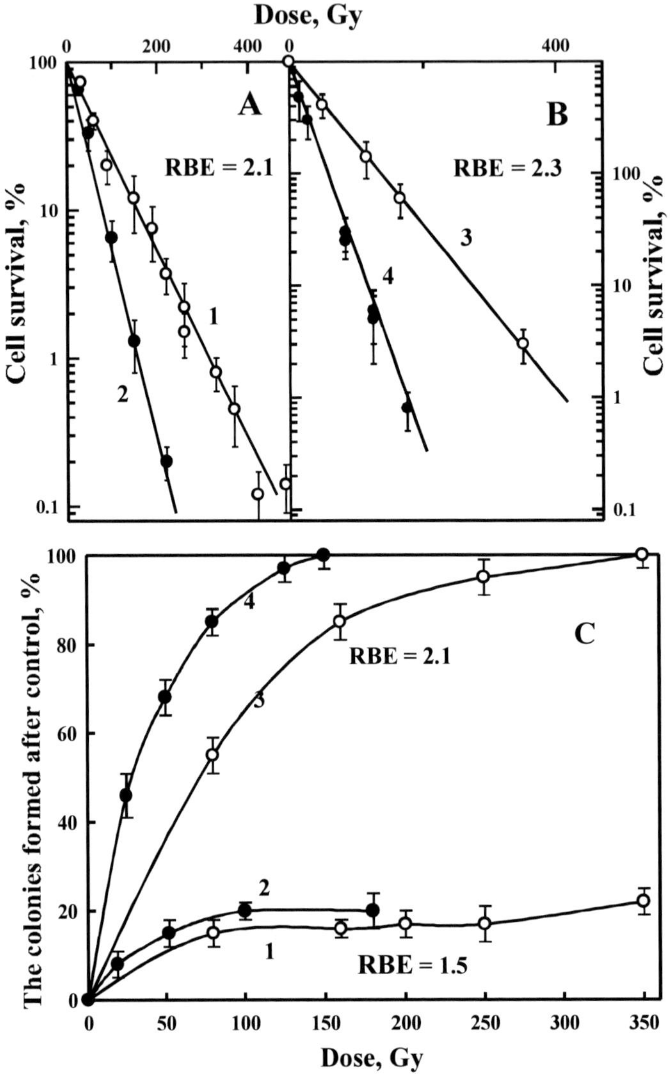

Figure 3.12. Radiobiological responses of radiosensitive mutants of haploid (strain XS774-4d, *rad*51) and diploid (strain XS806, *rad*51/*rad*51) *S. cerevisiae* yeast cells: survival curves for haploid (A) and diploid (B) yeast cells, the delayed appearance of colonies (C) by haploid (curves 1, 2) and diploid (curves 3, 4) yeast cells irradiated with γ-rays of ^{60}Co (open circles) and α-particles of ^{239}Pu (closed circles).

These data are consistent with previously published numerous results (Kapultsevich et al., 1972; Korogodin et al., 1974, 1977 b; Kapultsevich, 1978; Tolstorukov et al., 1979, 1982; Tolstorukov, Effremov, 1984; Korogodin et al., 2007). Fundamentally new data obtained in our study and presented in Figure 3.11, relate to the quantification of the RBE for genetic instability of haploid and diploid yeast cells and its comparison with the RBE for cell survival. It can be seen from this figure that even after exposure to high LET radiation, genetic instability remains low for a haploid strain ~ 25% and again reaches 100% for a diploid strain. These values are achieved at lower doses of α-particles indicating high RBE values of densely ionizing radiation for the genetic instability. The proportion of cells containing sub-damages ineffective for cell inactivation but responsible for the delayed appearance of cells survived irradiation increased with an increase in the radiation dose. As shown above, this effect reflects the genetic instability.

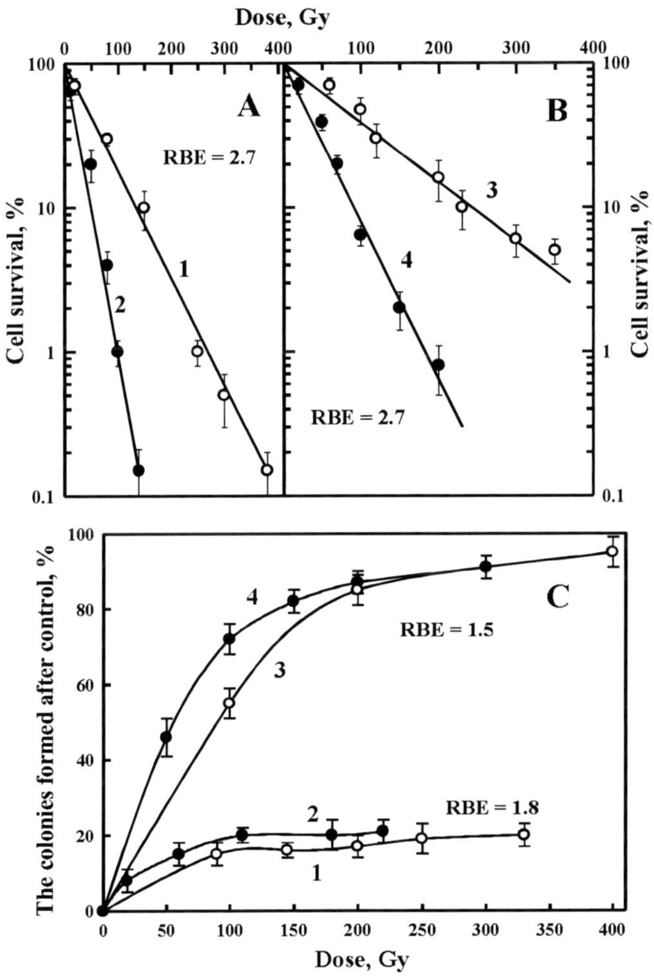

Figure 3.13. Radiobiological responses of radiosensitive mutants of haploid (strain g160/2d, *rad*52) and diploid (strain XS1898, *rad*52/*rad*52) yeast cells *S. cerevisiae*: survival curves of haploid (A) and diploid (B) yeast cells and the delayed appearance of colonies (C) by haploid (curves 1, 2) and diploid (curves 3, 4) yeast cells irradiated with ^{60}Co γ-rays (open circles) and ^{239}Pu α-particles (closed circles).

We quantitatively estimated the RBE of α-particles for genetic instability by the maximum ratio of isoeffective doses of γ- and α-radiation for colonies formed after control. This ratio was evaluated within the range of this parameter from 10 to 20% for haploid strain and from 40 to 70% for a diploid one. It was demonstrated that the RBE of α-particles is 2.3 for the haploid strain and 5.7 for the diploid strain that corresponds to the RBE of α-particles, assessed by the cell survival (Figure 3.11, Table 3.1). The same increased efficiency of α-particles for cell inactivation and late formation of colonies by irradiated cells may be associated with the greater efficiency of densely ionizing radiation to produce lethal radiation damage and accompanying sub-damage, which persist in the distant descendants of cells that survived after irradiation and thereby cause destabilization of the genome. To understand whether genetic instability is due to the ability of cells to recover from radiation damage or the shape of the survival curves, as was traditionally assumed, it was of interest to obtain similar dependences on the dose of cell survival and the delayed colony formation by cells surviving after irradiation with γ-rays and α-particles of radiosensitive mutants defective in some repair systems. The results are shown in Figures 3.12, 3.13 and 3.14.

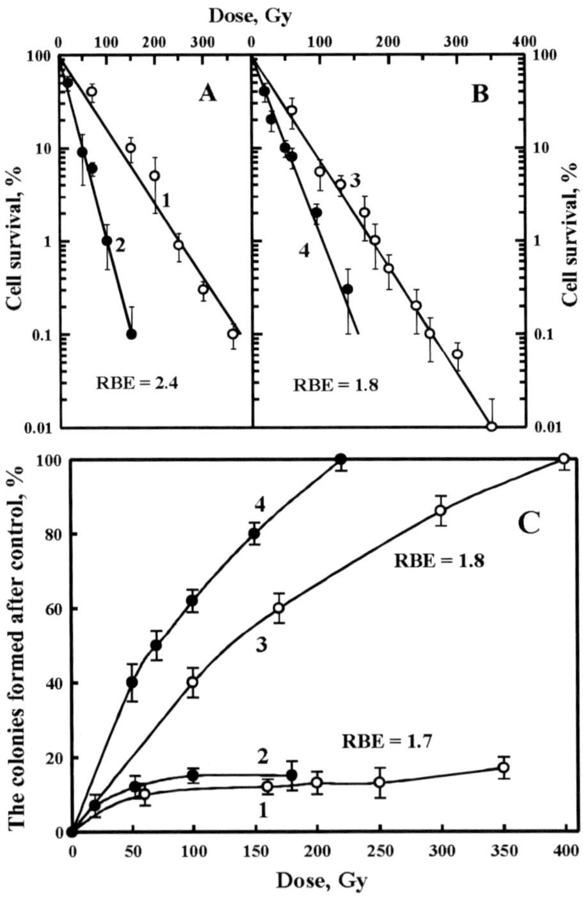

Figure 3.14. Radiobiological responses of radiosensitive mutants of haploid (strain g218/7c, *rad*54) and diploid (strain XS1879, *rad*54/*rad*54) *S. cerevisiae* yeast cells: survival curves of haploid (A) and diploid (B) yeast and the delayed appearance of colonies (C) for haploid (curves 1, 2) and diploid (curves 3, 4) yeast cells irradiated with ^{60}Co γ-rays (open circles) and ^{239}Pu α-particles (closed circles).

Radiobiological parameters of these data are summarized in Table 3.1. It can be seen that the relative radiosensitivity of mutant strains, assessed by the ratio LD_{90}(wild)/LD_{90}(mutant) for sparsely ionizing radiation, increased to a much greater extent for diploid strains than for haploids in comparison with the radiosensitivity of wild-type strains. Here LD_{90} is a dose that reduces cell survival to a 10% level.

Table 3.1. Radiobiological parameters of haploid and diploid yeast cells *S. cerevisiae* of various genotypes

Strain	Genotype	The greatest genetic instability, %	$\frac{LD_{90}(wild)}{LD_{90}(mutant)}$	RBE for cell survival	RBE for genetic instability
S288C	RAD	30	–	2.1 ± 0.3	2.3 ± 0.2
XS800	RAD/RAD	100	–	4.5 ± 0.4	5.7 ± 0.3
XS774-4d	*rad*51	20	1.07	2.1 ± 0.2	1.5 ± 0.1
XS806	*rad*51/*rad*51	100	6.00	2.5 ± 0.1	2.1 ± 0.3
g160/2d	*rad*52	20	1.14	2.7 ± 0.3	1.8 ± 0.4
XS1898	*rad*52/*rad*52	100	3.7	2.7 ± 0.4	1.5 ± 0.2
g218/7c	*rad*54	20	1.33	2.4 ± 0.1	1.7 ± 0.3
XS1879	*rad*54/*rad*54	100	10.4	1.8 ± 0.1	1.8 ± 0.3

The presented data show that mutant genes lead to only a slight increase in the sensitivity to the action of sparsely ionizing radiation of haploid cells in comparison with the parent wild-type cells. These ratios are 1.07, 1.14, and 1.33 for cells with the *rad*51, *rad*52, and *rad*54 mutations. On the contrary, these mutations in the homozygous state of diploid cells led to a significantly greater increase in cell radiosensitivity. The corresponding ratios are 6.0, 3.7 and 10.4. A greater increase in the radiosensitivity of diploid cells occurred because of the loss of their ability to recover from double-strand DNA breaks, mainly due to molecular defects in the process of recombination of homologous chromosomes (Friedberg et al., 2006).

The RBE of α-particles for cell survival of haploid and radiosensitive diploid mutants yeast cells with exponential survival curves was quantified by the ratio of the mean lethal doses D_0 after exposure to ionizing radiation with low and high LET radiation: RBE = $D_0(\gamma)/D_0(\alpha)$. In contrast to haploid cells, the survival curves for wild-type diploid cells were sigmoid after irradiation with both γ-rays and α-particles. For this case, RBE was assessed by the ratio of the corresponding D_0 values at the exponential parts of the survival curves. The high efficiency of densely ionizing radiation for genetic instability has been reported in many works by researchers working with cultured mammalian cells (Kadhim et al., 1994; Little et al., 1997; Dugan, Bedford, 2003; Ponnaiya et al., 2004). Due to the criteria used for genetic instability, the authors of these works were unable to assess RBE quantitatively. Our proposed test of genetic instability allowed us to quantify RBE values by the ratio of isoeffective doses after irradiation with γ-rays and α-particles.

The results presented in Figures 3.12, 3.13 and 3.14 and summarized in Table 3.1 show that exposure to high LET radiation was more effective not only for survival, but also for genetic instability. It is also clear that RBE values were close to each other for both studied tests, and the highest RBE values (4.5–5.7) were observed for wild-type diploid cells, while for wild-type haploid yeast cells and all radiosensitive mutants, regardless of their ploidy RBE varied within 1.8–2.7 without any systematic pattern. The data summarized in Table 3.1 shows that the greatest values of genetic instability reached 100% with increasing dose for all diploid yeast cells, regardless of the shape of the survival curve and cell ability to recover from radiation damage, while for all haploid strains, this effect was varied within the range of 20–30%. It was of interest to imagine the dependence of the delayed production of colonies not on the radiation dose, but on the cell survival. Such dependences for *Saccharomyces cerevisiae* yeast cells exposed to γ-rays and α-particles are presented in Figure 3.15.

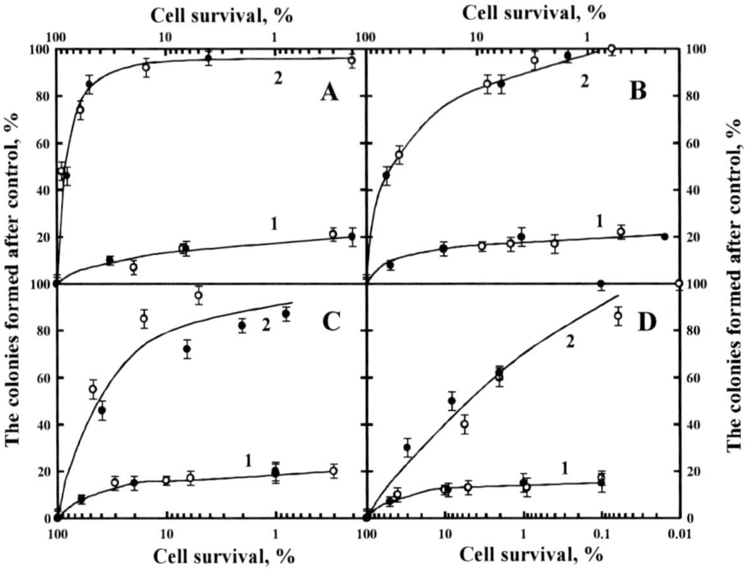

Figure 3.15. The dependence of the delayed appearance of colonies produced by haploid (curves 1) and diploid (curves 2) yeast cells of various genotypes that survived after irradiation with ^{60}Co γ-rays (open circles) and ^{239}Pu α-particles (closed circles) on cell survival. A: S288C (RAD), XS800 (RAD/RAD); B: XS774-4d (*rad*51), XS806 (*rad*51/*rad*51); C: g160/2d (*rad*52), XS1898 (*rad*52/*rad*52); D: g218/7c (*rad*54), XS1879 (*rad*54/*rad*54).

It is clear that such a presentation of the data shows that the survival of both haploid and diploid cells unambiguously determines the yield of colonies formed later than in control. It is also obvious that the results presented in Figure 3.15 demonstrate the independence of the delayed colony appearance from LET radiation for all strains. These data show that equally effective lethal damage, i.e., lesions leading to the same cell survival are accompanied by the formation of an equal number of sub-damages responsible for genetic instability. To compare the effectiveness of the effect of different types of radiation on genetic instability, we presented the dependence of the delayed colony formation on cell survival for wild-type haploid and diploid yeast cells (Figure 3.16).

Figure 3.16 contains data on the late formation of colonies by cells that survived after irradiation with γ-rays, α-particles and UV light. The results obtained after irradiation of yeast cells with UV light will be described in more detail in the next chapter. As a whole, the results shown in Figures 3.15 and 3.16 show that the delay in the formation of colonies by irradiated cells, depending on their survival, practically did not depend on the type of radiation for isoeffective irradiation. These data indicate that the equally effective lethal effect induced by different types of radiations is accompanied by an equal number of sub-damages responsible for genetic instability. In other words, genetic instability is associated with the accumulation of sub-damages that are ineffective for cell inactivation.

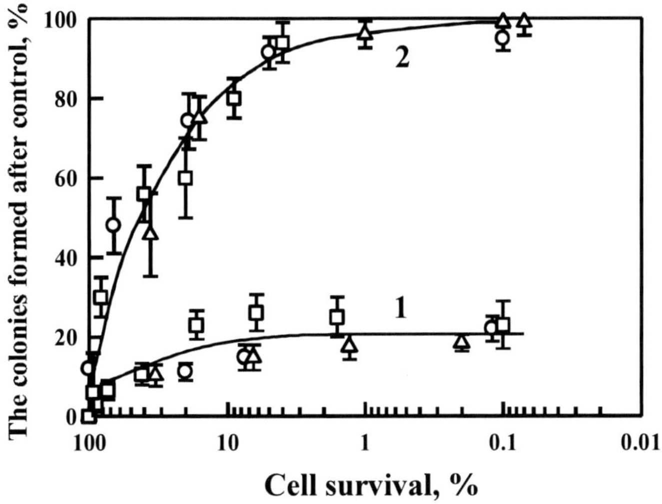

Figure 3.16. The relationship between cell survival and the delayed appearance of colonies formed by haploid (strain S288C, curve 1) and diploid (strain XS800, curve 2) yeast cells of *Saccharomyces cerevisiae* that survived after irradiation with ^{60}Co γ-rays (circles), ^{239}Pu α-particles (triangles) and UV light (squares).

Once again, we note that these sub-damages persist in the distant offspring of cells that survived after irradiation and can manifest themselves in a slowdown of their reproduction. This problem is important in connection with the use of the interaction theories of non-lethal sub-damages to explain the patterns of synergy effects (Petin et al., 2012, 2020; Petin, Morozov, 2015; Petin, Kim, 2016), which should be taken into account in the treatment of tumors with the combined use of ionizing radiation with hyperthermia and chemical sensitizers (Hahn, 1982; Yarmonenko et al., 1992; Lindenbraten, Korolyuk, 2000; Joiner, van der Kogel, 2009; Tsyb, Mardynskiy, 2010; Kaprin et al., 2017; Hall, Giaccia, 2018; Kaprin, Mardynskiy, 2018), as well as during photoradiation and photochemical therapy (Melloni et al., 1984; Castellani, 2012).

3.4. Cell Recovery after Repeated Irradiation with Ionizing Radiation

There are numerous data, which indicate both cell survival and RBE of ionizing radiation with high LET depend on genes that control cell recovery from radiation damage. These data were confirmed above for yeast cells of various genotypes. For unicellular organisms of various origins, RBE was higher for cells capable of recovery than for cells that do not have this ability (Haynes, 1966; Petin, 1977, 1987; Alper, 1979; Petin, Kabakova, 1981; Frankenberg-Schwager et al., 1984; Krasavin, 1989; Weyrather et al., 1999; Shadley et al., 1991; Petin, Petin, 1995; Petin et al., 2006, 2020; Petin, Kim, 2016). In other words, RBE is determined by both the probability of the formation of primary radiation damage (physical events) and the ability of cells to recover from radiation damage (biological events). There is general agreement that radiation with high LET produces a large proportion of damage that is considered less reversible (Ritter et al., 1977; Korogodin, Krasavin, 1982; Blöcher, 1988; Heilmann et al., 1993; Stenerlöw et al., 2000; Chausov et al., 2009; Nakajima et al., 2013; Okumura et al., 2013;

Zadnepryanets et al., 2018 a, b) compared to low LET radiation. Regularities of recovery have been well studied for yeast cells at the cellular level (Korogodin, 1966, 2010; Kapultsevich, 1978; Petin, 1987; Korogodin, 1993; Petin et al., 2020; Petin, Kim, 2016) and at the molecular level (Luchnik et al., 1977; Saeki et al., 1980; Glasunov et al., 1989; Frankenberg-Schwager et al., 1994). The ability of eukaryotic cells to recover from radiation damage was first discovered in experiments with yeast (Korogodin, 1958 a, b). If the irradiated yeast cells are stored after irradiation in a non-nutrient medium at 30°C before plating on a solid nutrient medium, their survival is increased. This phenomenon is now known as cell recovery in a non-nutritive environment or liquid holding recovery (LHR) (Korogodin 1958 a, b; Patrick et al., 1964; Korogodin, 1966). A quantitative approach describing the recovery process as a decrease in the effective dose is well known (Davidson, 1956; Korogodin, 1966; Kapultsevich, 1978), and it has been applied to yeast (Kim et al., 2004; Petin, Kim, 2004, 2016) and mammalian cells (Kim et al., 2005 a) exposed to ionizing or UV radiation in combination with various chemical and physical agents. This approach makes it possible to determine whether the recovery process is disrupted after combined actions, or whether a decrease in the rate and degree of recovery can be fully interpreted by an increase in the proportion of irreversible radiation damage.

It is known that during radiation treatment of tumors, the dose required to destroy all malignant cells cannot be applied once due to damage of normal cells (Yarmonenko et al., 1992; Lindenbraten, Korolyuk, 2000; Joiner, van der Kogel, 2009; Kaprin, Mardynsky, 2019). In this regard, fractionated irradiation is used in oncology, and the number of fractions can reach 10 or more. One of the main problems of fractionated irradiation is the cell recovery from sublethal radiation damage, which are responsible for the formation of the shoulder of the survival curves in the region of low doses of ionizing radiation. Phenomenologically, this type of recovery is manifested in the restoration of the sigmoid shape of the survival curves after each repeated irradiation. The patterns of this type of recovery have been sufficiently well studied quantitatively, and on this basis, various complex fractionation schemes have been proposed and introduced into practice (Ivanov, 1986; Kogl, 1986; Yarmonenko et al., 1992; Hall, Giaccia, 2018). To reduce this recovery, it is proposed to use densely ionizing radiation (Raju, 1980; Saha, 2006; Joiner, van der Kogel, 2009; Tsyb, Gulidov, 2010). It is known that the response of cells to radiation determines the response to radiation of the entire tumor (Weichselbaum, 1986; Steel, 2002). However, there is insufficient information in the literature on the change in radiosensitivity and the ability of cells to recover from potentially lethal cell damage after repeated exposure to densely ionizing radiation. Previously, it was shown that with repeated irradiation of yeast cells with γ-radiation, the ability of cells to recover from radiation damage decreased with an increase in the number of successive irradiation (Korogodin, 1966). However, in this work, the parameters of post-radiation recovery were not quantified. Therefore, this section of our work is devoted to the study of the ability of yeast cells to recover from potentially lethal radiation damage after repeated irradiation with ionizing radiation of various qualities. The study was carried out with diploid yeast cells capable to recovery in LHR, often serves as an indicator of the overall cell activity to recovery (Kumar et al., 1985). Therefore, our studies on the repeated irradiation and quantitative assessment of LHR parameters, presented below, may be relevant to other types of repairs. The main objectives of our study were (i) to study the radiosensitivity and recovery of yeast cells after repeated exposure to radiation with low and high LET radiation; (ii) to answer the question of whether radiation with high LET affects the recovery process itself, or whether such exposure leads to the formation of more serious irreversible radiation damage from which cells are unable to recover; (iii) to quantify the RBE of high LET radiation after repeated irradiation of yeast cells.

In the process of cell recovery in a non-nutritive environment, a number of primary radiation damage is eliminated that leads to a decrease in the effective dose of radiation and the corresponding increase in cell survival. For yeast cells, it was demonstrated (Korogodin, 1966; Kapultsevich, 1978; Petin, Kim, 2004; Kim et al., 2004, 2005 a) that LHR of irradiated cells can be described by a decrease in the effective dose (equation (2.3)). This equation contains two main parameters – the recovery constant, which characterizes the probability of recovery per unit time β (equation (2.7)) and the irreversible component of radiation damage K (equation (2.3)).

In this section of the work, the ability of cells to recover after repeated exposure to radiation with low and high LET radiation will be described. Diploid yeast cells were irradiated with ^{60}Co γ-rays and ^{239}Pu α-particles. The curves of the dependence of cell survival both on the radiation dose and the duration of recovery after exposure to radiation with different LET were obtained experimentally. When the irradiated cells were fully recovered, they were irradiated again and then recovered. The procedure was repeated three times. Figures 3.17 A, C, E shows the survival curves obtained immediately after γ-irradiation (closed circles) and after complete recovery in a non-nutrient medium at 30°C (open circles). Figures 3.17 B, D, F shows the dependence of cell survival on the duration of recovery in a non-nutritive medium after irradiation. The results were obtained after the first (A, B), second (C, D) and third (E, F) irradiation. Similar data after α-irradiation are shown in Figure 3.18.

It is seen that the shape of the survival curves was sigmoid after the first irradiation with both γ-rays and α-particles. After subsequent exposures, the survival curves became exponential, and the mean lethal doses characterizing the slope of the survival curves only slightly depended on the number of subsequent exposures. It is also clear that after each irradiation with γ-rays at a dose of 1200 Gy and complete recovery, the percentage of surviving cells decreased by an order of magnitude. Approximately such a decrease in survival occurred after irradiation with α-particles at a dose of 150 Gy. It is obvious that cell recovery gradually slowed down, reaching a plateau after 24–48 h of recovery (Figures 3.17 B, D, F and 3.18 B, D, F, Table 3.2).

Table 3.2. Parameters characterizing yeast cell recovery after exposure to sparsely and densely ionizing radiation

The type of radiation	The number of exposures	D_0, Gy; before recovery	RBE	D_0, Gy; after recovery	K	β, ч$^{-1}$
γ-rays	1	189	–	387	0.45	0.055
	2	172	–	366	0.46	0.055
	3	215	–	301	0.58	0.055
α-particles	1	65	2.9	107	0.47	0.041
	2	70	2.4	107	0.77	0.041
	3	81	2.6	81	1.00	–

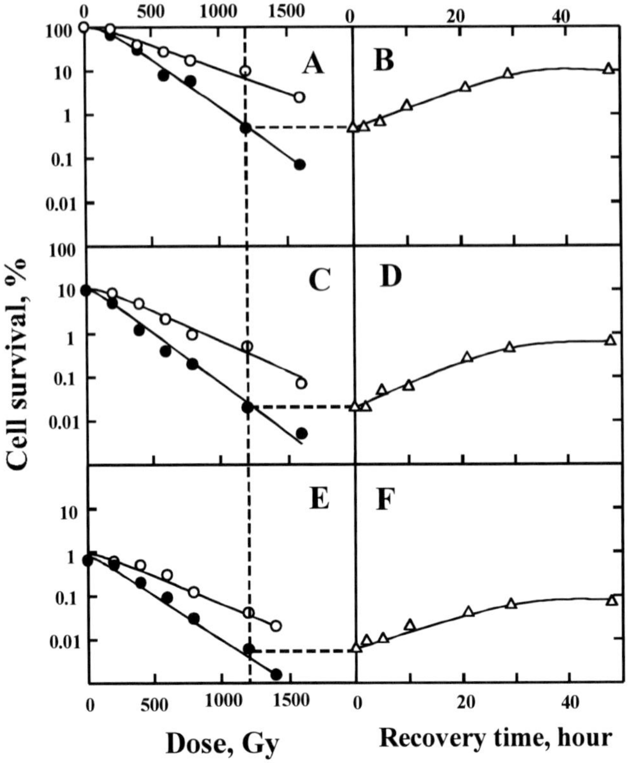

Figure 3.17. The dependence of the survival of diploid yeast cells *Saccharomyces cerevisiae* (strain XS800) on the dose of γ-rays (A, C, E), obtained immediately after irradiation (closed circles) and after complete cell recovery (open circles); cell survival (B, D, F) in the dependence on the duration of post-radiation recovery; A, B – the first exposure; C, D – the second exposure; E, F – the third exposure.

The data summarized in Table 3.2 include the following parameters of the curves shown in Figures 3.17 and 3.18: the mean lethal dose D_0 immediately after irradiation and after complete recovery of cells irradiated with γ-rays and α-particles. Table 3.2 also includes the main recovery parameters calculated on the basis of these figures: the irreversible component of radiation damage K and the recovery constant β after the action of sparsely and densely ionizing radiation, calculated by equations (2.3) and (2.7), respectively. The data presented in Table 3.2, allowed us to calculate the RBE values through the ratio $D_0(\gamma)/D_0(\alpha)$. After the first, second and third irradiation, these values were 2.9, 2.4 and 2.6. This means that, within the limits of possible experimental errors, the RBE of α-particles did not depend on the number of repeated irradiations. As can be seen from the Table 3.2, the degree of recovery after low LET radiation decreased after repeated exposure, as a result of which the irreversible component of radiation damage K increased from 0.45 after the first to 0.58 after the third exposure to γ-rays The increase in the proportion of the irreversible component was much more pronounced after irradiation with α-particles, when the irreversible component increased from 0.47 after the first exposure to 0.77 after the second and to 1.00 after the third. A value of K = 1.0 means that all potentially lethal damage become completely irreversible after the third exposure to densely ionizing radiation.

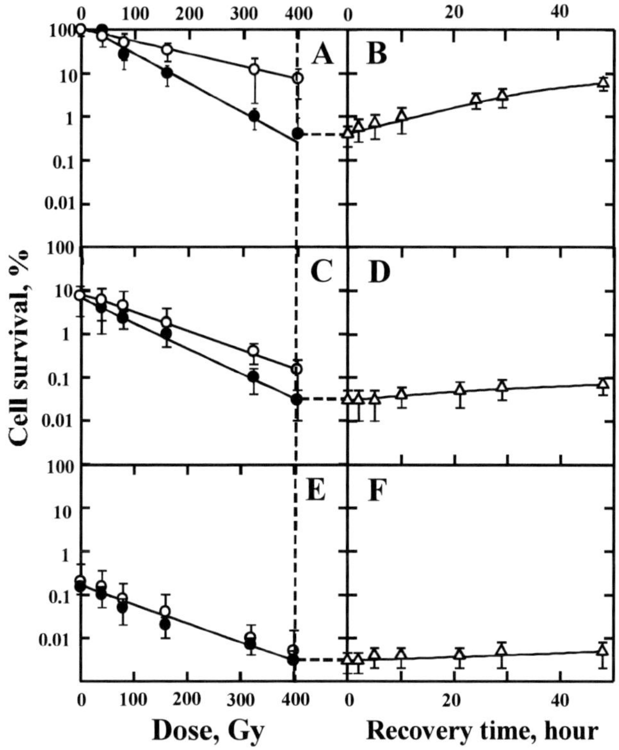

Figure 3.18. The dependence of the survival of diploid yeast cells *Saccharomyces cerevisiae* (strain XS800) on the dose of exposure to α-particles (A, C, E), obtained immediately after irradiation (*closed* circles) and after complete recovery (open circles); cell survival (B, D, F) depending on the duration of post-radiation recovery; A, B – the first exposure; C, D – the second exposure; E, F– the third irradiation.

It is also clear that for both types of radiation, regardless of the number of repeated exposures, the recovery constant β remains permanent for the successive repeated γ-irradiation. Exposure to α-particles resulted in somewhat reduced the recovery constant β. This means that the very process of recovery during irradiation is not damaged after repeated irradiation both with sparsely and densely ionizing radiation. This is consistent with previously published results (Petin, Kim, 2004) obtained for a large number of yeast cell strains. New results from this study indicate that repeated exposures to high LET radiation significantly reduce the ability of cells to recover from potentially lethal damage due to the formation of irreversible radiation damage, from which cells could not recover from radiation damage. These data show that a decrease in the ability of cells to recover after an increase in LET radiation is less associated with a violation of the repair systems but in a greater extent with an increase in the irreversible component of radiation damage, i.e., with the production of more "severe" damage, from which cells are unable to recover.

It is known that the main advantages of densely ionizing radiation used in the treatment of oncology diseases are the increased RBE values, due to both the higher probability of the formation of primary radiation damage and a decrease in the ability of cells to recover at the biochemical stage of radiation damage fixation (Raju, 1980; Kogl, 1986; Yarmonenko et al., 1992; Lindenbraten, Korolyuk, 2000; Joiner van der Kogel, 2009; Tsyb, Gulidov, 2010; Hall, Giaccia, 2018; Kaprin, Mardynskiy, 2018). From the presented data, it is clear the importance to know about a decrease in the degree of curvature of the initial part of the survival curve after repeated irradiation, which results in a significant decrease in the ability of cells to recover from sublethal damage. These new results indicate that repeated exposure to densely ionizing radiation sharply decreases the ability of cells to recover not only from sublethal, but also from potentially lethal damage due to the formation of irreversible damage, from which cells are unable to recover.

In general, the new results obtained in this section in conjunction with previously published data (Petin, 1977, 1987; Ritter et al., 1977; Blöcher, 1988; Krasavin, 1989; Heilmann et al., 1993; Stenerlöw et al., 2000) show that the mechanism of RBE increasing may be explained by two factors – an increased probability of the primary radiation damage formation and a decrease in the ability of cells to recover from damage caused by high LET radiation. The reason for the decrease in the ability of cells to recover after exposure to high LET radiation is an increase in the proportion of irreversible radiation damage, up to the complete cessation of recovery after the third α-irradiation. An increase in the relative proportion of irreversible radiation damage after repeated exposure can occur for two reasons: either due to the formation of damage that cannot be restored at all, or due to a violation of the recovery process itself. The presented data show that the irreversible component of radiation damage was increased after repeated irradiation with γ-rays from 0.45 after the first irradiation to 0.58 after the third exposure. This effect was more noticeable after irradiation with densely ionizing radiation, the corresponding values of the irreversible component were 0.55 and 1.0. The second possibility was also analyzed. Our results showed that regardless of the number of repeated irradiations, the recovery constant changes insignificantly for both types of radiation. The cessation of cell recovery after reaching the plateau of the recovery curve is unlikely due to depletion of the corresponding repair enzymes, since the recovery process in the non-nutritive medium continued not only after the first, but also after the second and third exposure to radiation with low LET. It was shown that the recovery constant slightly depends on the LET of radiation: it changed from 0.055 h^{-1} for γ-rays to 0.041 h^{-1} for α-particles. In other words, 5.5 and 4.1% of the lesions available for repair were repaired every hour after γ- and α-irradiation. It should be noted that in a detailed study of the kinetics of recovery in a non-nutritive medium in wild-type cells and radiosensitive mutants of *Saccharomyces* yeast exposed both to low- and high LET radiation, the recovery probability was identical after γ- and α-irradiation for 8 yeast strains and varied within 0.05–0.1 h^{-1} (Petin, Kim, 2005 a). The results obtained in present work (Table 3.2) are close to this range of values. This further confirms that the recovery process from potentially lethal radiation damage itself is not damaged by repeated exposures to low- and high LET radiation. The slowdown and decrease in the volume of cell recovery in a non-nutritive medium is fully explained by an increase in the proportion of irreversibly damaged cells.

Thus, the biological efficiency of radiation with different LET is determined by factors of physical and biological nature. The factor of physical nature affects the output of direct double-strand breaks as well as more complex DNA damage, which are difficult for cell repair. The factor of biological nature is aimed to correcting and eliminating DNA damage. This effect is under genetic control (Korogodin, Krasavin, 1982). It is known that single breaks constantly occur within the cell without any connection with radiation, but simply due to thermal instability of DNA, as well as a result of some oxidative and enzymatic processes. Moreover, single breaks are even necessary during DNA replication. Actually, the molecule must be unwound at the following synthesis, for which one strand must be able to rotate the other, which cannot be achieved without breaking it. It is assumed, however, that irradiation produces not only single breaks similar to those arising spontaneously, but also "complex" breaks appear, in which several broken bonds are located nearby in the DNA chain. Such breaks are repaired worse than those arising under normal conditions (Krasavin, 1989; Zadnepryanets et al., 2018 a, b). Double breaks can be produced as a result of random spatial coincidence of single breaks in opposite DNA strands, the probability of which increases with an increase in the dose of sparsely ionizing radiation. In such a case a large amount of energy is released in the microvolume of cell even after the action of sparsely ionizing radiation. The release of energy over the cell volume does not occur absolutely uniformly, but in a discrete portion of different sizes, so that in an approximately equal volume, several electron volts or several hundred electron volts may be transferred within the target (Korogodin, Krasavin, 1982; Krasavin, 2016). Under the action of densely ionizing radiation, the total number of lesions will be smaller, but the concentration of broken interatomic bonds per microvolume unit should be higher. Thus, the action of densely ionizing radiation produces a greater number of double strand and more complex breaks per dose unit than after the action of sparsely ionizing radiation. This is precisely the reason for higher RBE of densely ionizing radiation both for cell survival and its genetic instability.

Chapter 4

Genetic Instability of Yeast Cells after Irradiation with UV Light

Abstract

The general characteristics of UV-induced biological effects are discussed. New experimental data on the genetic instability of haploid and diploid yeast strains of wild type and their UV sensitive mutants are obtained. The main conclusion from the data presented here is that the genetic instability of the studied haploid and diploid strains, differing in UV sensitivity and in the shape of the survival curves is mainly determined by cell ploidy. The greatest value of genetic instability was 30–50% for haploid cells while it reached 100% with an increase in UV light dose for diploid cells regardless of their sensitivity to UV light, the shape of the survival curves and cell ability to recover from UV damage. These data show that the cells of all diploid strains that survived after irradiation retained some residual sub-damage emerging as many colonies produced after control, and these sub-damages preserved in the distant descendants of cells that survived after irradiation.

4.1. General Characteristics of UV-Induced Biological Effects

The new results, presented in Chapter 3 of this work, show that genetic instability, determined by the delayed colony forming ability by cells surviving after irradiation, is expressed to greater extent for diploid cells regardless of the shape of the survival curve, radiosensitivity and cell ability to recover from radiation damage. In order to confirm this main conclusion, it would be necessary to compare the genetic instability induced by UV radiation in wild-type haploid and diploid yeast cells and their UV sensitive mutants. The shape of the survival curves after irradiation with UV light, in contrast to the action of ionizing radiation, does not depend on cell ploidy and has a sigmoid shape for both haploid and diploid yeast cells. Then, a comparison of the genetic instability of haploid and diploid yeast cells that survived after exposure to ultraviolet radiation may contribute to a better understanding of the assumption whether this effect is related to the shape of the survival curve or it is mainly determined by cell ploidy.

UV radiation is electromagnetic radiation within the wavelength range from 380 to 10 nm. It is not visible to the eye. The sun is a natural source of UV radiation. Distinguish between near (to visible light) (380–200 nm) and far or vacuum parts of the UV spectrum (200 –100 nm). The near UV region is further subdivided into three ranges: UV-A (380–320 nm), UV-B (320–290 nm), UV-C (290–200 nm). Despite the fact that the radiation flux in the UV-A range is an order of magnitude higher than the intensity in the UV-B range, its effect is responsible only in the pigmentation (sunburn) of the skin. The UV-B range is most active, inducing a number of photobiological reactions. UV-C and shorter wavelength radiation does not hit the Earth's surface, it forms the ozone layer. The energy of quanta of short-wave UV radiation is sufficient to ionize atoms and molecules (20 nm – 60 eV; 100 nm – 12 eV; 200 nm – 6 eV).

Due to the ionization of air by short-wave UV radiation at an altitude of 50–80 km from the Earth's surface, the ionosphere is formed, which plays an important role in the propagation of radio waves. As stratospheric ozone depletes, humans and the environment will be exposed to increasing levels of UV radiation. Plain window glass, which is transparent to the visible part of the spectrum, provides protection against bioactive UV ranges, while quartz glass transmits the entire UV spectrum. About 45% of solar radiation comes from infrared rays, 48% from the visible part of the spectrum, and 7% from UV light.

Biological effects induced by UV radiation is similar to those from ionizing radiation – delayed cell division, chromosomal aberrations, cell inactivation, mutations, the ability of cells to recover from radiation damage, the heterogeneity of the dying population and colonies produced by cells surviving after irradiation, etc. UV radiation is fundamentally similar to the absorption of visible light. The work (Hoeijmakers, 1993) discusses a variety of structurally damages responsible for biological effects that occur after irradiation with UV light: cyclobutane-pyrimidine dimers, DNA cross-links, many types of chemical photoproducts and DNA adducts. The absorption and emission of UV light quanta are depicted in Figure 4.1 by solid lines. The wavy lines represent transitions from excited levels to lower levels, which are characterized by lower energy, while the excess energy of the electron is converted into heat without the emission of light. Upon absorption of UV light quanta, the molecule passes from the ground state (S_0) to the excited state S_1 (lower singlet state) or S_2 (upper singlet state). The energy of an excited electron from the upper singlet state extremely quickly (10^{-13}–10^{-11} s) is partially converted into heat through molecular vibrations or the electron passes to a more long-lived level (lower singlet state), where the electron is 3-4 orders of magnitude longer (10^{-9}–10^{-8} s). There are three possible options for the further fate of this electron. First, it can go to the ground level without radiation by converting the energy into heat through the vibrations of the molecule. Second, it can go to the ground level by emitting a fluorescence quantum. Since the duration of the residence of the electron in the lower singlet state (S_1) is short, fluorescence is observed directly during the illumination of the object. In addition, from this singlet state, the electron can go to a longer-lived triplet level (T), which is located slightly below S_1, and the duration of stay in which is much longer. From this state, an electron can go to the ground state S_0 with the emission of a phosphorescence quantum. The slowed down phosphorescence is due to the longer lifetime of the triplet state. This also explains the high probability of a molecule in this state to enter into photobiological reactions by transferring an electron from the triplet state to an acceptor.

The absorption spectrum is the dependence of the amount of absorption on the wavelength (or frequency) of the radiation incident on the object. The absorption spectra of individual atoms are linear and discontinuous, since the transition between the levels occurs during the absorption of quanta, the energy of which exactly corresponds to the energy difference between the two electronic levels. Molecules responsible for photobiological reactions consist of tens or even hundreds of atoms, so their absorption spectra are continuous with the presence of one or several maxima. This is because absorption of a quantum of UV light, the molecule receives additional vibrational energy associated with a change in conformation. Since the vibration energy is also quantized, there are a number of vibrational sublevels for each electronic level of the molecule. In Figure 4.1 schematically depicts (thin lines) vibrational sublevels of the ground (S_o), excited (S_1, S_2) and triplet (T) levels and possible transitions between them.

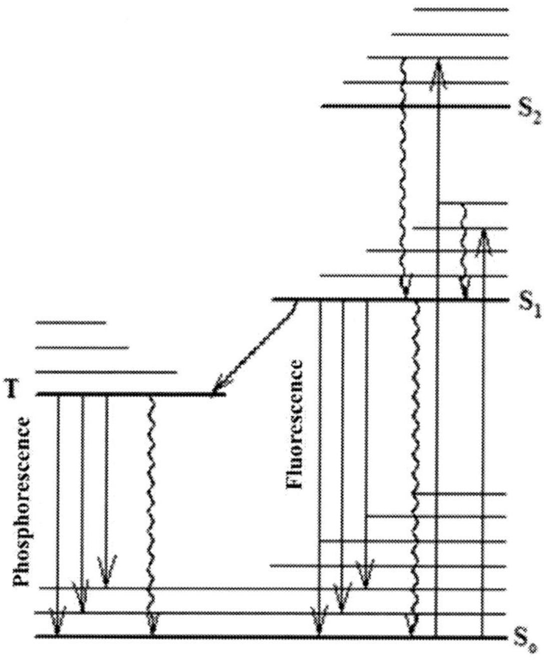

Figure 4.1. Energy levels, vibrational sublevels of molecules and possible transitions between them (solid lines – UV light, wavy lines – transitions due to thermal energy loss).

Not all of the electron binding energy is transformed in fluorescence emission, i.e., the transition of an electron is carried out from S_1 not to S_o (Figure 4.1), but ends at some higher vibrational sublevel. This means that part of the electron energy remains in the molecule, increasing its vibrational energy. Ultimately, additional vibrational energy is converted to heat (without radiation), or emitted by infrared radiation, characterized by low quantum energies.

The coincidence of the spectrum of UV radiation action with the absorption spectrum of nucleic acids and some proteins indicates that UV radiation is characterized by selective absorption and a corresponding action on nuclear and cytoplasmic structures responsible for the synthesis of macromolecular cellular components – DNA, RNA, proteins (Samoilova, 1967; Smith, Hanawalt, 1969; Castellani, 2012). A quantitative comparison of the action of ionizing and UV radiation (Lea, 1946) showed that the excitation of molecules is less biologically effective than their ionization, although even in those years it has been noted that this position requires further study and clarifications (Fritz-Neagley, 1961).

Biological effects caused by UV radiation depend on the radiation wavelength. According to the law known in photochemistry as the Einstein-Stark law of equivalence, only one activated molecule is formed in a result of the absorption of one quantum of light. Following the primary reaction, secondary reactions occur too. For example, the absorbed energy may be re-emitted at the same wavelength (resonance) or at longer wavelengths (fluorescence), or is converted into heat. An activated molecule can also acquire the ability to react with other molecules or change: a water molecule, for example, can decompose into free radicals (Konev, Volotovsky, 1979). Periodic intense UV irradiation is an important etiological factor in the development of malignant melanoma (Bald et al., 2014). In humans, UV radiation is capable of inducing two types of malignant skin tumors – melanoma and non-melanoma basal cell and squamous cell carcinomas (Shore, 1990). People with pale skin, red hair, freckles, and those unable to sunburn are at a higher risk of developing melanoma. Melanoma, the most lethal form of skin cancer, has a high growth rate worldwide and is responsible for more than 7,000 deaths each year in the United States.

The effect of DNA damage repair disorders on the UV sensitivity of *Saccharomyces cerevisiae* yeast cells at various stages of the cell cycle has been studied (Siede, Friedberg, 1990). The authors investigated the sensitivity of mutants from the RAD3 and RAD6 groups. The RAD3 group of mutants is defective in excisional nucleotide repair, while the mutants in the RAD6 group are defective in postreplicative repair processes and in UV-induced mutagenesis. Genes in the epistatic group RAD52 are mainly involved in the recombination repair of DNA strand breaks and do not play a major role in the repair of damage caused by ultraviolet radiation. In the work (Ferguson, Cox, 1980), it was shown that haploid yeast, in contrast to the action of ionizing radiation, is able to recover from UV-induced damage when kept in a non-nutritive environment during the post-radiation period. It was found that this occurs as a result of excision of dimers.

The genetic instability of UV-sensitive mutants (*rad*9) of the yeast *Saccharomyces cerevisiae* was studied (Fasullo et al., 1998, 2001). The *rad*9 mutation have a 5-fold higher rate of spontaneous mitotic recombination and a more than 10-fold increase in the number of recombinants containing translocations. The induction of mutations by UV light was investigated in UV-sensitive (*uvs*) and wild (UVS) strains of the yeast *Saccharomyces cerevisiae* (Resnick, 1969). It was shown that the reversion to prototrophy in both strains occurred regardless of whether the initial auxotrophic mutation arose by replacing DNA base pairs or as a result of deletion formation. DNA is the most important molecule in the defeat of cells by UV-B and UV-C radiation (Samoilova, 1967; Bochkov, 1971; Johnson et al., 1973; Troshin, 1979; Castellani, 2012). The biological effect of UV radiation on cells is based on the processes of interaction of UV quanta with the genetic apparatus of cells, as a result of which cells can be inactivated or cell division and repair processes can be stimulated (Resnick, Setlow, 1972; Johnson et al., 1973; Saha, 2006). At the same time, the probability of the formation of new forms that are not typical for a given cell population due to mutagenesis induced by UV radiation increases.

The absorption spectrum of UV radiation by nucleic acids has a maximum at a wavelength of 260 nm, and the spectrum of action of UV radiation for inactivating *Escherichia coli* bacteria resembles in its shape the absorption spectrum of nucleic acids and has a maximum near 265 nm (Konev, Volotovsky, 1979; Castellani, 2012). It is concluded that in the case of cell death under the influence of UV radiation, the main target is a DNA molecule. UV photons with a wavelength of more than 200 nm can be selectively absorbed by various molecules, transferring their valence electrons to a higher energy level. Such excited molecules readily enter into photochemical reactions, which ultimately produce the effects observed at the biological level.

UV light is one of the naturally occurring mutagenic factors. UV light-induced mutations occur in viruses, microorganisms, unicellular plants and animals, as well as in cells and tissue cultures of higher plants and animals (Konev, Volotovsky, 1979). UV irradiation activates many genes and some viruses, including HIV (WHO, 1995). Genes activated by UV-B and UV-C radiation are different from genes activated by UV-A radiation. Research on DNA repair defects has clearly established a link between UV-induced damage to skin DNA and various types of cancer (Smith, Hanawalt, 1969; Letavet, Taylor, 1975; Castellani, 2012). Along with mutations, both ionizing and UV radiation cause various types of chromosomal aberrations (Troshin, 1979; Sevankayev, 1987; Kudryashov, 2004).

As a result of the evolutionary process, several mechanisms have emerged that protect the cell from the damaging effects of UV radiation. These are photoreparation (photoreactivation) and dark repair. Photoreactivation consists in the fact that as a result of illumination of cells with long-wave UV radiation or visible light, damage induced by UV light is repaired in cells. The spectrum of photoreactivation action in bacteria is in the range of 300–500 nm with a maximum at 380 nm (Konev, Volodarsky, 1979). The enzyme DNA photolyase is involved in the process of photoreactivation. Photolyase repairs damage caused by exposure to ultraviolet light. For this enzyme to work, visible light is needed. Photolyase is a phylogenetically ancient enzyme that is presents and functions in many species, from bacteria to fungi, plants, and animals (McEachern, Haber, 2006).

For a long time, it is believed that under the action of ionizing radiation, the excitations of molecules and atoms can be neglected, and the biological effect is completely due to ionizations. However, in the 70s and 80s of the last century, a series of publications appeared (Myasnik, 1974; Morozov, Myasnik, 1980; Moss, Smith, 1980; Myasnik et al., 1980, 1985; Munson, 1981; Skvortzov et al., 1981; Morgan et al., 1982; Petin, Komarov, 1989; Petin, Morozov, 2015), which demonstrated the photoreactivation of mutant strains of bacteria *E. coli* and yeast *Saccharomyces cerevisiae* of a certain genotype, irradiated with high-energy ionizing radiation. Data were obtained proving that such photoreactivation is related with the flux of UV light accompanying the action of ionizing radiation. The source of UV light was the Čerenkov emission, which occurs when fast electrons move in a medium with a speed exceeding the speed of light in this medium. The energy spectrum of this radiation includes not only the visible light, but also the ultraviolet part of the spectrum. It is known that this radiation appears only at electron energies exceeding 250 keV, and its value increases with an increase in the radiation energy (Table 2.2) and the volume of the irradiated object (Table 2.3).

It follows from the data described that Čerenkov radiation can be considered as a means of delivering UV light to deeply located tissues, which are known to be capable of enzymatic reparation of UV damage in DNA (Smith, Hanawalt, 1969). The greatest energies of radionuclides contained in humans and animals significantly exceed the threshold energy for the appearance of Čerenkov radiation, for example, this energy for ^{40}K is 1.31 and 1.50 MeV, for ^{137}Cs – 0.51 and 1.17 MeV. The human body weighing 70 kg contains 0.0164 g of ^{40}K, which ensures the decay of about 10^{11} atoms per year.

It is also possible to hypothesize about the possible participation of the UV component of ionizing radiation in the mechanism of radiation hormesis or adaptive response, which could be observed after exposure to ionizing radiation at low doses, when charged particles crossed only a limited part of the cell population (Little, 2007). At the same time, the entire population was characterized by the positive effect of ionizing radiation or changed its response to subsequent irradiation with ionizing radiation at higher doses. Since the number of quanta of Čerenkov radiation at a radiation dose of 100 Gy (^{60}Co) is about 10^{13} per 1 cm^3 of irradiated water (Morozov, 1996 b; Morozov, Myasnik, 1980), a huge number of these quanta took place even at low doses of ionizing radiation, for example, 10^9 quanta per 1 cm^3 for a dose of 1 cGy. Since the quanta of UV light are distributed over the irradiated volume more uniformly than ionizations, it is possible that ionizing radiation even at small doses, including natural background, could trigger repair processes in cells (similar to SOS repair) due to the accompanying UV light, thereby increasing the radioresistance of biological objects to subsequent irradiation.

Mutations in critical genes are thought to be a necessary part of cancer induction. The traditional view of radiation mutagenesis is that radiation induces many mutations in cells shortly after irradiation due to erroneous repair or lack of DNA repair before or during DNA replication. It was shown that the delayed mutations were observed at the *hprt* gene locus of Chinese hamster fibroblasts (V79), which appeared in many cell generations after three types of carcinogenic irradiation: UV-A and UV-B radiation or X-rays (Dahle, Kvam, 2003; Dahle et al., 2005). The frequency of mutations was measured in clones 14 days after irradiation with doses that inactivate 80% of cells. The proportion of unstable clones was 10–7500 times higher than the background and was slightly higher for cells irradiated with UV-A radiation. On the contrary, this component of UV radiation induced fewer direct mutations, but significantly more delayed mutations that determine genetic instability.

It is believed that natural UV radiation is a more effective carcinogen than ionizing radiation. UV radiation is a major cause of non-melanoma skin cancers such as squamous cell carcinoma (WHO, 1995). UV-A radiation plays a role in the induction of genomic unstable malignant melanoma. Both UV-A and UV-B radiation induce genome instability in mammalian cells in the form of the delayed gene mutations (Federico et al., 2016). UV-A radiation acts primarily in the presence of oxygen and endogenous photosensitizers, generating reactive oxygen species. In other words, UV-A radiation induces a type of oxidative stress that resembles endogenous oxidative stress caused by inflammation or other endogenous sources. However, UV radiation subsequently also induces oxidative stress, especially in the form of hydrogen peroxide and lipid peroxides, and induces pyrimidine dimers. Reactive oxygen species are involved in many aspects of biology such as atherosclerosis, aging, mutagenesis, and carcinogenesis. Antioxidants reduce the oxidation of biomolecules by reactive oxygen species and other oxidants and may be a factor in reducing the incidence of many types of human cancers, including skin cancer, for people with high consumption of tea, fruits and vegetables (WHO, 1995). It is assumed that antioxidants added after UV-A and UV-B radiation are potent inhibitors of genome instability, which manifests itself in the form of the delayed mutations.

Irradiation of cells with UV-A radiation causes oxidative damage in DNA, single-strand breaks, cross-linking of DNA-protein through reactions with reactive oxygen species. Urushibara et al. (2014) investigated whether lesions other than double-stranded DNA breaks play a role in the induction of chromosomal instability. The authors exposed human chromosome 21 to UV- and transferred this chromosome into non-irradiated mouse recipient cells. Chromosomal instability was investigated using chromosome staining and fluorescent *in situ* hybridization. The ploidy of murine recipient cells increased, and chromosomal aberrations arose not only in the human chromosome irradiated with UV-A, but also in non-irradiated mouse chromosomes. The frequency of these disorders increased with an increase in the radiation dose transmitted by the carried human chromosome. In contrast, in a control experiment, in which an unirradiated human chromosome was transferred, cell hybrids remained diploid, and the frequency of chromosomal aberrations in both the transferred human chromosome and the chromosomes of recipient mice remained low. This study indicates that the chromosome lacking double-strand DNA breaks caused by UV-A radiation may be is unstable and transmits the instability to the chromosomes of recipient cells in non-irradiated mice.

The study of the combined action of UV radiation with various physical and chemical factors that change the sensitivity of cells to UV light is relevant not only from a fundamental point of view, but also has real practical significance. Indeed, the increased background of UV radiation due to the depletion of the ozone layer in the stratosphere is potentially hazardous to the environment (Shore, 1990; WHO, 1995). In addition, with the advent of new therapies using UV radiation (Waldow, Dougherti, 1984), it becomes important to study the mechanisms of its synergistic interaction with other agents. The synergistic interaction of UV and ionizing radiation was shown in bacteria *Escherichia coli* (Han, Elkind, 1978). It has been demonstrated that with an increase in the preliminary dose of UV radiation, an increase in the slope of the survival curves of bacteria irradiated with ionizing radiation is observed. For bacterial cells, the limiting dose was about 120 J/m^2. A further increase in the UV dose did not lead to an increase in the radiosensitivity of bacteria. This demonstrates the limitation of the maximum synergism with the sequential action of damaging agents.

The works (Calkins et al., 1978; Petin, Komarov, 1989) describe the results of the combined action of densely ionizing radiation (α-particles) and UV light. The effect might be additive, synergistic or antagonistic, depending on the doses of UV radiation and α-particles. Interestingly, in these studies, the dose-effect curves after UV exposure to cells pre-irradiated with α-particles were characterized by an unusual shape – an oscillation of survival with significantly different maxima and minima. There are data showing that UV radiation of different energies can synergistically interact with each other (Tyrrell, Peak, 1978), with different chemical agents (Borchers et al., 1990) and hyperthermia (Mitchell et al., 1985; Petin et al., 2012, 2020; Petin, Kim, 2016).

The study of cell recovery after UV radiation, used in combination with certain chemical agents and hyperthermia, showed that under the conditions the rate of recovery slows down and the volume of recovery decreases, i.e., the proportion of irreversibly damaged cells increases (Averbeck, Moustacchi, 1975; Petin, Kim, 2004). For example, the rate and volume of yeast cell recovery from sublethal damage during dose fractionation decreased when caffeine was added to the incubation medium between subsequent dose fractions (Kiefer, 1975). The slowdown in the recovery rate can be associated both with the suppression or disruption of the recovery process itself and with the formation of more "severe" irreversible damage after combined actions of UV light with other modalities. Analysis of this problem showed that the synergistic effects observed with the simultaneous combined effect of hyperthermia with both UV light and ionizing radiation are due to the formation of irreversible damage, from which cells were unable to recover, while the recovery process itself is not disturbed (Petin, Kim, 2004; Komarova, Petin, 2007; Petin et al., 2012, 2020).

UV radiation with a wavelength of less than 400 nm penetrates the skin to a depth of about 2 mm, without reaching the subcutaneous layers. The observed differences between the UV sensitivity of light and dark skin are due to differences in the depth of penetration of UV radiation, since the reflectivity of light and dark skin in relation to UV light is approximately the same. Studies in experimental animals have shown that UV irradiation can cause both acute and long-term effects, such as cataracts, photokeratitis, damage to the corneal epithelium and various changes in the retina. Studies in monkeys have shown that the retina is six times more vulnerable to UV light than it is to visible light (WHO, 1995). The effect of UV radiation on human skin leads to two well-known effects – erythema, considered as the first stage of suntan and manifested in reddening of the skin, and pigmentation (second stage of suntan). The maximum erythema efficiency occurs at 295 nm while the maximum pigmentation efficiency takes place at the wavelength of 345 nm (Figure 4.2).

The observed differences in the maximum efficiency of erythema and pigmentation effects indicate the distinction in the main photochemical reactions underlying the two stages of suntan. Pigmentation appears after a longer latency period than erythema and persists longer after irradiation. In addition, for pigmentation, a decrease in the effectiveness of UV radiation is observed with repeated exposures. The mechanism of occurrence of ultraviolet erythema is due to the expansion of blood vessels deeper than UV radiation penetrates. It is believed that UV radiation in the superficial layers of the skin releases significant amounts of histamine and similar vasodilators.

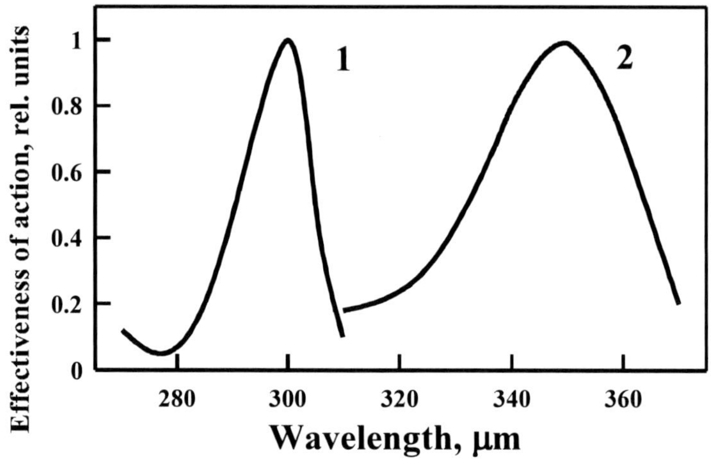

Figure 4.2. Relative efficiency of erythema (curve 1) and pigmentation (curve 2) emergence on human skin.

Low levels of both γ- and UV radiation cause stochastic effects and health risks, which are defined as the likelihood of cancer and genetic damage. High doses result in deterministic effects that can undoubtedly lead to serious tissue damage (Tsyb et al., 2005). UV radiation can induce two types of malignant skin tumors – melanoma and non-melanoma cancers (Ferlay et al., 2015). The number of new cases is expected to increase by about 70% over the next 20 years. It is known that the processes of tumor induction are preceded by an increase in genetic instability (Hanahan, Weinberg, 2000). The maximum in the spectrum of skin carcinogenesis occurs at a wavelength of 310 nm. The incidence of skin cancers increases with age and with the intensity of UV radiation. In the latitudinal zone 30–50 °N. The incidence of cancer in the skin for white population makes up 160–400 per 100,000 people/year. This magnitude is close to the total tumor incidence of all other types of cancer – about 300 per 100,000 people/year. It is known that people with pigmented skin are less susceptible to this disease. For example, the incidence of tumors in representatives of the Negroid race is more than 60 times lower, and in the Hispanic race, it is 7–10 times lower than in the white population in the same area, with almost the same frequency of tumors other than skin cancer. The annual increase in the incidence of skin cancer in the United States is 6-7%. Mortality from melanoma in European countries is increasing by 4-5% per year, in Canada and the United States – by 5-6%, in

Australia – by 8%. This constant increase in skin oncology is directly related to the increase in the flow of UV radiation to the Earth.

Ultraviolet photophthalmia is caused by overexposure of eyes to UV radiation. It includes photokeratitis, photoconjunctivitis, and cataractogenesis. Photokeratitis is a clouding of the lens, the maximum efficiency of which observed at a wavelength of 270–275 nm. The maximum in the spectrum of UV radiation action for the induction of photoconjunctivitis (inflammation of the mucous membranes and the anterior part of the eyeball) is observed at a wavelength of 260 nm. It is known that most UV radiation shorter than 300 nm is completely absorbed when passing through the cornea and the anterior chamber of the eye, i.e., before hitting the anterior surface of the lens. So, the effect of UV radiation appears only on the surface tissues of the eyes. The maximum efficiency of cataractogenesis is observed at a wavelength of 305 nm. Unlike the skin, the eye and its surrounding tissues do not develop resistance to repeated exposure of UV light.

UV radiation plays an important role in providing the body with vitamin D, which is involved in the regulation of phosphorus-calcium metabolism. Vitamin D deficiency causes rickets, tooth decay and possibly prostate cancer, which is also characterized by latitudinal dependence – an increase in mortality from prostate cancer as we move north. The spectrum of action of photochemical reactions leading to the synthesis of vitamin D is close to erythemal, with the maximum yield of this reaction at 300 nm. The dose of UV radiation required to compensate for vitamin D corresponds to daily exposure to the open sun for 0.5–1.0 hours at the latitude of Moscow. It is important to note that the role of UV radiation in providing the body with vitamin D cannot be compensated for by its consumption with food, since the process of vitamin D synthesis under the influence of UV radiation is self-regulating and excludes the possibility of hypervitaminosis. At the same time, the increased content of vitamin D causes the deposition of calcium in various tissues, accompanied by their necrotic degeneration. The intensity of vitamin D photobiosynthesis decreases with an increase in skin pigmentation. As a consequence, skin pigmentation can be the cause of vitamin D deficiency in immigrants of color living in low temperate at northern latitudes. Along with the photobiosynthesis of vitamin D, UV radiation plays a certain role as a general health remedy, for example, in the treatment of bacterial skin diseases. Hence, it is concluded that in addition to the maximum permissible levels of exposure to UV radiation, there is also a minimum required dose of UV exposure. Therefore, optimal UV radiation standards should allow a maximum of health-improving effects with a minimum of negative effects (WHO, 1995).

Since more than 30% of the world's supply of protein for human nutrition comes from the sea, a significant decrease in biomass production would lead to a decrease in fish stocks with the continued growth of the world's population. The contribution of aquatic ecosystems to the total biomass of the Earth is greater than that of all terrestrial ecosystems combined. UV exposure also affects these systems (Letavet, Taylor, 1975; WHO, 1995). It has been shown that marine ecosystems, which in a number of countries are the main food source for humans, are more sensitive to UV-B radiation than terrestrial plants. A decrease in leaf area, wet and dry weight, lipid content, and photosynthetic activity are typical reactions of plants sensitive to UV-B radiation (Troshin, 1979; WHO, 1995). Many plant species are slowed down by elevated levels of UV-B radiation. A 17% decrease in photosynthesis in UV-B sensitive soybeans has been described when exposed to radiation equivalent to an 18% decrease in ozone (Castellani, 2012). Exposure to growth of 12 varieties of rice for 12 weeks with additional UV-B irradiation levels equivalent to a 20% decrease in ozone (15.7 kJ/m^2) revealed changes in rice biomass, morphology and photosynthesis. A three-year field study (WHO, 1995) showed an overall decrease in photosynthetic capacity in pine trees that received additional doses of UV-B radiation, simulating a decrease in ozone concentration by 16 and 25% (11.5 and 13.6 kJ/m^2).

UV radiation is of great environmental importance. If all UV radiation falling on the upper atmosphere reached the surface of the Earth, then all animals and plants would die. Even a small flux of UV radiation (about 1%) reaching the Earth induces a significant incidence of skin cancer. We are protected from the harmful effects of UV radiation, since most of it (over 99%) is absorbed by the layer of ozone (O_3) located in the stratosphere at an altitude of about 10-50 km from the Earth's surface, with a maximum concentration at an altitude of about 25 km. This layer is commonly referred to as the ozone shield. Ozone is a blue gas with a pungent odor, a strong oxidizing agent. The deodorant properties of ozone are known to be used to eliminate odors. Ozone has a bactericidal effect; therefore, it is used to disinfect water in swimming pools. Under natural conditions, ozone is generated by an electrical discharge (for example, during a thunderstorm). The ozone layer effectively absorbs the main flow of solar UV radiation, with the maximum in the absorption spectrum of the ozone layer at a wavelength of 320 nm, which is very close to the optimal wavelength for skin carcinogenesis (310 nm). The ozone layer is formed due to the dissociation of oxygen molecules into two atoms and their subsequent attachment to an oxygen molecule ($O + O_2 = O_3$). Dissociation of oxygen molecules occurs upon absorption of high-energy short-wave UV radiation at a wavelength of less than 240 nm. The maximum efficiency of this process is observed at 185–190 nm. An oxygen atom can combine with an ozone molecule to form two oxygen molecules. Therefore, a certain equilibrium is established and maintained between oxygen and ozone, since the lower layers of the atmosphere and stratosphere are poorly mixed.

The ozone layer has been in dynamic equilibrium for a long time. However, since the 70s of the last century, decrease in the thickness of the ozone layer and the formation of ozone holes have been constantly recorded. In 1985, satellites detected a 50% decrease in the thickness of the ozone layer over Antarctica over an area equal to the area of Brazil. The decrease in the ozone layer has received the figurative name of ozone holes. Since 1985, their formation has been relatively regularly observed not only over Antarctica and the Arctic, but also over South America and Australia. In fact, it has turned into one of the global environmental problems due to the increased frequency of cataractogenesis and carcinogenesis. According to the United Nations (Handbook, 2020), a 1% reduction in the ozone layer could lead to 100,000 additional cases of cataracts and 10,000 cases of skin cancer worldwide. The mechanism for reducing the ozone layer is associated with the catalytic effect of chlorine and some other elements on the ozone decomposition process. A relatively small number of chlorine atoms can cause significant damage to the ozone shield, and this effect will continue indefinitely, as chlorine atoms leave the stratosphere very slowly. Its water-soluble compounds represent most of the chlorine used on Earth, so they do not reach the stratosphere, being washed out of the atmosphere by precipitation. The exception is chlorofluorocarbons, i.e., hydrocarbon molecules in which some of the hydrogen atoms are replaced by chlorine and fluorine. They are very volatile and insoluble in water, so they easily reach the stratosphere, where they can decompose with the release of chlorine atoms. Chlorofluorocarbons are gases at room temperature, chemically inert and non-toxic, but they turn into a liquid at low pressure with the release of heat. On the contrary, when they evaporate, they absorb heat and give off cold. This property is widely used in refrigerators and air conditioners. Many other applications of chlorofluorocarbons are known, where they ultimately escape into the atmosphere. Several

International Meetings have been held recommending the reduction or even complete cessation of their production (Handbook, 2020). Another group of chemicals that deplete the ozone layer are called halons and are used by fire departments in many countries. It should be noted that another source of chlorine entering the atmosphere is also known – tectonic processes in the earth's crust, in which chlorine can be released at the faults of the earth's crust. Emissions from supersonic aircraft, as well as launches of spacecraft and satellites, also contribute to the destruction of the ozone layer.

4.2. Patterns of Yeast Cell Inactivation and Recovery after Exposure to UV Radiation

One of the differences between the effects induced by ionizing radiation and UV light is the different response of haploid and diploid cells surviving after irradiation. As demonstrated in previous chapters, the survival curves of wild-type diploid yeast cells are sigmoid after irradiation with both γ-rays and α-particles. Haploid cells are characterized by exponential survival curves regardless of the quality of radiation, while they are 5 times more radiosensitive than diploid cells at the level of 10% survival after exposure to sparsely ionizing radiation and 2.1 times after exposure to radiation with high LET (Figure 3.4). This pattern was due to the different ability of haploid and diploid cells to recover from damage induced by ionizing radiation. However, the survival curves of haploid and diploid cells were sigmoid after irradiation with UV light and their sensitivity to UV exposure did not differ much, which is apparently due to the greater similarity of these cells to recovery.

Another fundamental difference between the effect of UV light and ionizing radiation on biological objects, including yeast cells, is the property of molecular oxygen present in cells and tissues to enhance the biological effect of ionizing radiation. The greatest value of the oxygen enhancement ratio is about 3 after exposure to ionizing radiation, not varying much for various biological effects (Eidus, Korystov, 1984). Under the action of UV radiation on biological objects, oxygen did not affect the sensitivity of cells to the action of UV radiation in a great extent. This significant difference between the two types of radiation may be a consequence of the fundamental difference in the biological consequences of excitations and ionizations of atoms and molecules of biological objects.

Among the biological effects of UV radiation on yeast cells, the release of various organic compounds into the environment by damaged cells, such as vitamins, amino acids and nucleotides, which promote cell proliferation, deserves attention (Kabakov, 1963). This work notes that some authors, using labeled phosphorus, observed the leakage of organophosphorus compounds from UV exposed yeast cells. The accumulation of nucleotides occurred not only in the environment, but also inside the irradiated cells. Based on spectrophotometric studies, it was concluded that the synthesis of nucleotide-like substances increases in UV-irradiated cells. However, it remains unclear whether the loss of biological material by yeast cells is one of the stages in the development of radiation damage or whether this process plays the role of cleansing the intracellular environment from toxic radiation products. A change in UV sensitivity also occurred during prolonged exposure of the yeast cells to radiation in a non-nutritive medium (Kabakov, 1963). This effect was found upon repeated irradiation after 24 h of the yeast suspension incubated at 30ºC. The main condition for the manifestation of this effect is long-term aeration at a physiologically optimal temperature. UV sensitivity decreased both as a result of incubation of cells in a non-nutrient medium and active aeration with air. This effect is by no means related to the classical oxygen effect, which is recorded when comparing the results of cell irradiation under aerated and anoxic conditions.

Experiments with yeast cells on photoreactivation and dark repair have shown that the repaired lesions are pyrimidine dimers in DNA (Johnson et al., 1973). It was shown (Kabakov, 1963) that the survival curves observed immediately after irradiation and after complete photoreactivation of cells remain similar to each other, while the efficiency of photoreactivation, assessed by the ratio of isoeffective doses on the dose-effect curves with and without photoreactivation is equal to 2.4. A comparative study of light and dark recovery of yeast cells of different ploidy, affected by ultraviolet radiation, was carried out (Korogodin et al., 1962). The experiments used genetically related strains of the yeast *Saccharomyces cerevisiae*, from haploids to hexaploids, obtained from R. K. Mortimer (USA). Irradiation was carried out in a glass bottle covered with black lacquer on the outside. A small amount of suspension was used for uniform irradiation with UV light – the height of which in the glass cuvette was 2–5 mm. Photoreaction was provided by a 300 W incandescent lamp, the infrared part of which was absorbed by a water filter. Dark recovery took place in a thermostat. It was shown that a significant proportion of UV affected cells (50–70%) died without dividing. According to this indicator, the effect of UV radiation significantly differed from the effect of ionizing radiation, when such a form of death was not observed in the range of doses that reduce cell survival to 0.1–0.01% (Korogodin, 1966; Petin et al., 2006). A detailed study of death forms after exposure to ionizing radiation at very high doses showed that when irradiated cells were kept in a non-nutritive medium, the cells did not recover, but died without division (interphase death) (Petin et al., 2006). However, after exposure to UV radiation, this form of death immediately after irradiation was 50–70%, decreased to 0.1–2.0% after photoreactivation of diploid and other polyploid cells (Korogodin et al., 1962). It was shown that haploid yeast, in contrast to the action of ionizing radiation, is able to recover from UV-induced damage when kept in a non-nutritive environment during the post-radiation period (Ferguson, Cox, 1980). It was found that this occurs as a result of excision of dimers.

UV light has a powerful bactericidal, antimitotic, mutagenic, carcinogenic and therapeutic effect (Konev, Volotovsky, 1979; Troshin, 1979; WHO, 1995; Castellani, 2012). The specificity of the action of UV radiation, as well as the relatively high reversibility of its effects, the possibility of accurate dosing and local irradiation determines its use in the treatment of malignant tumors (Melloni et al., 1984; Waldow, Dougherti, 1984). With the advent of new methods of therapy using UV radiation (photodynamic therapy), the problem of enhancing the UV light action when it is used with other environmental factors becomes urgent. Insufficient knowledge of the effects after combined actions UV light with other factors could be due to the fact that the effects of combined action, for example, hyperthermia with UV light, were studied by various authors in a relatively narrow temperature range.

Experimental study of the elevated temperature effect on the UV sensitivity of yeast cells differing in ploidy and sensitivity to UV irradiation showed the influence of UV light intensity on the synergistic effects observed after the simultaneous action of ultraviolet radiation with hyperthermia (Zhurakovskaya, 2001; Petin et al., 2012; Petin, Kim, 2016). Original new results were obtained in these works, which can be briefly summarized as follows. It is demonstrated for the first time that there was a certain temperature range within which the synergistic effect is displayed for the simultaneous action of UV radiation and hyperthermia. For a fixed UV intensity, there is an optimum temperature, at which the synergistic enhancement ratio reaches the greatest value. At temperatures below this optimal value, cell inactivation is mainly determined by the action of UV radiation. On the contrary, at temperatures that are higher of this value, the biological effect of the combined action is largely due to the effect of elevated temperatures.

The results of UV light interaction with hyperthermia, obtained in a wide range of applied temperatures, showed the existence of two extrema: at some temperatures, small antagonism was observed, at others, the greatest synergistic effect was obtained (Zhurakovskaya, 2001). In other words, in contrast to thermoradiation interaction, there is a temperature range, within which the interaction of heat and UV radiation is antagonistic, when the final result of the simultaneous action of these agents was less than the amount expected for independent addition of the effects from each agent. The possibility of antagonistic interaction of these factors was also noted (Vilenchik, 1991).

An interesting information has been obtained confirming the pattern of synergistic interaction of the simultaneous application of hyperthermia and both the temperature, at which the irradiation occurs, and the intensity of UV light determine UV radiation the final effect. For all UV light intensities investigated, the optimum temperature was obtained at which the synergistic effect reaches it's the greatest value. With an increase in the intensity of UV radiation, it was necessary to increase the temperature at which the exposure occurs, and, conversely, with a decrease in the intensity, the temperature of the irradiated suspension should be lowered to achieve maximum synergy.

Comparing the just noted patterns of synergistic effects after the simultaneous use of UV radiation and hyperthermia with those for the combined action of hyperthermia with ionizing radiation (Petin, Zhurakovskaya, 1995; Petin, Kim, 2004; Petin et al., 2012), we can note some general characteristics. The synergism of the simultaneous action of hyperthermia with ionizing radiation or UV light depends on the intensity of the agents used and is observed only in a certain temperature range, within which there is an optimal temperature, at which the synergistic effect is maximal. Similar patterns were noted for the simultaneous use of hyperthermia with ultrasound, microwave radiation, and various chemical agents (Petin et al., 2012, 2020; Petin, Kim, 2016). The totality of these data indicates the universality of the revealed patterns and their general biological significance.

For experimental study of recombinations after the action of UV light (254 nm) we used diploid yeast cells *Saccharomyces cerevisiae* (strain T1) at the stationary stage of growth This strain was specially designed for the induction of mitotic crossing over and mitotic segregation (Zakharov et al., 1980, 1984). Due to segregation among the descendants of cells that have undergone various physical or chemical influences, cell mutations appeared. Therefore, this strain is convenient for quantifying the yield of recombinant mutations after action of various modalities. The strain has the following genetic characteristics: name – T1 (PG-154); genotype – RAD/RAD, a/α, *ade*2-192, *ade*2-G45. This strain is highly sensitive to damaging agents of chemical and radiation nature (Zakharov et al., 1980, 1984; Pakhomova, Tsyb, 1994). The T1 strain is a prototroph and forms white colonies. The first mutation, *ade*2-192, causes the colony to be red, and the second, *ade*2-G45, – pink. If the resulting colonies have twin sectors – red and pink, this was considered as a mitotic crossing over event. All other changes in coloration of colonies, namely, completely red or completely pink colonies or colonies with one colored sector, can be attributed to the consequences of mitotic crossing over when one of the crossover cells dies. The frequency of induced recombinant cells was quantitatively expressed as a percentage relative to the total number of colonies examined. The experiments were carried out in 4 replicates (Korogodin et al., 2007; Petin et al., 2008). For each duration of irradiation, about 1000 colonies were counted. The results are presented in Table 4.1. It can be seen that the yield of induced recombinant cells at a low intensity of UV radiation increased to 2.8% with an increase in the duration of irradiation to 45 min and a decrease in cell survival to 20%. A more intense yield of recombinant cells was observed at a higher intensity of UV light – up to 6.9% with an irradiation duration of 4 min, which reduced cell survival to 8%.

Table 4.1. The yield of the induced recombinant yeast cells *Saccharomyces cerevisiae*, strain T1, after exposure to UV radiation (254 nm)

UV radiation intensity, W/m^2	Duration of exposure, min	Cell survival, %	Yield of recombinant cells, %
0.15	0	100	0
	10	90	0.7
	15	80	1.3
	30	40	2.0
	35	34	2.5
	45	20	2.8
1.5	0	100	0
	1	70	1.8
	2	45	3.8
	3	20	5.7
	4	8	6.9

We also note that the simultaneous effect of UV radiation and hyperthermia on cell survival and yield of recombinants of diploid yeast cells of *Saccharomyces cerevisiae* was characterized by a synergistic interaction, i.e., mutual strengthening of the agents applied (Petin et al., 2012; 2020; Petin, Kim, 2016). Since the presence of recombinant yeast cells after UV irradiation indicates their genetic instability, it is of interest to obtain experimental data on the delay in the formation of colonies by cells that survived after UV irradiation and quantitatively determine the genetic instability of the cells.

4.3. Genetic Instability of Haploid Yeast Cells of Different Genotypes after UV Irradiation

It was noted above that UV radiation is a type of non-ionizing electromagnetic radiation including the shortest wavelength, which does not reach the Earth's surface and their energy is sufficient for ionization. In small doses, UV radiation is beneficial for humans and is essential for the production of vitamin D (IARC, 2009). Wavelengths 320–400 nm (UV-A) and 290–320 nm (UV-B) can induce specific DNA damage leading to malignant tumors of the human skin (Shore, 1990; WHO, 1995; Kielbassa et al., 1997), therefore, these UV wavelengths are classified as Group I carcinogens by the International Agency for Research on Cancer (IARC, 2009). The maximum in the spectrum of skin carcinogenesis induction occurs at a wavelength of 310 nm. Intense UV radiation may be responsible for the development of malignant melanoma (Bald et al., 2014). People with fair skin, red hair, freckles, and those prone to sunburn are at a higher risk of developing melanoma than dark-skinned individuals, due to the protective effect of melanin, which is enriched in the dark-skinned population. Melanoma, the most lethal form of skin cancer, is characterized by a high growth rate worldwide. The incidence of melanoma is comparable to that of other cancers (WHO, 1995). There are data indicating that the onset of cancer and aging of the organism is often preceded by genetic instability (Marder, Morgan, 1993; Vorobtsova, 2008; Urushibara et al., 2014).

It was shown in previous chapters and in our recent publications (Evstratova et al., 2016; Evstratova, Petin, 2018a, b, c) that the delay in the formation of colonies by cells that survived after irradiation with γ-rays or α-particles does not depend on the shape of the survival curves and cell ability to recover from radiation damage, but are completely determined by cell ploidy. Experiments have been performed with haploid and isogenic diploid wild-type strains and their radiosensitive mutants. The effect of the delayed colony formation with increasing dose reached 100% for both radioresistant and radiosensitive diploid strains, but was less pronounced for haploid cells (20–25%). Based on these results, it was concluded that the manifestation of long-term radiation consequences (genetic instability) is determined mainly by cell ploidy, and not by their radiosensitivity and cell ability to recover from radiation damage. Therefore, it was of interest to obtain similar data on the delayed colony forming ability by surviving UV-irradiated wild-type haploid and diploid yeast cells and their UV-sensitive mutants. It is well known that the shape of the survival curves after irradiation with UV light does not depend on ploidy and has a sigmoid shape for both haploid and diploid wild-type yeast cells (Petin et al., 2012; Petin, Kim, 2016). Considering these data and information on the increased yield of skin cancer under the influence of UV light (Shore, 1990; WHO, 1995; Douglas et al., 2014), it was of interest to compare the effect of genetic instability in cells of different genotypes that survived after exposure to UV radiation. Such data would help to understand whether this effect is due to the ability of cells to recover from radiation damage and the shape of the survival curve or largely is determined by cell ploidy, regardless of their sensitivity to damaging factors.

In experiments with wild-type haploid yeast cells of *Saccharomyces cerevisiae* (strain S288C, RAD) and the most UV-sensitive mutant strains LMG318 (*rad2*), XC6 (*rad6*), and 21-LMG-3031 (*rad18*) were used. Strains S288C and XC6 were initially obtained from Dr. R.K. Mortimer (USA), strains LMG318 and 21-LMG-3031 were obtained from V.G. Korolev (Gatchina, Russia). The cells were grown on a solid nutrient medium (peptone – 1%, glucose – 2%, yeast extract – 1%, agar-agar – 2%) for 5–8 days. The duration of pre-radiation cultivation was determined by the minimum content of budding cells. In our experiments the irradiated suspension contained 1–2% of budding cells.

The cell suspension (concentration 10^6 cells/ml, volume 2 ml) was irradiated in an open quartz vessel with UV light at a wavelength of 254 nm at an intensity of 1.5 W/m^2 (installation Transilluminator UVstar 15, Biometra, Germany). At this wavelength, the greatest absorption is observed for DNA, mutagenesis, bactericidal and fungicidal action (IARC, 2009). To avoid photoreactivation, irradiation, dilution of suspensions, and other procedures were performed under red light, while post-radiation incubation of cells took place under dark conditions. Irradiated and control cells were placed in Petri dishes with a solid nutrient medium so that, as a result, 150–200 colonies were formed. Cell survival was traditionally assessed by the ratio of the number of colonies formed by the cells that survived after irradiation to the number of colonies formed in the control after 5 days of incubation on a solid nutrient medium. To quantify genetic instability, the percentage of colonies formed later than control by cells surviving after exposure to UV light was calculated. Each experiment was repeated 2–5 times. Results are presented as a mean and standard error.

Figure 4.3 A shows the dependence of cell survival of wild-type haploid yeast cells of *S. cerevisiae* (RAD) on the dose of UV radiation. Figure 4.3 B exhibits similar dose-effect curves for three UV-sensitive haploid yeast strains (*rad*2, *rad*6 and *rad*18), differing in sensitivity to UV light due to the reduced ability of cells to recover from induced potentially lethal damage. It can be seen that the cells of the original wild-type strain (RAD) were characterized by a sigmoid survival curve with a characteristic shoulder in the region of low doses. The shape of the survival curves was exponential for *rad*2, *rad*6, and *rad*18 mutants, which were 33, 27, and 14 times more sensitive to UV light irradiation compared to the original parent strain at the 1% of survival (Table 4.2). The lethal dose for this survival, is indicated in Table 4.2 as LD 1%, and the ratio of parent strain (RAD) UV sensitivity to that of UV mutants as LD 1% (RAD)/LD 1% (*rad*). Despite significant inactivation of cells (99%), one percent of cells that survived after irradiation retained some residual damage that was ineffective for lethal effects and may be observed in distant offspring of cells that survived after irradiation, which led to a delay in the formation of colonies by survivors after UV irradiation of cells.

Figure 4.3 B and 4.3 D show the dependence on the UV light dose of the delayed production of colonies by cells surviving after UV irradiation, which characterizes genetic instability.

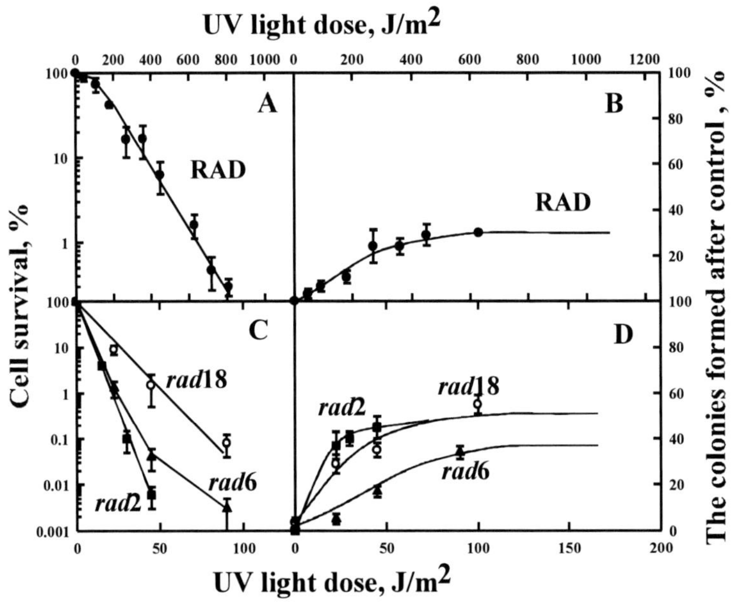

Figure 4.3. The dependences of survival (A, C) and the effect of the delayed appearance of colonies (B, D) on the dose of UV light for wild-type *S. cerevisiae* haploid yeast cells, strain S288C (RAD, closed circles) and the most UV-sensitive mutants, strains LMG318 (*rad*2, closed squares), XC6 (*rad*6, closed triangles) and 21-LMG-3031 (*rad*18, open circles).

It is clear that the delay in colony formation increased with increasing UV light dose, reaching a plateau at about 30% for wild-type cells and 40–55% for UV-sensitive mutants. It is also seen that this effect was displayed in the region of lower doses of UV light for UV-sensitive mutants than for wild-type strain. As a criterion for sensitivity to UV radiation of genetic instability, we used a 50% effective dose (*ED* 50%), i.e., the dose at which a 50% effect of the delay in the formation of colonies by cells surviving after irradiation is obtained (Table

4.2). For this test, the ratio of the sensitivities of the wild-type strain and UV mutants sensitive to UV light is indicated in Table 4.2 as *ED* 50%(RAD)/*ED* 50%(*rad*). For strains *rad*2, *rad*6, and *rad*18, this ratio was 15.0; 5.4 and 9.3.

These data show that the quantitative parameters characterizing the increase in UV sensitivity of haploid yeast cells according to cell survival and the delay in the formation of colonies by the cells surviving after irradiation are somewhat comparable with each other – the ratio of UV light fluxes providing 1% survival and 50% of the delayed cell formation is 33 and 15 for *rad*2, 27 and 5.4 for *rad*6, 14 and 9.3 for *rad*18.

Table 4.2. Radiobiological parameters of *S. cerevisiae* haploid yeast cells of different genotypes after UV irradiation

UV-sensitive locus	*LD* 1%, J/m²	*LD* 1% (RAD) / *LD* 1% (*rad*)	*ED* 50%, J/m²	*ED* 50% (RAD) / *ED* 50%(rad)
RAD	2100	–	700	1,0
*rad*2	64.2	33	47	15
*rad*6	76.3	27	130	5.4
*rad*18	155.0	14	75	9.3

An increase in the sensitivity of mutant cells both in cell survival and in the delayed colony forming ability by the irradiated cells may be associated with sub-damage accompanying lethal damage and responsible for the genetic instability emerging as a result of the delayed appearance of colonies by cells surviving UV irradiation. These sub-damages are ineffective for cell inactivation and persist in the distant offspring of cells that survived after irradiation.

The main conclusion from the presented data is that the genetic instability of the studied haploid strains, differing in UV sensitivity and in the shape of the survival curves, in no case reached the values observed in the previous chapters for diploid strains (100%), regardless of their radiosensitivity, the shape of the survival curves and cell ability to recover from radiation damage.

4.4. Delayed Colony Forming Ability by Diploid Yeast Cells of Different Genotype after UV Light Irradiation

To make sure that our conclusion that genetic instability of yeast cells is mainly determined by cell ploidy rather than by their sensitivity to damaging agents and cell ability to recover from radiation damage, it is of interest to quantify the delayed colony forming ability in the diploid yeast cells of different genotypes after irradiation with UV light. It has already been noted above that the delayed appearance of colonies produced by cells surviving after irradiation, along with many other delayed effects initiated by ionizing radiation, are examples of genetic instability or distant display of inherited damage (Kapultsevich et al., 1972; Tolstorukov et al., 1979; Korogodin et al., 2007; Petin et al., 2008). In these studies, it was concluded that the sigmoid (diploid strains) rather than exponential (haploid strains of wild type) shape of the survival curve mainly determines the delayed effects produced by irradiated cells, i.e., genetic instability. However, a common drawback of the cited works is the use of wild-type haploid and diploid yeasts rather than their radiosensitive mutants. We took this drawback into account when analyzing the relationship of genetic instability for radiosensitive mutants that survived after irradiation with sparsely and densely ionizing radiation (Chapters 2 and 3). The general conclusion of these studies was the assertion that genetic instability is completely determined by cell ploidy, and not related to cell radiosensitivity, the shape of survival curves, or cell ability to recover from radiation damage. In the previous section of this chapter, it was shown that after irradiation of wild-type haploid yeast and UV-sensitive mutants, genetic instability with increasing UV radiation dose reached 30% for the wild-type strain and 40-50% for UV-sensitive mutants. It was of interest to obtain similar quantitative data for cell survival and the delayed colony formation by wild-type *S. cerevisiae* diploid yeast cells and their homozygous UV-sensitive mutants survived after UV irradiation.

We used in our experiments wild-type *Saccharomyces cerevisiae* diploid yeast cells (RAD/RAD, strains XS800 and T1) and the most UV sensitive homozygous diploid mutants – *rad2*/*rad2* (strain T2); *rad6*/*rad6* (strain XS1956); *rad18*/*rad18* (strain XS1924). Strains XS800, XS1956 and XS1924 were originally obtained from Nakai (Japan), strains T1 and T2 from Zakharov (Russia).

Figure 4.4 A shows the survival curves of wild-type diploid yeast cells of *S. cerevisiae* (RAD/RAD, strain XS800) and UV sensitive mutants *rad6*/*rad6* (strain XS1956) and *rad18*/*rad18* (strain XS1924). Figure 4.4 C exhibits survival curves for another wild-type cells (RAD/RAD, strain T1) and its homozygous UV sensitive mutant (*rad2*/*rad2*, strain T2). Figures 4.4 B and 4.4 D show the curves characterizing the delay in the formation of colonies by cells surviving after UV irradiation.

It is clear from Figure 4.4 A that all three strains were characterized by sigmoid forms of survival curves. The mutant strains were more sensitive to UV irradiation compared to the cells of the parent wild-type strain – 2.4 times for *rad6*/*rad6*, 1.5 times for *rad18*/*rad18* at the level of 1% cell survival (Table 4.3). For these strains, the delay in the formation of colonies by irradiated cells, regardless of their sensitivity to UV light, reached 100% (Figure 4.4 B), although this effect was manifested for mutants in the region of lower UV irradiation doses than for the wild- type strain. Figures 4.4 C and 4.4 D show similar data for the diploid wild-type strain T1 (RAD/RAD) and its UV-sensitive mutant T2 (*rad2*/*rad2*), differing not only in the ability to recover from radiation damage, but also in the shape of the dose-effect curves. It can be seen that cells of T1 strain are characterized by a sigmoid dose-response curve, while that of T2 strain showed an exponential dose-dependent survival curve and was 10.5 times more sensitive to UV irradiation as compared with the original T1 strain at the level of 1% cell survival (Table 4.3). Despite such large differences, the delay in the formation of colonies by cells that survived after irradiation reached 100% with an increase in the UV light dose.

These data show that the cells of all diploid strains that survived after irradiation retained some residual sub-damage emerging as a lot of colonies produced after control, and these sub-damages preserved in the distant descendants of cells that survived after irradiation. This effect, in contrast to haploid strains, reached 100% with an increase in UV light dose regardless of cell sensitivity to UV light, the shape of the survival curves and cell ability to recover from UV damage.

The delayed formation of colonies by cells of UV sensitive mutants that survived after UV irradiation was manifested in the region of lower doses of UV irradiation than for the wild-type strain. The ratio of these doses at the level of about 50% of the delayed effect was 7.4 for the *rad2*/*rad2* mutant, 2.0 for the *rad6*/*rad6* mutant and 2.8 for *rad18*/*rad18* mutant (Table 4.3).

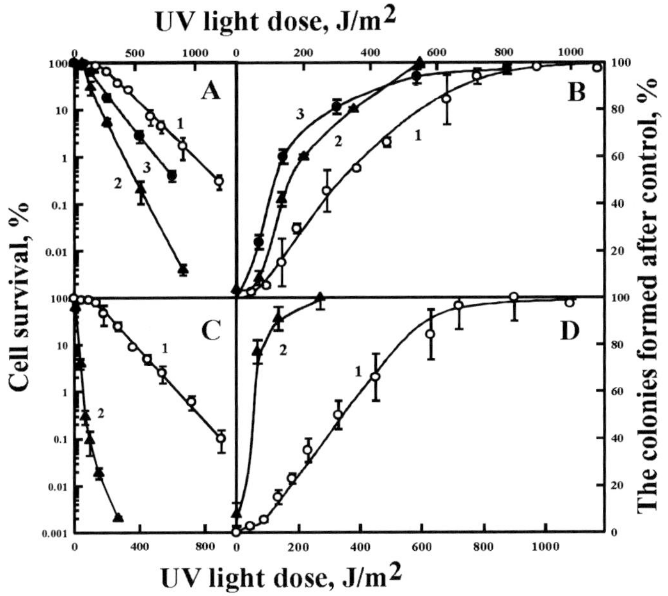

Figure 4.4. The dependence of cell survival (A, C) and the delayed colony forming ability (B, D) on the dose of UV light for wild-type *S cerevisiae* diploid yeast cells and UV-sensitive mutants. A, B: XS800 (RAD/RAD, curves 1), XS1956 (*rad6/rad6*, curves 2), strain XS1924 (*rad18/rad18*, curves 3). C, D: T1 (RAD/RAD, curves 1), T2 (*rad2/rad2*, curves 2).

Table 4.3. Radiobiological parameters of *S. cerevisiae* diploid yeast cells of different genotypes after UV irradiation

Strain	UV-sensitive locus`	*LD* 1%, J/m²	*LD* 1% (RAD) / *LD* 1% (*rad*)	*ED* 50%, J/m²	*ED* 50% (RAD) / *ED* 50% (*rad*)
T1	RAD/RAD	630	–	340	–
T2	*rad2/rad2*	60	10.5	46	7.4
XS800	RAD/RAD	985	–	310	–
XS1956	*rad6/rad6*	405	2.4	154	2.0
XS1924	*rad18/rad18*	645	1.5	110	2.8

The data presented show that the quantitative parameters characterizing the increase in the UV sensitivity of cells according to the cell survival test and the delay in the formation of colonies by irradiated were 10.5 and 7.4 for *rad2/rad2*, 2.4 and 2.0 for *rad6/rad6*, 1.5 and 2.8 for *rad18/rad18* mutants.

An increase in the UV sensitivity of mutant cells in terms of both survival and the delay in the formation of colonies by irradiated cells may be associated with the greater efficiency of UV radiation to produce not only effective lethal damage, but also sub-damage that are ineffective for cell inactivation and persist in the distant offspring of cells surviving irradiation. As a result, one can observe a delay in the formation of colonies, which determines genetic instability.

To compare the effectiveness of ionizing radiation action for genetic instability with the influence of UV light on this effect, the dependence of the delay in colony formation was presented not on the dose, but on the survival of wild-type haploid and diploid yeast cells

(Figure 3.16). This figure also shows the data presented above and published by us earlier for the late appearance of colonies by cells of the same strains that survived after irradiation with γ-rays and α-particles (Evstratova et al., 2016; Evstratova, Petin, 2018 a, b, c). The results obtained show that the delay in the formation of colonies by the surviving yeast cells in the dependence of their survival, was practically the same after irradiation with UV light, sparsely and densely ionizing radiation. These new data indicate that isoeffective irradiation, i.e., irradiation with different agents, leading to the same survival, induces the same number of effective lethal damage as well as the accompanying sub-damage, which are responsible for genetic instability displayed by a delay in the formation of colonies by irradiated cells. These sub-damages persist in the distant descendants of cells surviving after irradiation and thereby cause destabilization of the genome, which manifests itself in the delay in colony formation and in the appearance of phenotypically different clones (Korogodin et al., 2007; Petin et al., 2008).

We presented above (section 2.4) the radiobiological evidences of yeast cell genetic instability. It was shown that cells from late formed colonies after exposure to ionizing radiation are characterized by reduced growth efficiency and increased sensitivity to various environmental factors. The later the colonies are produced in comparison with the control, the more the delayed effects were displayed both immediately and in the long term after irradiation. A common property of long-term radiation effects is the transmission and preservation of some sub-damage ineffective for cell inactivation and responsible for genetic instability. The idea of interaction of sub-damage produced by various agents is used to interpret and optimize synergistic effects (Petin et al., 2012, 2020; Petin, Kim, 2016), an urgent problem of using the combined effects of chemical and physical factors that may increase cell sensitivity to the agents applied. This idea is also important when cells exposed to UV light combined with other physical and chemical agents resulting in the synergistic effects at the treatment of tumors in photoradiation and photochemotherapy (Melloni et al., 1984; Waldow, Dougherti, 1984).

4.5. Mechanisms of Genetic Instability Emerging in Yeast Cells after UV Light, γ-Rays and α-Particles Irradiation

Many events that can cause some changes within the genome can be called genetic instability. This term encompasses various genetic changes that can be classified in different ways (Yuen et al., 2007; Zhu et al., 2012; Skoneczna et al., 2015). In particular, the delayed formation of colonies on a solid nutrient medium by cells that survived after exposure to ionizing radiation or UV light, as compared with the control, similar to other effects, such as reproductive cell death, genome rearrangement, malignant transformation, chromosomal aberrations, decreased cloning efficiency, formTtion of micronuclei and heterogeneity among the descendants of cells that survived after irradiation are examples of the genetic instability of an irradiated cell (Klein et al., 1990; Chang, Little, 1992 a, b; Morgan, Murnane, 1995; Little, 1998, 2000; Burlakova et al., 2001; Mazurik, Mikhailov, 2001; Morgan, 2003; McMurray , Goettschling, 2004; Korogodin et al., 2007; Yuen et al., 2007; Vorobtsova, 2008; Shen, 2011; Stirling et al., 2011; Aggarwal, Brosh, 2012; Kumaran et al., 2013; Lee et al., 2015). A common property of long-term radiation effects is the transmission and preservation in offspring surviving after irradiation some sub-damage that is ineffective for cell inactivation (Kim et al., 2006; Petin et al., 2008; Zhu et al., 2012). These sub-damages are not effective for lethal events. The authors of the work (Skoneczna et al., 2015) discussed various factors that can affect the stability of the genome, such as cell ploidy, the phase of the cell cycle, the transcriptional activity of a specific DNA region, the ability of cells to control the repair of DNA breaks, the metabolic stage of the cell, its respiratory capacity, and finally, the potential impact of endogenous or exogenous stress.

Genetic instability concerns such important consequences of exposure to ionizing radiation and UV light as mutagenesis, carcinogenesis and aging, which are the main long-term effects of radiation (Cheng, Loeb, 1993; Kalb et al., 2006; Shen, 2011; Gundem et al., 2015). Genetic instability is one of the main signs of cancer, which include six biological features acquired by a tumor during its multi-stage development. These include maintenance of proliferative signaling, avoidance of tumor suppressors, inhibition of apoptosis (which makes replicative immortality possible), induction of angiogenesis, and activation of invasion and metastasis (Hanahan, Weinberg, 2000). An insufficiently studied aspect of this problem is a quantitative comparison of the survival and genetic instability of haploid and diploid cells that survived after exposure to many factors, for example, γ-rays, α-particles and UV light. The most convenient model for this purpose is yeast cells differing in sensitivity to the action of various factors (Korogodin, 1966, 2010; Kapultsevich et al., 1972; Petin et al., 2012; Petin, Kim, 2016). The role of genetic instability in cell radioresistance has been widely discussed in many works (Gatti et al., 1991; Morgan, Murnane, 1995; Kaplan, Morgan, 1998; Kalb et al., 2006; Petin et al., 2008; Sugasawa, 2008).

Yeast cells, as the simplest example of eukaryotic cells, are a convenient object for comparing the lethal effect and genetic instability of haploid and diploid cells, showing the delayed appearance of colonies by cells that survived after irradiation, which makes it possible to obtain new fundamental data. Many authors investigated genetic instability using *Saccharomyces cerevisiae* as the model of unicellular organisms (McMurray, Gottchchling, 2004; Yuen et al., 2007; Sheltzer et al., 2011; Duffy et al., 2016). Over the past decade, yeast has served as a good model for a number of discoveries related to the genetic instability. For example, molecular aspects of aneuploidy have been carefully investigated. The budding yeast *Saccharomyces cerevisiae* has become a reliable and versatile model system for studying genetic changes. These unicellular eukaryotes are particularly suitable for studying the effect of aneuploidy and cellular physiology due to their small genome containing 16 chromosomes and due to the fact that they tolerate aneuploidy well (Mulla et al., 2014). To investigate the mechanism by which aneuploidy could promote tumorigenesis, Sheltzer et al. (2011) investigated the effect of aneuploidy on genetic stability. The authors analyzed 13 budding yeast strains and found that all aneuploidy strains exhibited instability. Various strains of yeast cells provide an excellent model for studying the role of ploidy in the observed radiobiological reactions. The yeast *Saccharomyces cerevisiae* was used to study the mechanisms associated with genome instability (Duffy et al., 2016) and to determine the relationship between stress and genetic instability (Degtyareva et al., 2008).

The late appearance of colonies by cells that survived after exposure to various types of radiations is used in this monograph as a test of genetic instability. The delayed appearance of colonies, along with other delayed effects induced in individual cells surviving after exposure to ionizing radiation of various quality or UV light, subsequently manifests itself as genetic instability of the offspring. This effect occurs in cultured cells of various origins (Klein et al., 1990; Chang, Little, 1992 a, b; Morgan, Murnane, 1995; Little, 1998; Dugan et al., 2003;

Korogodin et al., 2007). This effect reflects genetic instability. The radiobiological evidence of the genetic instability of yeast cells surviving after irradiation were presented above. It was shown that in the distant offspring of cells that survived after irradiation, an increased yield of various mutations and morphological changes is observed – the content of respiratory mutants, the proportion of cells recombinant for the *ade*1 and *ade*2 mutations, the number of morphologically altered colonies, and a significant content of nonviable cells. These effects were expressed largely for clones, the formation of which was delayed for several days in comparison with the control. In addition to this radiobiological evidence of genetic instability, data on re-seeding of late-formed colonies are added. These experiments showed that cells from different colonies that grew later than the control again formed colonies that were heterogeneous in size. The aggregate of these data allowed us to attribute the delay in the formation of colonies by irradiated cells to genetic instability. Cells from late formed colonies were characterized by increased radio- and thermal sensitivity (Korogodin, 2010; Korogodin et al., 2007; Petin et al., 2008). This problem is important in connection with the use of synergetic ideas in the treatment of tumors not only with sparsely and densely ionizing radiation (Lindenbraten, Korolyuk, 2000; Hall, Giaccia, 2018; Kaprin, Mardynsky, 2019), but also with the use of UV radiation in photoradiation and photochemotherapy (Melloni et al., 1984; Waldow, Dougherti, 1984).

It was shown in many works (Kapultsevich et al., 1972; Korogodin et al., 1974, 1977 a, b; Kapultsevich, 1978; Korogodin et al., 2007; Petin et al., 2008) that genetic instability is a natural property of yeast cells with a sigmoid shape of the survival curves due to the ability of cells to recover from radiation damage, although this effect was negligible or even not observed for haploid strains with exponential survival curves. We assumed it was time to change this paradigm (Evstratova, Petin, 2018 a, b, c). It is possible that genetic instability can be determined mainly by cell ploidy and is not unambiguously associated with the shape of the survival curve and cell ability to recover from radiation damage. To test this hypothesis, we compared the effect of slower growth of irradiated cells, which determines genetic instability in wild-type haploid and homozygous diploid cells, as well as in their radiosensitive and UV-sensitive mutants that survive after irradiation with γ-rays, α-particles, and UV light.

The set of data presented in this chapter for radiosensitive mutants means that, due to the defects that block the repair of DNA double-strand breaks, cells of diploid strains are characterized by exponential forms of survival curves. Nevertheless, their genetic instability, quantified by the delay in the formation of colonies by cells that survived after irradiation with γ-rays, reaches 100% with increasing dose. These data show that cells of radiosensitive strains, regardless of the shape of the survival curves and cell ability to repair radiation damage, inherit some sub-damage responsible for the delay in division. This effect did not strongly depend on the radiosensitivity of haploid strains and was about 20%. Consequently, the effect of the late appearance of colonies by irradiated yeast cells is not unambiguously associated with the dynamics of survival, characterized by a certain shape of the survival curve or with the ability of cells to recover from double-strand DNA breaks. In other words, the presence of diploid-specific repair (Saeki et al., 1980) is not the main determinant of genetic instability of irradiated cells. As a result, the data presented show that the delay in the formation of macrocolonies by cells surviving after irradiation with sparsely ionizing radiation is strongly determined by the cell ploidy and not by the sigmoid shape of the survival curve of yeast cells and their ability to post-radiation recovery, as has been repeatedly postulated in earlier publications (Kapultsevich et al., 1972; Korogodin et al., 1974, 1977 a, b; Kapultsevich, 1978; Glasunov et al., 1989; Korogodin et al., 2007; Petin et al., 2008).

The practical use of ionizing radiation with high LET seems promising for several reasons. Such radiation is currently used to treat cancer. This is due to their high efficiency, especially in the field of low doses usually used in fractionated irradiation of malignant tumors, as well as due to the suppression of cell ability to recover from sublethal and potentially lethal damage (Frankenberg et al., 1981; Petin, Kabakova, 1981; Frankenberg-Schwager et al., 1984; Petin, Kapultcevich, 2014; Hall, Giaccia, 2018). Although genetic instability is well known after exposure to high LET radiation (Kadhim et al., 1994; Little et al., 1997; Ponnaiya et al., 2004), the RBE of densely ionizing radiation for this effect has never been quantified before this work. Most of the results on genetic instability have been obtained with cultured mammalian cells in a diploid state, so there is no work comparing RBE of densely ionizing radiation for haploid and diploid cell inactivation and genetic instability. The quantitative assessment of the delay in the formation of colonies by cells that survived after irradiation, presented in this work, makes it possible to compare the RBE of high-LET radiation both for lethal effects and for the delayed appearance of colonies by haploid and homozygous diploid cells.

Data on the relationship between RBE of densely ionizing radiation for inactivation of various cells with their ability to recover from radiation damage have been published by many authors (Petin, 1977, 1979, 1987; Alper, 1979; Frankenberg et al., 1981; Petin, Kabakova, 1981; Korogodin, Krasavin, 1982; Frankenberg-Schwager et al., 1984; Tobias et al., 1984; Blöcher, 1988; Krasavin, 1989; Petin, Petin, 1995; Petin, Kim, 2005, 2016; Petin, Kapultcevich, 2014; Zadnepryanets et al., 2018 a, b) and were confirmed in this work for cell survival. The correlation between the RBE of α-particles and the ability of cells to recover radiation damage was confirmed in this work – the strains that were most sensitive to the action of ionizing radiation, i.e., the most "deficient" in terms of their ability to recover, had lower RBE values – in the range 2–4, and for most of these strains RBE values were in the range between 2 and 3. On the contrary, for strains that were more resistant to ionizing radiation, had a greater ability to recover from damage induced by ionizing radiation, RBE varied being in the range from 3 to 5. These differences are even more pronounced when assessing RBE after cell recovery in a non-nutrient medium (Figure 3.10), in this case RBE varied from 3 to 10 or more. The highest RBE value was observed for wild-type diploid cells. For wild-type haploid yeast cells and all mutant cells, regardless of their ploidy, RBE values randomly varied in the range 1.6–2.7 without any systematic regularity. These data clearly confirm that cell survival and RBE of densely ionizing radiation is determined not only by the increased probability to produce primary radiation damage at the physicochemical stage of interaction of ionizing radiation, but also by the ability of cells to recover from radiation damage at the biochemical stage of such damage fixation.

An important new fact obtained in our studies is that the RBE for the survival of wild-type diploid yeast cells capable of recovering from radiation damage is 4.6 ± 0.2 and does not differ significantly from the RBE for genetic instability, quantified by the delayed formation of colonies by irradiated cells (4.7 ± 0.4). RBE for cells with significantly reduced ability to recover from radiation damage was significantly lower for both survival and genetic instability. These RBE values randomly varied in the range 1.8–2.7 without any systematic pattern for wild-type haploid yeast cells and all mutant cells regardless of their ploidy. This indicates that RBE values for both studied tests are mainly determined by the ability of cells to recover from radiation damage. The results obtained are not new for cell survival, while they are fundamentally new for genetic instability. The identity of the RBE values or cell survival and their genetic instability may be associated with the greater efficiency of densely ionizing radiation to produce lethal radiation damage and accompanying sub-damage, which are responsible for the genetic instability, i.e., for the delay in the appearance of colonies by cells that survived after irradiation. This sub-damage keeps in distant descendants of cells and thereby cause destabilization of the genome, which is manifested in the delayed appearance of colonies after irradiation.

Large RBE values for high LET radiation may be due to the fact that high-LET radiations create clusters of ionizations and excitations in DNA (Goodhead, 1994; Zadnepryants et al., 2018 a, b), which lead to more severe and complex damage than reference radiation with low LET. This is also consistent with our previous results that densely ionizing radiation inhibits cell repair due to the formation of a larger proportion of irreversible radiation damage, while the recovery constant is stayed unchanged after irradiation with both sparsely and densely ionizing radiation (Petin, Kim, 2005 b; Petin, Kim, 2016). These data mean that the recovery process itself is not disturbed by radiation with high LET.

Other new and interesting results of this study are related to the effect of the delayed colony forming ability of haploid and diploid yeast cells of various genotypes that survived after exposure to low- and high-LET radiation. Both resistant and radiosensitive diploid strains, in contrast to haploid ones, demonstrate a greater degree of genetic instability (100% versus 20%) regardless of the quality of the radiation. Despite the defects of the DNA double-strand breaks repair in radiosensitive diploid mutants, cells that survived irradiation inherit some residual sub-damage, which is responsible for the genetic instability display to the same extent as diploid yeast cells of wild-type capable of repairing double-strand breaks in DNA. This means that the genetic instability of yeast cells caused by exposure to ionizing radiation is determined by both the quality of radiation and the ploidy of the cells, but not by the shape of the survival curve and cell ability to recover from radiation damage.

According to many of the works cited above, the maximum RBE values of densely ionizing radiation, estimated in the region of high doses of ionizing radiation for yeast cells, cultured mammalian cells and plants, are in the range of 3–5. In many cases, when extrapolating RBE values to the region of low doses, RBE increases and is in the range of 5–10. In accordance with the Russian radiation safety standards (NRB-99/2009), an RBE value of 20 was taken for α-radiation, obtained not only as a result of extrapolation of data obtained on various cellular systems to man, but also in connection with the introduction of a safety factor that takes into account various uncertainties. Since the assessment of genetic instability in the region of low doses of ionizing radiation seems to be rather difficult at present, we did not discuss the problem of estimating the RBE for genetic instability in the region of low doses.

Comparison of survival and genetic instability depending on the dose of UV light for haploid and isogenic diploid wild-type cells was carried out. The survival curves for these strains were characterized by a pronounced shoulder in the region of low doses of UV radiation, reflecting the lower efficiency of UV light action in this region and cell ability to recover from sublethal damage. Haploid cells were slightly more sensitive to UV radiation compared to diploid cells. For wild-type haploid cells that survived after UV irradiation, the maximum genetic instability was 20–25%, and for diploid cells, 100%. Qualitatively similar results were obtained for UV-sensitive mutants. The maximum delay in the formation of colonies by the surviving haploid cells after irradiation was 40–50%, and by diploid cells, 100%, regardless of their UV sensitivity and the shape of the survival curves. This indicates that the delay in colony formation is determined by the number of chromosomes sets in the cell, and not by the exponential or sigmoid shape of the survival curves or the repair of damage induced by UV light. Once again, we note that, after exposure to ionizing or UV radiation, the survival curves of diploid cells of wild-type strains have a sigmoid shape, and they are much more resistant to influencing agents than haploid cells characterized by an exponential survival curve. This is due to the diploid-specific ability of wild-type cells to recover from double-stranded DNA breaks (Luchnik et al., 1977; Saeki et al., 1980; Frankenberg et al., 1981). The ability of haploid cells to repair double-stranded DNA breaks is now also known due to the discovery of non-homologous repair (Lewis, Resnick, 2000; Mao et al., 2008), the efficiency of which is lower than that of restoration due to the recombination of homologous chromosomes. However, it cannot be concluded from this data that it is precisely the different degree of restoration of haploid and diploid cells that determines the various effect of the late formation of colonies by haploid and diploid cells, since both radio- and UV-sensitive diploid, rather than haploid strains, demonstrated the same genetic instability as theirs initial wild-type cells, which reached this effect 100% with increasing dose of the acting agent. This indicates that the genetic instability determined by the delayed formation of colonies by cells that survived after UV irradiation is mainly determined by the number of chromosome sets in the cell, and not by the shape of the survival curves and/or the repair of damage induced by UV light. This conclusion is fully consistent with our studies with wild-type yeast cells and radiosensitive mutants defective in various repair pathways and irradiated with γ-rays or α-particles.

The dependences of the delayed colony formation on cell survival show that this effect does not depend on the type and quality of ionizing radiation. Since cell survival after action of various agents is unambiguously determined by the number of lethal events, these results indicate that the number of lesions effective to cell inactivation determines the number of sub-damage responsible for the delayed colony forming ability by irradiated cells. Interestingly, the effect of the late formation of colonies by both haploid and diploid yeast cells, depending on their survival, was the same after irradiation with UV light, sparsely and densely ionizing radiation. These data indicate that equal effect of radiation of different quality induces the same number of effective (lethal) damage and an equal or proportional number of accompanying sub-damage, which are responsible for genetic instability, manifested by a delay in the formation of colonies by irradiated cells. In other words, this effect is associated with the accumulation of sub-damage that are ineffective for cell inactivation. This sub-damage keeps in the distant offspring of cells that survived after irradiation and thereby cause destabilization of the genome, which is manifested in the delayed production of colonies and in the appearance of phenotypically different clones.

Many mechanisms have been proposed for how radiation exposure can lead to genetic instability (McMurray, Gottschling, 2004; Yuen et al., 2007; Korogodin, 2010; Sheltzer et al., 2011; Shen, 2011; Zhu et al., 2012; Skoneczna et al., 2015; Duffy et al., 2016). At present, the study of the role of DNA damage in genetic instability, aging and cancer formation is being actively continued (Cheng, Loeb, 1993; McMurray, Gottschling, 2004; Korogodin et al., 2007; Petin et al., 2008; Shen, 2011). There is evidence that double-stranded DNA breaks are important for the induction of the delayed reproductive cell death and, consequently, the delayed colony formation by cells that survived exposure to ionizing radiation (Klein et al., 1990; Chang, Little, 1992 a; Little, 1998; Morgan, Murnane, 1995; Brosh, Bohr, 2007; Korogodin et al., 2007; Korogodin, 2010). It was suggested that double-stranded DNA breaks initiate the phenotype of the delayed colony formation (Chang, Little, 1992 b). As was mentioned, double-strand DNA breaks requires two homologous chromosomes. Due to this ability, diploid yeast cells are more resistant to the action of ionizing radiation than haploid ones, the ability of which to repair such breaks is significantly reduced and can be realized through less efficient non-homologous reunification of DNA breaks (Lewis, Resnick, 2000; Mao et al., 2008). The results of our work show that genetic instability determined by the delayed appearance of colonies was well expressed and reached 100% for diploid cells, regardless of the shape of the survival curve and cell ability to recover from radiation damage. This effect was weakly expressed (20–25%) for all haploid strains characterized by an exponential shape of the survival curve. Consequently, genetic instability is largely determined by the ploidy of yeast cells and is not related to the shape of the dose-survival curve and the ability of cells to recover from radiation damage. We compared the delayed colony formation by haploid and homozygous wild-type diploid strains and radiosensitive mutants that survive exposure to radiation with different LETs. It was shown that both resistant and radiosensitive diploid strains, in contrast to haploid ones, demonstrate a high degree of genetic instability (100% versus 20–25%) regardless of the quality of the radiation, radiosensitivity of cells, and their ability to recover from radiation damage. Consequently, the genetic instability caused by ionizing radiation is largely determined by cell ploidy, and not by the shape of the dose-response curve and the ability of cells to recover from radiation damage.

The study of the role of DNA damage in genetic instability, aging and cancer formation is being actively continued (Cheng, Loeb, 1993; McMurray, Gottschling, 2004; Korogodin et al., 2007; Yuen et al., 2007; Petin et al., 2008; Sheltzer et al., 2011; Shen, 2011; Duffy et al., 2016). There is evidence that double-stranded DNA breaks are important for the induction of delayed reproductive cell death and, as a consequence, the delayed colony formation by cells that survived exposure to ionizing radiation (Klein et al., 1990; Chang, Little, 1992 a; Little, 1998; Morgan, Murnane, 1995; Brosh, Bohr, 2007; Korogodin et al., 2007; Korogodin, 2010). There is evidence that double-stranded DNA breaks initiate the phenotype of the delayed colony formation (Chang, Little, 1992 b). The results of our work show that genetic instability determined by the delayed appearance of colonies was well expressed and reached 100% for diploid cells, regardless of the shape of the survival curve and cell ability to recover from radiation damage. This effect was weakly expressed after ionizing exposure (20–25%) for all haploid strains characterized by an exponential shape of the survival curve. Consequently, genetic instability is largely determined by the ploidy of yeast cells and is not related to the shape of the dose-survival curve and the ability of cells to recover from radiation damage. To test this hypothesis, we compared the delayed colony formation by haploid and homozygous wild-type diploid strains and radiosensitive mutants that survive exposure to radiation with different LETs. It was shown that both resistant and radiosensitive diploid strains, in contrast to haploid ones, demonstrate a high degree of genetic instability (100% versus 20–25%) regardless of the quality of the radiation. Consequently, the genetic instability caused by ionizing radiation is largely determined by cell ploidy, but not by the shape of the dose-response curve and the ability of cells to recover from radiation damage. In this regard, it is interesting that recent studies of cancer cells and budding yeast have demonstrated that aneuploidy, a condition of abnormal chromosome numbers, is correlated with disordered genetic instability. It was shown that genetic instability correlates with cellular ploidy in an unexpected way: cells with a chromosomal content close to the haploid state are much more stable than cells with a large number of chromosomes (Zhu et al., 2012; Zheng et al., 2016). Published evidence suggests that increasing the number of chromosome sets increases genomic instability, while improving the cellular ability to mitigate the deleterious effects of gene mutations or chromosome loss. The double set of chromosomes allows the resulting aneuploid cells to survive, mitigating the harmful effects of chromosome loss or gene mutation. These unique functions can generate cells with variable karyotype, thereby providing material for natural selection (Lee et al., 2015).

We have obtained other additional confirmations about the dominant role of ploidy, rather than the sensitivity of cells and their ability to recover from radiation damage, in the genetic instability of cells that survived after irradiation. It has long been known that the shape of the survival curves after irradiation with UV light, in contrast to the action of ionizing radiation, does not depend on ploidy and is sigmoid for both haploid and diploid yeast cells (Korogodin, 1966; Petin, 1987; Petin, Komarov, 1989). Then, a comparison of the genetic instability of haploid and diploid yeast cells that survived after exposure to ultraviolet radiation may contribute to a better understanding of the assumption whether this effect is really related to the shape of the survival curve or is mainly determined by cell ploidy. Despite the sigmoid shape of the survival curves of both haploid and diploid yeasts after UV irradiation, the delayed appearance of colonies is more pronounced for diploid rather than haploid cells (100% versus 30–50%). Consequently, the effect of the delayed colony formation by diploid yeast cells surviving after irradiation is not related to the shape of the survival curve and the ability of cells to recover from radiation damage.

Despite the fact that the biological effects associated with radiation-induced genetic instability are widely discussed in the literature, the specific molecular nature of the events that initiate and maintain instability in the offspring of irradiated cells remains unclear. A thorough discussion of the molecular mechanisms underlying the genetic stability is beyond the mainstream of this work. However, there are some points worth mentioning. Many mechanisms have been proposed on how exposure to ionizing radiation can lead to genetic instability (Aggarwal, Brosh, 2012; Skoneczna et al., 2015). Although it has been shown that the nucleus is a target for radiation-induced genetic instability, it is very likely that a single gene or even a group of genes, or a single radiation mutation or DNA double breaks, is not sufficient to induce genetic instability (Kaplan, Morgan, 1998). It is believed that an unrepaired or mistakenly repaired DNA double-strand break can lead to both cell inactivation and genetic instability. Defects in the process of DNA repair can initiate genomic instability and carcinogenesis (Klein et al., 1990; Chang, Little, 1992 a, b; Cheng, Loeb, 1993), cause cell cycle arrest and apoptosis. Homologous recombination is an important metabolic pathway found in all organisms and may be implicated in genetic instability (Lee et al., 2015).

In other works, it was also noted that mitotic recombination or initiation of double DNA breaks might be responsible for genome instability (Korogodin et al., 1974; Tolstorukov et al., 1979; Chang, Little, 1992 a; Duffy et al., 2016). Mutations (Chang, Little, 1998) and DNA repair errors (Marder, Morgan, 1993; Mazurik, Mikhailov, 2001) can lead to genetic instability. It was demonstrated that the formation of double-strand DNA breaks, as well as damage to chromosomes, can initiate delayed reproductive death of both yeast cells and cultured mammalian cells (Chang, Little, 1992 b; McMurray, Gottschling, 2004; Urushibara et al., 2014). Rad51 and Rad52 are key proteins involved in the repair of DNA double-strand breaks through homologous recombination in diploid and polyploid cells of various origins (Lee et al., 2015). In our work, genetic instability was demonstrated for yeast strains defective in recombination (*rad*51/*rad*51, *rad*52/*rad*52). For these strains, as for other radiosensitive mutants, the delay in the formation of colonies by cells that survived after irradiation reached

100%, as in wild-type cells capable of repair double-stranded DNA breaks. On this basis, it can be assumed that homologous recombination cannot be considered as the main reason for the slow formation of colonies by irradiated cells. Consequently, the mechanism of different manifestations of genetic instability by haploid and diploid yeast cells surviving after irradiation does not depend on the ability of cells to recover from radiation damage and may be associated with some chromosomal damage, which can be largely fatal for haploid rather than diploid cells.

Genetic analysis of unstable and normal clones was carried out using artificially obtained haplonts diploid yeast *Pichia pinus*, which have four chromosomes in a haploid set (Tolstorukov, Benevolensky, 1978; Tolstorukov, Effremov, 1984). It has been shown that cells that have lost one or several nonhomologous chromosomes give rise to colonies that grow on a nutrient medium later than control (Tolstorukov et al., 1979). All four chromosomes in the analyzed diploid cells could be easily identified by auxotrophic mutations in a heterozygous state (Tolstorukov et al., 1979, 1982). In control cells of this strain, as in unirradiated cells of the yeast *Saccharomyces cerevisia*e, unstable clones arise spontaneously with frequencies of 10^{-2}–10^{-3}. An analysis of the segregation of heterozygous markers during divisions of mitotically unstable clones showed that the loss of individual chromosomes occurs in unstable clones. These chromosomes, initially represented by two homologues, could be lost after irradiation up to complete haploidization. The frequency of chromosome loss in unstable clones, calculated in relation to mitotic segregation of one marker, was estimated as $1.5 \cdot 10^{-3}$ per cell per division. It was concluded that cells that have lost one or several chromosomes are not absolutely lethal and give rise to unstable clones of the yeast *Pichia pinus*, which grow on a nutrient medium later than control (Tolstorukov et al., 1979, 1982). It is possible that the delayed colony forming ability in *Saccharomyces cerevisiae* yeast may be also associated with a violation of mitotic divisions. This, in particular, is indicated by the data on significant chromosomal abnormalities in unstable clones of *Saccharomyces cerevisiae*, which were observed during the analysis of karyotypes using pulsed field electrophoresis (Korogodin et al., 2007). The works (Sheltzer et al., 2011; Urushibara et al., 2014) also provide data demonstrating that aneuploidy can lead to genetic instability in yeast. The variety of possible mechanisms for triggering genetic instability indicates the need for further studies of this phenomenon.

Among other possible mechanisms for radiation-induced genetic instability, the following issues can be noted. It is assumed that damage to DNA, cell membranes or increased production of reactive oxygen species (Myasnik, 1974; Pelevina et al., 1985; Bruskov, 1990; Polivoda et al., 1990) are the necessary conditions for instability. It is known that genome stability is provided at the cellular level by three interconnected systems (Pelevina et al., 1985). The first of these is a system for monitoring the cell cycle at checkpoints. It creates conditions for the removal of cells with altered DNA structure. By stopping the movement of cells along the cycle, this control system provides the time required for repair, and if it is ineffective, it induces the processes of apoptosis. The central component of this system is a tumor suppressor protein *p53*. In general, the system for monitoring the cell cycle at checkpoints is effective when exposed to ionizing radiation. However, the *p53* gene is highly mutable. The mutant *p53* protein may lose the ability to fully interact with the chromosome and, as a consequence, lose effective control over other genes directly involved in the supervision of cell passage through the cycle and the elimination of the defective part of the cell population through apoptosis. As a result, the frequency of spontaneous homologous recombination between chromosomal sequences increases by a factor of 5–20, which indicates an increase in genetic instability.

The second system consists of DNA repair mechanisms aimed at eliminating damage arising during cellular metabolism, thermal fluctuations of polymer atoms and the action of genotoxic agents, including ionizing radiation. These mechanisms include base excisional repair, nucleotide excisional repair, repair of base pairing disorders, and recombination repair of DNA breaks by a homologous mechanism and non-homologous DNA end joining. The efficiency of repair is determined by the nature of damage, the actual conditions for its implementation, the activity of enzymes of the repair system, the state of DNA packing of chromatin filaments, and the functional usefulness of the *p53* protein involved in the control of repair. Changes in each of these factors have a significant effect on maintaining the stability of the genome and can lead to the emergence of its instability. These facts make it clear why the participation of DNA repair systems in the development genome instability is a subject of intensive study.

The third system that has a significant impact on maintaining the genome stability is the redox homeostasis system. Reactive oxygen species and antioxidants form a single prooxidant-antioxidant system. Until recently, only protective functions were associated with this system, but recent studies show that, in addition to damaging functions, reactive oxygen species can play an important role in intracellular signaling. The role of this system in response to ionizing radiation is traditionally considered extremely important, since the main role in eliminating the consequences of the indirect action of ionizing radiation is assigned to antioxidant compounds and antioxidant enzymes. The question of how this mechanism can participate in the genetic instability caused by UV irradiation is still open.

Another possible cause of genetic instability, manifested in the form of delayed reproductive cell death, is the loss of telomeric DNA in cells (Artandi, DePinho, 2010). There is abundant evidence that telomeres that protect the ends of chromosomes are directly involved in the ability of cells to proliferate indefinitely (Blasko, 2005; Shay, Wright, 2000). Telomeres gradually shorten after each cycle of cell division, ultimately losing the ability to protect DNA located at the ends of the chromosome from joining each other. Thus, the length of telomere DNA in a cell determines how many successive divisions a cell can undergo before telomeres are severely damaged and lose their protective functions (Chen et al., 2009). The bare ends of the chromosomes are recognized as breaks in both strands of DNA. Damage of this kind is usually repaired by connecting the broken ends. However, in this case, the ends of different chromosomes may be connected, since they are no longer protected by telomeres. This temporarily solves the problem of the absence of telomeres, however, during anaphase of cell division, linked chromosomes are torn apart randomly, which leads to a large number of mutations and chromosomal abnormalities.

Genome rearrangements lead to mutations that underlie many human diseases, and continued genomic instability is likely to contribute to the development of many types of cancer. The tools for studying genome instability in mammalian cells are limited, while model organisms such as *Saccharomyces cerevisiae* are convenient for such studies. There are many genetic analyzes designed to measure the frequency of chromosomal rearrangements in yeasts. These genetic analyzes have been used to identify many types of chromosomal rearrangements, including translocations, deletions, and damaged chromosomes repaired by *de novo* telomere addition, and genes have been identified that act to suppress and produce chromosomal rearrangements. The results of these studies have contributed to understanding the pathways and mechanisms that suppress genome instability and how these pathways interact with each other (Putnam, Richard, 2017).

Based on the presented data, it can be concluded that the mechanism of different manifestations of genetic instability by haploid and diploid yeast cells surviving after irradiation may be associated with some chromosomal damage, for example, deletions or even loss of chromosomes, – events that are more fatal for haploid, not diploid cells. That is why the genetic instability of haploid cells is manifested to a much lesser extent than for diploid cells, regardless of their radiosensitivity or cell ability to recover from radiation damage.

The final conclusion is that the mechanism of different manifestations of genetic instability by haploid and diploid yeast cells surviving after irradiation does not depend on the ability of cells to recover from radiation damage and may be associated with some chromosomal damage (deletions, aneuploidy), which are largely fatal to haploid but not diploid cells.

Part II. Unusual Radiobiological Effects

Chapter 5

Universal Patterns of Synergistic Interactions

Abstract

The regularities of the synergistic interaction of physical and chemical environmental factors were described in several monographs based on the results obtained by the staff of Biophysical laboratory of the A.F. Tsyb Medical Radiological Research Center, Obninsk (Petin, Komarov, 1989; Petin et al., 2012, 2020; Petin, Morozov, 2015; Petin, Kim, 2016). New results are summarized in our recent publications (Pronkevich, Evstratova et al., 2017; Evstratova, Petin, 2018 a, 2020; Evstratova et al., 2018 b), on the basis of which this chapter was prepared. It begins from mathematical description and quantitative prediction of synergistic effects after combined application of two damaging agents. The model is based on the suggestion that synergism is due to the production of additional effective damage resulting from interaction of sub-damage that are non-effective after separate application of agents. Potential significance of synergy at low intensities of physical and chemical environmental factors is proved. Biophysical interpretation of synergy dependence on the intensity of the applied agents is presented. General biological and clinical significance of synergy ideas are discussed. A number of universal patterns of synergic interactions are demonstrated, which did not depend on the used biological objects and test as well as on the interacting physical and chemical factors.

5.1. Description and Prediction of Synergistic Effects

In general, synergetic is defined as the science of self-organization or the emergence of a new quality, including the ordered structures arising from chaos (Haken, 1980; Berthoud, 2013). Synergistic effect in the economy is an increase in the efficiency of activities as a result of combination, connection, integration, merging of separate parts into a single system, due to the appearance of new qualities (emergence) of the resulting system (Knyazeva, 2013). The term "synergism" used in radiation biology means "joint action" (from the Greek "sin" – jointly and "ergos" – action) of different factors. In this case, two fundamentally different phenomena may arise – synergistic and antagonistic interactions (Petin et al., 2012, 2020; Petin, Morozov, 2015; Petin, Kim, 2016). In the first case, the resulting biological effect is greater than the expected sum of the effects after independent action of each factor separately, in the second case – *vice versa*. The synergistic interaction of physical and chemical agents of natural and anthropogenic origin, as well as biological and social factors affecting living objects and the biosphere as a whole is a reality of the modern world. For example, a synergistic interaction of ionizing radiation with a number of chemical radiosensitizers can be used in the treatment of cancer patients (Wang et al., 2007; Zhang et al., 2011; Segova-Mendoza et al., 2015; Patties et al., 2016). The term synergistic interaction can be used not only to enhance the harmful, but also positive effect with combined effects. Thus, the authors of works (Suganuma et al., 1999; Fujiki et al., 2015) paid special attention to patients with cancer in Japan, who daily consumed green tea together with anticancer drugs. The findings show a synergistic interaction of green tea catechins with anti-cancer drugs.

Figure 5.1 shows an example of calculating the thermal enhancement ratio (TER) and the synergistic enhancement ratio (k) after simultaneous and separate action of hyperthermia (52°C) and ionizing radiation (10 Gy/min) on haploid yeast cells of *Saccharomyces cerevisiae* (strain 6a). Curve 1 in this figure reflects the dependence of cell survival on the duration of exposure to hyperthermia (upper abscissa), and curve 2 – on the dose of ionizing radiation (lower abscissa). Curve 4 shows a decrease in yeast cell survival for the case of simultaneous application of agents. For any experimental point of this curve, one can indicate both the absorbed dose (on the lower abscissa) and the duration of exposure to the agents (on the upper abscissa). For example, a dose of 120 Gy was reached in 12 minutes, a dose of 200 Gy – in 20 minutes, i.e., the upper and lower abscissas are equally acceptable for curve 4. Curve 3 is the theoretical curve expected for the independent action of the effects produced by heat and radiation. It is important to note that for the independent action of agents, the probability of the final effect, in accordance with the probability theory (Feller, 1969; Gmurman, 2014; Vasina et al., 2018; Olkin et al., 2019), is determined not by a simple sum of the effects, but by the product of probabilities of the effects induced by each agent applied separately. So, for cell survival, we have

$$S = S_1 \cdot S_2, \tag{5.1}$$

where S is the probability of cell survival after combined exposure, S_1 and S_2 are the probabilities of cell survival after the action of each agent applied separately. Then, after taking the logarithm of equation (5.1), we have

$$lnS = lnS_1 + lnS_2 \text{ or } lgS = lgS_1 + lgS_2. \tag{5.2}$$

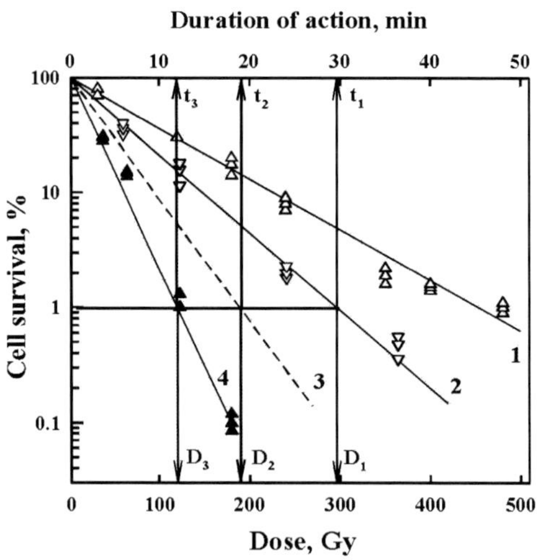

Figure 5.1. Survival curves of *Saccharomyces cerevisiae* haploid yeast cells, strain 6a: 1 – the effect of the elevated temperature (52°C); 2 – the effect of ionizing radiation (γ-rays of ^{60}Co, 10 Gy/min) at room temperature; 3 – calculated curve for independent summation of the effects of ionizing radiation and hyperthermia; 4 – experimentally obtained survival curve after the simultaneous action of these agents.

Consequently, the independent action of two agents is characterized by the addition of the logarithms of the probabilities of cell survival.

According to Poisson statistics (Zimmer, 1962; Lea, 1963; Timoffeev-Resovsky, Zimmer, 1947; Olkin et al., 2019), the number of lesions induced by a factor causing cell inactivation can be defined as

$$N = -lnS. \tag{5.3}$$

Then equation (5.2) means that the total number of damages caused by the independent action of two factors is determined by the sum of the damage generated by each of the acting agents. In accordance with Figure 5.1, after combined action of ionizing action and hyperthermia, the thermal enhancement ratio (TER), often used to quantitatively describe the increase in cell radiosensitivity in medical radiology and oncology (Sreffer et al., 1990; Lindenbraten, Korolyuk, 2000), for a survival rate of 1% (Figure 5.1) can be written as

$$TER = D_1/D_3 = t_1/t_3, \tag{5.4}$$

where D_1 is the absorbed dose after the action of ionizing radiation, D_3 – after the simultaneous use of agents. This ratio increases with an increase in the operating temperature (due to a decrease in D_3 and does not in any way reflect the nature of the interaction of the two agents, i.e., does not answer the question of whether the observed effects is independent or synergistic. Curve 3 in Figure 5.1 is constructed using equation (5.1). To quantify the synergistic effect, we used the synergistic enhancement ratio k (Petin, Berdnikova, 1979), which for 1% of cell survival is described by the expression

$$k = D_2/D_3 = t_2/t_3, \tag{5.5}$$

where D_2 is the absorbed dose in the case of independent action of agents. In accordance with this equation, at low temperatures, when the biological effect is mainly determined by ionizing radiation and the contribution of thermal damage is insignificant, there is no synergistic interaction, i.e., $k = 1$. With a further increase in the effective temperature, k increases. At temperatures when cell inactivation is mainly due to thermal damage, k can tend to unity. In fact, this coefficient shows how many times the registered effect has increased in comparison with the independent action of agents. In the case of sigmoid survival curves, this ratio can also be used, meaning the ratio of the mean lethal durations of exposure that reduce survival by a factor of e on the straight-line portion of the dose-response curves presented on a semi-logarithmic scale, where e is the base of natural logarithms. Note that for each experimental point in the following figures, 4 survival curves were used to calculate the synergistic enhancement ratio (equation (5.5)) and 3 curves to estimate the TER (equation (5.4)). The arrows in Figure 5.1 shows an example of finding the doses required to quantify these parameters. The results shown in the following figures are based on more than 400 curves describing the dependence of cell survival on the duration of exposure to ionizing radiation, hyperthermia, cisplatin, and other factors, as well as after the combined application of hyperthermia with physical or chemical agents. These data were published in works (Ben-Hur, Elkind, 1974; Ben-Hur, 1976; Hahn, 1982; Dewey, Holahan, 1984; Urano et al., 1990;

Zhurakovskaya, 2001; Petin et al., 2012. 2020; Petin, Kim, 2016; Hall, Giaccia, 2018). It is not possible to give all these graphic data, but they are used to construct the subsequent figures.

The increased efficiency of the combined effects of ionizing radiation with other agents is relevant for medical radiology, sterilization of medical equipment, food products, etc. (Kuzin, Kaushanskiy, 1981; Yarmonenko et al., 1992; Tsyb, Mardynskiy, 2010; Kaprin et al., 2014; Hall, Giaccia, 2018). The use of ionizing radiation with hyperthermia or chemical radiosensitizers in radiation therapy has become widespread (Dietzel, 1975; Streffer et al., 1978; Hill, Bellami, 1984; Sugahara, 1984; Iliakis, Kurtzman, 1988; El-Awady et al., 2001; Kurpeshev, Mardynsky, 2010; Tsyb, Gulidov, 2010). A synergistic interaction of agents has been observed when their effect exceeds the independent sum of the effects produced after their separate application (Dethlefsen, Dewey, 1982; Kuzin, 1983; Sugahara, Saito, 1988, 1989; Streffer et al., 1990; Buhtoiarov et al., 2011). One of the common points of view on the mechanism of synergistic interaction is the suppression of cell ability to recover from radiation damage. However, experimental data have shown that after combined exposures, the process of cell recovery from potentially lethal damage is not disturbed, and the observed inhibition of recovery is completely due to the formation of additional irreversible damage because of the interaction of some sub-damage that are not effective when the agents are used separately (Zhurakovskaya, 2001; Komarova, Petin, 2007; Petin et al., 2012, 2020). On this basis, mathematical models of synergism have been proposed (Petin, Komarov, 1989, 1997; Petin, Kim, 2016), which make it possible to optimize the parameters of influencing agents to achieve the greatest synergy and explain its dependence on the intensity of the factors.

The variety of physical factors and chemical agents necessitates the development of new conceptual foundations of synergistic interaction, which would not depend on the agents, biological objects and tests used in combinations. Many regularities in the manifestation of synergistic effects (Petin et al., 2012, 2020; Evstratova et al., 2018) indicate the existence of some general idea regarding the mechanism of the synergistic interaction, which would not depend on the biological object, test, and agents applied in combination. It was suggested that the mechanism of synergistic interaction might be due to the formation of additional effective damage resulting from the interaction of sub-damage induced by both agents (Petin, Komarov, 1997). It should be noted that the idea concerning the interaction of radiation damage is widely used in the construction of mathematical models of the biological action of ionizing radiation and other agents including hyperthermia (Alper, 1970; Kellerer, Rossi, 1972; Loshek et al., 1977; Murthy et al., 1979; Zaider, Rossi, 1980; Tobias, 1985; Ivanov, 1986; Jain, Pohlit, 1986; Jung, 1986; Ager, Haynes, 1987; Harder et al., 1994). In some of these works mathematical models of synergism were described, but none of them allows to describe the general patterns of synergistic interaction.

The hypothesis of the interaction of sub-damage was used to develop several of the mathematical model of synergy, which are described in previous publications (Petin, Komarov, 1989, 1997; Petin et al., 2012, 2020; Petin, Kim, 2016). Let us briefly describe the essence of the simplest version of this model. A synergistic interaction means that the average number of damages per cell is greater than the sum of the damage from each agent. Consequently, in conditions of synergistic interaction, additional effective damage appears. Let us assume that they are formed due to the interaction of some sub-damage produced after the action of each agent and which are not effective when the agents are used separately. Suppose that in order to form additional effective damage, a sub-damage from one agent interacts with a sub-damage from another. Let each effective damage from one agent correspond to p_1 sub-damage induced

by this factor. Similarly, we introduce the second parameter p_2 – the number of sub-damages induced per one effective damage from another factor. Then, if we assume that the number of effective (for example, lethal) damage induced by the first factor is N_1
many of our experiments it was hyperthermia) – N_2
formed by the first agent will be equal to p_1N_1, and by the second one – p_2N_2. Then the additional number of effective damages N_3

$$N_3 = \min\{p_1N_1;\ p_2N_2\}. \tag{5.6}$$

The synergistic enhancement ratio k, determined by the ratio of the number of effective damages in the case of combined impacts to the number of damages expected from independent sum of damage generated by each agent. Taking into account expression (5.6), it can be written as

$$k = \frac{N_1+N_2+N_3}{N_1+N_2} = \frac{N_1+N_2+min\{p_1N_1;p_2N_2\}}{N_1+N_2} = 1 + \frac{min\{p_1;p_2N_2/N_1\}}{1+N_2/N_1}. \tag{5.7}$$

If $p_1N_1 < p_2N_2$, i.e., cells die mainly from the action of hyperthermia, then $\min\{p_1N_1;p_2N_2\} = p_1N_1$, and from equation (5.7) we obtain

$$k_1 = 1 + \frac{p_1}{1+N_2/N_1}. \tag{5.8}$$

From here

$$p_1 = (k_1 - 1)\cdot(1 + N_2/N_1). \tag{5.9}$$

Similarly, if the cells inactivated mainly from the action of the first factor, i.e. $p_1N_1 > p_2N_2$ and $\min\{p_1N_1;p_2N_2\} = p_2N_2$, we have

$$p_2 = (k_2 - 1)\cdot(1 + N_1/N_2). \tag{5.10}$$

From equation (5.7), it is easy to obtain the condition for achieving the greatest synergistic interaction

$$p_1N_1 = p_2N_2. \tag{5.11}$$

This means that the greatest synergistic effect is achieved when both agents produce equal numbers of sub-damage. Equation (5.11) might be rewritten as

$$N_2/N_1 = p_1/p_2. \tag{5.12}$$

The ratio N_2/N_1 can be estimated from the ratio of the survival curve slopes after exposure to ionizing radiation or another agent $(1/t_{01})$ with hyperthermia $(1/t_{02})$, where t_{01} and t_{02} are the duration of the separate action of these factors, which reduce cell survival by e times on the exponential parts of the survival curves. Then, based on the experimental results, the ratio N_2/N_1 can be estimated by the formula

$$N_2/N_1 = t_{01}/t_{02}. \tag{5.13}$$

Equations (5.11) and (5.12) determine the conditions for achieving the maximum synergistic interaction. Taking into account equations (5.7) and (5.11), the value of the greatest synergistic enhancement ratio can be represented as

$$k_{max} = 1 + \frac{p_1 p_2}{p_1 + p_2}. \tag{5.14}$$

As can be seen from this equation, k_{max} is compleyely determined by the values of p_1 and p_2 but does not depend on N_1 and N_2
synergistic enhancement ratio (equation (5.14)), the ratio N_2/N_1 (equations (5.12) and (5.13)), at which it is achieved.

Let us analyze the applicability of the considered model for the sequential action of agents. We present the example of the sequential action of ionizing radiation and hyperthermia. The total effect of the agents is determined by the sum of the effective damage caused by the action of each agent N_1 and N_2 as well as the number of additional effective damage N_3
the interaction (equation (5.6)). The degree of radiosensitivity modification after sequential action of ionizing radiation and hyperthermia is often characterized by thermal enhancement ratio (TER) or by the dose modifying factor (DMF) (Petin et al., 2012) determined by the ratio

$$TER = DMF = (N_1 + N_2)/N_1. \tag{5.15}$$

Taking into account equation (5.6) we have

$$TER = 1 + \min\{p_1 N_1;\ p_2 N_2\}/N_1. \tag{5.16}$$

If the predominant number of effective damages is due to the action of ionizing radiation, i.e., $p_1 N_1 > p_2 N_2$, then the degree of modification of the biological effect of ionizing radiation by hyperthermia, taking into account equations (5.16), will be determined by the equation

$$TER = 1 + p_2 N_2/N_1. \tag{5.17}$$

It can be seen that the thermal enhancement ratio for a fixed dose of ionizing radiation (N_1 = const) increases linearly with an increase in the number of effective damage from hyperthermia N_2 $p_1 N_1 < p_2 N_2$, then

$$TER = 1 + p_1, \tag{5.18}$$

i.e., the thermal enhancement ratio will be a constant value independent of the modifying agent dose. The modification limit occurs under the condition $p_1 N_1 = p_2 N_2$, when both agents cause the same number of sub-damages. The total number of effective (in our case, lethal) damage N_Σ under the combined action of two factors will be described by the expression

$$N_{\Sigma} = N_1 + N_2 + N_3 = N_1 + N_2 + \min\{p_1 N_1;\ p_2 N_2\}. \quad (5.19)$$

The main parameters of the model p_1 and p_2 in this case can be determined using equation (5.19). So, with the predominant contribution of ionizing radiation to the total effect during the combined action $p_1 N_1 > p_2 N_2$ we obtain

$$p_2 = (N_{\Sigma} - N_1 - N_2)/N_2, \quad (5.20)$$

where N_{Σ}

In a situation where the leading factor in the process of cell inactivation during combined thermoradiation exposure is hyperthermia ($p_1 N_1 < p_2 N_2$), the value of p_1 is determined by the equation

$$p_1 = (N_{\Sigma} - N_1 - N_2)/N_1. \quad (5.21)$$

If in experiment the limit value of TER is observed, then its value can be used to estimate the value of p_1. In accordance with equation (5.18) p_1 will be determined by the expression

$$p_1 = TER_{max} - 1. \quad (5.22)$$

Much data on the synergistic interaction of various environmental factors have been obtained. The works (Petin et al., 2012, 2020; Petin, Morozov, 2015; Petin, Kim, 2016; Zhavoronkov, Petin, 2018) summarize a lot of such results. These works describe a large complex of previously unknown universal regularities of synergism, the manifestation of which during the simultaneous action of hyperthermia with various physical and chemical agents does not depend on the interacting factors used in combination, the analyzed effect and the biological objects used. Briefly, these patterns and the consequences arising from them can be summarized as follows.

1. The effectiveness of synergistic interaction is maximal after the simultaneous action of agents. An increase in the time interval between the agent application for the sequential action leads to a decrease in the effectiveness of synergism.
2. Synergism might not be observed at any randomly selected ratio of the agent parameters used in combination.
3. For any constant dose rate of ionizing radiation or intensity of another physical factor, synergism, when applied simultaneously with hyperthermia, may be observed only within a certain temperature range, within which there is a specific temperature providing the greatest synergistic interaction.
4. The range of temperatures synergistically interacting with physical and chemical agent shifts to lower temperatures with a decrease in the intensity of the agents applied, which indicates the fundamental importance of synergistic effects at low intensities of factors that actually occur in the biosphere.
5. The rate and the volume of cell recovery after combined exposure to ionizing radiation or UV light, synergistically interacting with hyperthermia or chemical agents,

significantly decreases, while the proportion of irreversible damage increases with increasing thermal load.

6. The recovery constant characterizing the probability of cell recovery from radiation damage per unit time does not depend on the temperature, at which the cells were irradiated, as well as on the concentration or duration of chemical compounds used in combination with ionizing radiation
7. The synergistic effect depends on the dose rate of ionizing radiation and the intensity of other physical factors or the concentration of chemical agents used in combination with hyperthermia. It was demonstrated that the lower the intensity of the physical factor or the concentration of the chemical agent, the lower temperature provides the greatest synergistic effect.
8. These universal patterns, obtained mainly on prokaryotes and yeast cells of various genotypes, have been partially published for mammalian cells (Ben-Hur, Elkind, 1974; Ben-Hur, 1976; Hahn, 1982; Dewey, Holahan, 1984; Urano et al., 1990; Hall, Giaccia, 2018). However, none of these works gave a complete understanding of synergistic effect patterns for mammalian cells. To assess the possibility of using the ideas of synergy in medical radiology, the examples given below are devoted to the comparison of the synergistic effects of yeast cells and cultured mammalian cells.

In the experiments described below, *Saccharomyces cerevisiae* yeast cells of diploid (strain XS800) and haploid (strain 6a) type were used in the stationary stage of growth, cells were cultured on a standard solid nutrient medium. Cell viability was determined by the counting of macrocolonies after 3–5 days of growing at 30°C. Acute exposure to ionizing radiation took place at room temperature (22°C) or at temperatures of 47–58°C, exposure duration 0–9 h. Pulsed electron accelerator LUE-25 with an electron energy of 25 MeV was used as source of ionizing radiation, the average dose rates were 5 and 25 Gy/min, as well as γ-rays of ^{60}Co at the installations: "Gammacell-220" (2 Gy/min), LUCH-2 (10 Gy/min) and "Researcher" (80 Gy/min). Each irradiation with electrons was accompanied by measurement of the absorbed dose by the ferrosulfate method. Dosimetry of cobalt sources was carried out with a calibrated ionization chamber (Siemens).

The effectiveness of the inactivating effect of the antitumor drug cisplatin (concentration 0.05 or 0.25 mg/ml, duration of exposure 0–3 h) on yeast cells was tested at temperatures of 47–55°C. Cisplatin (manufactured by LENS-PHARM LLC, Russia), another name for cis-DDP (cis-Diamminedichloroplatin) is a drug that is currently widely used in medical radiology as an antitumor agent (Yarmonenko et al., 1992; Kurpeshev, Mardynsky, 2010; Hall, Giaccia, 2018). Experiments for each investigated regime were repeated 3–5 times. The details of the methods of cultivation, determination of cell survival, and statistical processing were described earlier (Zhurukovskaya, 2001; Petin et al., 2012; Petin, Kim, 2016).

Extensive experimental data on cell survival after exposure to separate ionizing radiation and hyperthermia, as well as after their simultaneous application on cultured mammalian cells, have been published by other authors (Ben-Hur, Elkind, 1974; Ben-Hur, 1976; Joshi et al., 1978; Hahn, 1982; Sugahara, Saito, 1989; Urano et al., 1990; Hall, Giaccia, 2018). Using the averaged results presented in these works, we calculated the dependence of the synergistic enhancement ratio on the X-ray dose (250 kV) and the temperature, at which the irradiation took place. The work (Urano et al., 1990) presented a large set of experimental data reflecting the processes of cell death in the Chinese hamster CHO under the action of cisplatin (at concentrations of 3 and 6 μmol/L), hyperthermia, and the simultaneous use of these agents. Using these data, we performed a quantitative analysis of the effect of cisplatin and elevated temperature in terms of their synergistic interaction. It should be noted that the authors of the works cited above did not carry out such an assessment themselves. Figure 5.2 shows the dependence of the synergistic enhancement ratio (equation (5.5)) after the simultaneous action of ionizing radiation and hyperthermia on the dose rate of ionizing radiation and the temperature, at which irradiation of diploid yeast cells (Figure 5.2 A) and Chinese hamster cells (Figure 5.2 B) was occurred.

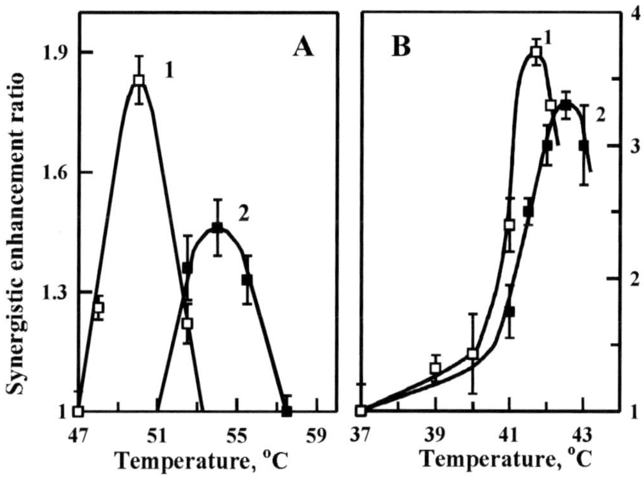

Figure 5.2. The dependence of the synergistic enhancement ratio for inactivation of diploid yeast cells of *Saccharomyces cerevisiae* (A) and cultured mammalian cells CHO (B) after the simultaneous action of hyperthermia and ionizing radiation (A – 25 MeV bremsstrahlung, B – 250 kV X-ray) on the temperature, at which the irradiation took place, and the dose rate of ionizing radiation: A – 10 and 250 Gy/min (curves 1 and 2, respectively), B – 0.033 and 0.12 Gy/min (curves 1 and 2, respectively).

The values of the synergistic enhancement ratio for yeast cells presented in this figure were calculated on the basis of our previously published survival curves (Zhurakovskaya, 2001; Petin et al., 2012; Kim et al., 2005 b, 2007 a, b; Petin, Kim, 2016). The survival of cultured mammalian cells was calculated according to the data published by other authors (Ben-Hur, Elkind, 1974; Ben-Hur, 1976; Dewey, Holahan, 1984; Sugahara, Saito, 1988, 1989; Streffer et al., 1990; Hall, Giaccia, 2018), who themselves did not assess the magnitude of the synergistic interaction.

It is obvious from Figure 5.2 that for each of the studied dose rates there is a certain temperature range that synergistically enhances the effect of ionizing radiation, within which there is a temperature maximizing the synergistic effect. It is also clear that the effectiveness of the synergistic interaction is determined both by the value of the temperature, at which the irradiation occurs, and the dose rate of ionizing radiation. It can be seen that with a decrease in the dose rate, in order to ensure the greatest synergistic interaction for both yeast and mammalian cells, it is necessary to decrease the temperature, at which the irradiation took place. This pattern indicates the fundamental possibility of synergistic interactions at low intensities of therapeutic factors. The greater value of the synergistic enhancement ratio for mammalian cells from the point of view of our concepts indicates the formation of a larger number of sub-damage responsible for synergism.

In the literature, there are various points of view on the dependence of the synergy on the intensity of the agents used. Some authors argue that with an increase in the operating temperature the effectiveness of synergistic interaction does not change (Kiefer et al., 1976) or even decreases (Kuzin, 1983; Streffer et al., 1990), others state that in some cases, with an increase in the operating temperature, the efficiency of synergistic interaction can increase (Petin, Komarov, 1997). The bell-shaped dependences of the synergistic enhancement ratio on temperature show that both of these situations can actually be realized in the case of combined actions. Indeed, as can be seen from Figure 5.2, for each fixed dose rate, this coefficient can both increase (left parts of curves 1 and 2) and decrease (right parts of curves 1 and 2) with increasing temperature, at which irradiation took place. Moreover, the intersection of curves 1 and 2 in each of the panels indicates the temperature, at which the effect may not depend on the dose rate.

Figure 5.3 presents the dependences of the thermal enhancement ratio (TER) on the duration of exposure to hyperthermia after its sequential application with ionizing radiation on diploid yeast cells of *Saccharomyces cerevisiae* (Figure 5.3 A) and Chinese hamster cells CHO (Figure 5.3 B). TER was quantitatively calculated on the basis of experimental survival curves and equation (5.4). Yeast cells were subjected to a sequential action of ionizing radiation at a dose rate of 2 or 80 Gy/min, and then a temperature of 50°C was applied (curve 1). Curve 2 reflects similar data for the reverse sequence of the agents used. It can be seen from this figure that the efficiency of thermal enhancement of acute irradiation does not depend on its dose rate, but the sequence of application of the agents is very important – irradiation of yeast cells before heating was more effective than irradiation after heating. So, the maximum value of TER is 3.0 in the first case and only 1.7 in the second. In other words, the data presented show that the degree of thermal radiosensitization was maximal when prolonged heat treatment (50°C) was applied after irradiation, and was less for the reverse sequence. The efficiency of thermal radio-sensitization at a high dose rate was the same (TER = 3) as at a dose rate reduced by a factor of 40.

The obtained data, from the point of view of the above presented mathematical model of synergy, indicate a longer preservation of sub-damage produced by ionizing radiation than sub-damage originated from hyperthermic exposure. Therefore, elimination and/or restoration of thermal sub-damage during prolonged heating did not seem to affect the efficiency of synergistic interaction when cells were irradiated with ionizing radiation, followed by prolonged thermal exposure. In this case, the resulting thermal sub-damage could immediately interact with the relatively longer-lived sub-damage from ionizing radiation.

For Chinese hamster cells (Figure 5.3 B), thermal exposure (45.5°C) was performed before irradiation; the results were published by other authors (Dethlefsen, Dewey, 1982; Dewey, Holahan, 1984; Mitchel et al., 1985; Sugahara, Saito, 1988, 1989). An increase in the TER value means a progressive increase in radiosensitivity with an increase in the duration of thermal exposure. It is also seen that in both cases (Figure 5.3 A, B) there is a certain value of the duration of heat exposure (2–3 h for yeast cells and about 20 min for mammalian cells), after which a further increase in the duration of exposure does not lead to increased cell radiosensitivity. It is these patterns that were predicted by the mathematical model of synergy described above.

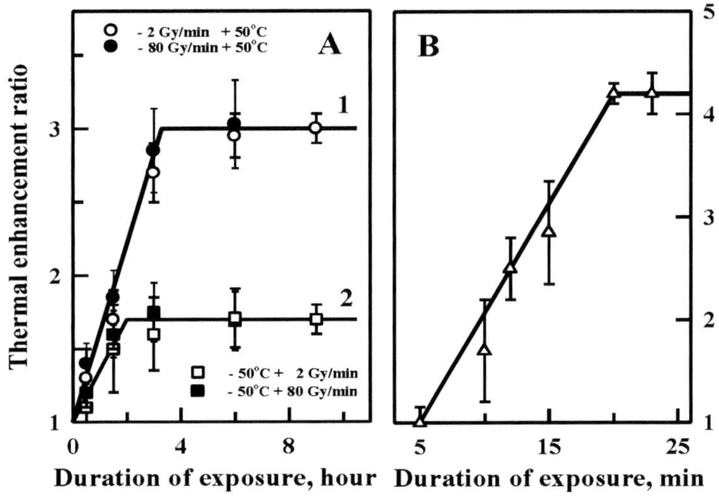

Figurer 5.3. The dependence of the thermal enhancement ratio on the duration of hyperthermia action on yeast cells *Saccharomyces cerevisiae* (A) and the mammalian CHO cells (B). A – various combinations of sequential exposure to hyperthermia (50°C) and ^{60}Co γ-rays, 2 and 80 Gy/min (details in the text); B – the sequential action of hyperthermia (45.5°C) and 250 kV X-ray radiation, 2 Gy/min.

It was of interest to obtain similar data for the synergistic interaction of chemical toxicants and hyperthermia. As an example, we have chosen the combination of cisplatin and hyperthermia, which is often used in medical radiology for tumor treatment. Above, we have already given arguments indicating that, in the first approximation, for chemical agents, as a dose the duration of action can be used while the analogue of the intensity will be the concentration of chemical agents. Then it can be expected that a decrease or increase in the concentration of cisplatin should be accompanied by a corresponding decrease or increase in the operating temperature in order to provide the greatest synergistic effect. We received additional data confirming this pattern. Figure 5.4 shows the dependence of the synergistic enhancement ratio on the operating temperature for yeast cells (Figure 5.4 A) and for mammalian cells (Figure 5.4 B).

To calculate this dependence in the first case, we used our own experimental data, in the second – the data published by other authors (Urano et al., 1990). It can be seen that in both cases the synergistic effect for fixed drug concentrations is observed only in a certain temperature range, within which the greatest synergistic enhancement ratio can be observed after the simultaneous action of the anti-tumor drug and hyperthermia. Our new results obtained are associated with the dependence of synergism on the concentration of the studied compounds – with a decrease in the concentration of the drug, the temperature, at which the greatest synergistic enhancement ratio is observed, also decreases. In this case, the entire range of operating temperatures, synergistically interacting with chemical agents, is shifted to the region of lower values. Note again that yeast cells are more heat-resistant, therefore, for them the range of temperatures synergistically enhancing the effect of cisplatin shifts to higher temperatures, since at lower temperatures no inactivation of yeast cells was observed.

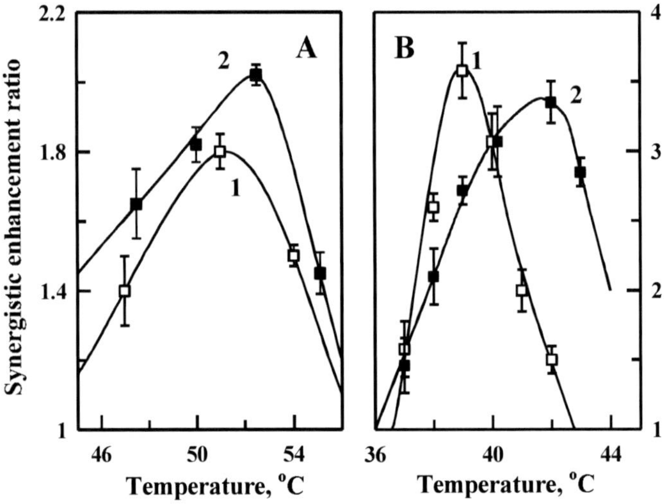

Figure 5.4. The dependences of the synergistic enhancement ratio on the temperature, at which the cisplatin acted on the inactivation of diploid yeast cells *Saccharomyces cerevisiae* (A) and the Chinese hamster CHO cells (B): A – 167 and 835 µmol/L, curves 1 and 2, respectively, B – 3 and 6 µmol/L, curves 1 and 2, respectively.

The results presented in this section and published earlier (Zhurakovskaya, 2001; Petin et al., 2012, 2020; Evstratova et al., 2016; Petin, Kim, 2016) demonstrate the existence of some general patterns of synergistic interaction for yeast and mammalian cells and do not depend on the nature of the physical and chemical factors used. It can be assumed that the knowledge and consideration of the main patterns of combined application of various agents described here and in our previous publications (Petin et al., 2012, 2020; Petin, Morozov, 2015; Petin, Kim, 2016; Zhavoronkov, Petin, 2018) will be useful for specialists using such effects in various practical directions.

5.2. Potential Significance of Synergy at Low Intensities of Physical and Chemical Environmental Factors

The purpose of this section is to present new experimental data demonstrating the versatility of the synergistic effects display for various biological objects, tests and interacting agents, as well as to provide convincing evidence of a possible synergistic interaction of agents of different origins at a natural intensity of environmental factors.

The combined effects of various physical and chemical substances are significant attribute of modern life (Streffer et al., 1990; Woodcock et al., 2011; Berthoud, 2013; Cain et al., 2014; Sucher, 2014; Petin, Kim, 2016). We have already noted that combined action can lead to synergistic interactions when the final biological effect exceeds the independent addition of the effects induced by each agent when used separately. In other words, the combined action of two agents can enhance the damaging effect of each other.

The nature of the synergistic interaction has been well studied for high doses, intensities and concentrations of chemical agents (Li et al., 1976; Leith et al., 1977; Petin, Zhurakovskaya, 2009, 2015; Petin, Morozov, 2015; Petin et al., 2012, 2020). These data are taken into account in the development of standards for hazardous substances in food, water and air in many countries. The safety standards in some countries for the simultaneous action of harmful agents simply summarize the adverse effects. It is clear that any risk assessment of a combined action must take into account the possibility of synergistic interaction.

Assessing the potential significance of the synergistic interaction of adverse environmental factors acting at the level of intensities found in the biosphere remains an intriguing and unsolved problem (Kim et al., 2001, 2004, 2006; Kolganova et al., 2010; Petin, Kim, 2016). Real experiments at low intensities of various agents found in occupational and environmental conditions are subjected to great uncertainty. A possible approach to solving this problem is to analyze the dependence of the synergistic interaction on the intensity of the agents used. The case where synergism decreases with decreasing intensity of the harmful agents is irrelevant. The same applies to a situation where a decrease in the intensity of one factor must be accompanied by an increase in the intensity of another in order to keep their synergistic interaction at the same level. The only possibility that is important for the natural intensity of environmental factors is as follows. The maximum or some fixed level of synergy can be achieved by using a lower intensity of the agents, if a lower intensity of another agent is used. In other words, the lower the intensity of one of the agents applied, the lower the intensity of another factor can provide the same synergistic enhancement ratio. In this case, it can be expected that even low intensities of hazardous agents can interact with each other in a synergistic manner and thereby increase their effect.

There are some contradictions in the literature devoted to this problem (Kuzin, 1983; Kiefer et al., 1976; Cebulska-Wasilewska et al., 1981; Little, 1990; Maehara et al., 2005). It has been shown that hyperthermia, applied simultaneously with ionizing radiation or UV light, significantly enhances the radiation response of the diploid yeast *Saccharomyces cerevisiae* (Petin, Zhurakovskaya, 1995; Kim et al., 2001; Petin et al., 2016). These results differ from those of other studies (Kiefer et al., 1976). In this work, only a small effect of temperature was noted for surviving yeast cells irradiated with X-rays. One of the possible reasons for this discrepancy, in addition to the strains used, is the use of different dose rates (480 and 10 Gy/min) and, consequently, the duration of the combined effects. The existence of optimal intensities to achieve maximum synergism has been shown for combined action of ethyl methanesulfonate and X-ray radiation to induce somatic mutations in Tradescantia (Cebulska-Wasileaska et al., 1981). Ethyl methanesulfonate is an organic compound with the molecular formula $C_3H_8SO_3$ and strong mutagen with teratogenic and possibly carcinogenic effects. In the work (Kuzin, 1983), it was suggested that the manifestation of synergism can be observed only at high intensities of active agents. The possibility of observing synergism at low radiation doses was discussed in (Little, 1990). Therefore, it is of interest to study the dependence of the synergistic interaction of various physical and chemical agents when changing in a wide range of intensities of physical agents or concentrations of chemical compounds takes place.

In our laboratory, a lot of experimental data were obtained for the simultaneous action of hyperthermia with ionizing radiation (Petin, Zhurakovskaya, 1995), UV light (Petin et al., 2012, 2020), ultrasound (Petin et al., 1999) and chemical agents (Petin et al., 2012; Omelchenko et al., 2014) on diploid yeast cells. In these studies it was shown, that with an increase in temperature at a simultaneous action of these agents, the synergistic effect first increases, reaches the greatest value, and then decreases. When using a lower intensity of physical or chemical agents, to ensure maximum synergistic interaction, a lower temperature should be applied.

To gain a deeper understanding of synergistic interactions display at low intensities of environmental factors, this section of the work was assigned the following tasks: (i) to prove the universality of the manifestation of synergistic effects for various biological objects, tests and interacting agents; (ii) provide data indicating the potential significance for synergi at low intensities of environmental factors.

The following experimental results were obtained on diploid yeast cells *Saccharomyces cerevisiae* (strain XS800). Homogeneous cell populations in the stationary stage of growth were used. The cells were incubated before irradiation for 3–5 days at 30°C on solid nutrient medium. Suspensions with a concentration of 10^6 cell/ml were exposed to various doses of ionizing radiation. Each of the independent experiments (ionizing radiation, hyperthermia, UV radiation, ultrasound, chemicals and their combinations) was performed in 3–5 replicates in each series.

To demonstrate the general patterns of synergistic effects, we used the data obtained in our own studies, as well as the results published by other authors for *Streptococcus sobrinus* (Maehara et al., 2005), *Bacillus subtilis* spores (Reynolds, Garst, 1970; Reynolds, Brannen, 1973), seeds of *Arabidopsis thaliana* (Dineva et al., 1993) and cultured mammalian cells (Johnson, Pavelec, 1973; Ben-Hur, Elkind, 1974; Ben-Hur, 1976; Urano et al., 1990). The authors of these works themselves did not estimate the values of the synergistic enhancement ratio. The source of ionizing radiation was ^{60}Co γ-rays ("Gammacell 220," Atomic Energy of Canada LTD, 10 Gy/min, and "Researcher," Radon, Russia, 25 Gy/min). In our experiments, we also used an electron beam from a 25 MeV pulsed linear accelerator. The pulse duration was 2.7 μs, the pulse repetition rate was 50 Hz, and the average dose rate was 5, 10, 25, and 250 Gy/min (Petin, Zhurakovskaya, 1995). A cell suspension (1.5 ml, 10^6 cells/ml) in a quartz vessel was irradiated with a bactericidal lamp, which emitted mainly UV radiation with a wavelength of 254 nm at 0.033, 0.25, and 1.5 W/m^2 at various temperatures (5–60.0°C). The UV irradiation technique is described above.

Continuous exposure to ultrasound was performed on a Fisher dismembrator with a frequency of 20 kHz at dose rates of 0.05 and 0.2 W/cm^2, which were measured by the calorimetric method. For sonication, the cell suspension (volume 0.1 ml, concentration 10^8 cells/ml) was placed in sterile water (volume 9.9 ml) at room temperature in one case or preheated to the required temperature in the other case. The suspension was placed in a metal vessel designed together with a source of ultrasound. To determine the effect caused by ultrasound alone, the absorbed ultrasonic heat was completely removed by cooling with water.

For the simultaneous action of hyperthermia and other agents, the time interval between the introduction of cells into pre-heated water and the beginning of the exposure was about 0.1–0.3 min, which was significantly less than the total duration of exposure. At the end of the irradiation, the samples were quickly cooled to room temperature. Therefore, the duration of the use of physical and chemical agents and heat exposure were practically the same.

The chemical agents used were cyclophosphamide (Baxter Oncology GmbH, Germany) and cisplatin. Both of these drugs are cytostatic alkylating agents. The concentrations of cyclophosphamide in our experiments with yeasts were 2 and 6 mg/ml. Original results obtained under the combined action of hyperthermia with chemical agents on mammalian cells were published by other authors (Johnson, Pavelec, 1973; Urano et al., 1990). We used these data to calculate the synergistic enhancement ratio.

To send any apparatus to distant space a very high degree of sterilization is required. The best effect was obtained with the simultaneous combined action of high temperatures (95 or

105°C) and ionizing radiation (γ-rays of ^{60}Co) with a dose rate varying over a wide range on bacterial spores of *Bacillus subtilis*. Survival curves were obtained for various irradiation conditions and published by other authors (Reynolds, Garst, 1970; Dugan, 1971; Reynolds, Brannen, 1973). We averaged these data and used them in this section of the work to assess the dependence of the synergistic enhancement ratio on the dose rate of ionizing radiation (Figure 5.5 A) at a constant exposure temperature (95°C). The authors of the cited works themselves did not carry out such calculations. In our own experimental studies with diploid yeast cells, the dependences of the synergistic enhancement ratio on the operating temperature and the intensity of ionizing radiation (Petin, Zhurakovskaya, 1995) and ultraviolet light (Petin, Zhurakovskaya, 2009) were obtained.

The results shown in Figure 5.5 A, B, C demonstrate an unexpected dependence of the synergistic interaction on the intensity of the agents used. Why unexpected? Indeed, it is well known that biological effects depend on the dose rate of ionizing radiation unidirectionally, and low dose rates in the overwhelming majority of cases are less effective due to recovery processes under protracted irradiation, which results in a reduced probability to produce primary radiation damage (Yarmonenko et al., 1992; Hall, Giaccia, 2018). The same is true for UV radiation.

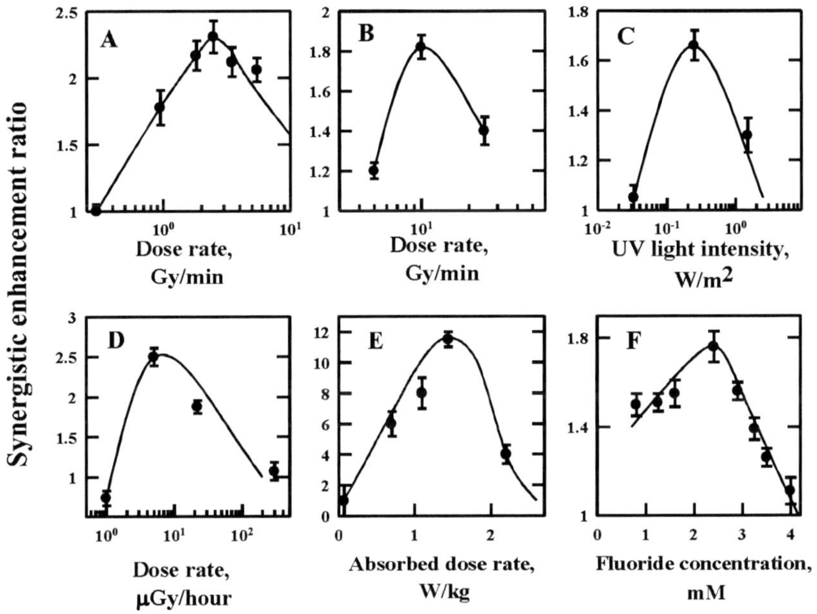

Figure 5.5. The dependence of the synergistic enhancement on the intensity of damaging factors: A – spores of *Bacillus subtilis* after simultaneous exposure to heat and ionizing radiation; B, C – diploid yeast *Saccharomyces cerevisiae*, strain XS800, after simultaneous exposure to temperature 50°C and ionizing radiation (B) or 52.5°C with UV light (C); D – seeds of *Arabidopsis thaliana* after successive exposure to ionizing radiation and lead nitrate; E – *Shinella* rabbits after simultaneous exposure to microwave radiation (7 GHz) and increased ambient temperature (30°C); F – bacteria *Streptococcus sobrinus* after the simultaneous action of fluoride and xylitol.

An impressive feature of the data presented is that the synergistic enhancement ratio first increases with increasing dose rate, then reaches its highest value, followed by a decrease. In other words, at a constant temperature at which irradiation occurs, synergism can be observed within a certain range of the intensities of the studied agents. It is clear that the optimal intensity of ionizing radiation or ultraviolet light within this range might be indicated, which provides the greatest synergy. To emphasize the general importance of this rule, we analyzed various interacting agents. Figure 5.5 D shows the relationship between the synergistic enhancement ratio and the dose rate of ionizing radiation for a sequential combination of chronic influence of ionizing radiation and lead nitrate on *Arabidopsis thaliana* seeds. This relationship, demonstrating the existence of an optimal dose rate to achieve the greatest synergy, was calculated using experimental results published by other authors (Dineva et al., 1993). The seeds were selected from populations of wildlife that had been growing for five years in places with varying levels of radioactive contamination in a 30-kilometer zone around the Chernobyl nuclear power plant, and then they were treated with lead nitrate (3.39 g/L). For this case, the coefficient of synergistic enhancement ratio was defined as the ratio of the increment of mutant or lethal embryos after the combined action to the sum of these increments after the separate action of each agent.

The synergistic effect has a pronounced maximum at a certain intensity of 7 GHz microwave radiation (Kolganova et al., 2010; Petin et al., 2011; Petin, Kim, 2016). Using the results of these works, we calculated the dependence of the synergistic enhancement ratio on the absorbed dose rate of microwave radiation (Figure 5.5 E).

An extensive quantitative study of the simultaneous combined action of fluoride and xylitol on streptococcal glycolysis showed (Maehara et al., 2005) that the formation of acid from glucose by the bacterial strain *Streptococus sobrinus* 6715 was inhibited in the presence of fluoride. The decrease in acid production was enhanced synergistically after the simultaneous action of fluoride and xylitol. We used these results to assess the dependence of the synergistic enhancement ratio on the fluorine concentration. These data are presented in Figure 5.5 E. More detailed data for the synergistic interaction of these two drugs are shown in Figure 5.6. Black circles in this figure denote experimental data (Maehara et al., 2005), light ones were obtained by us by extrapolation.

The set of data presented in Figures 5.5 and 5.6, demonstrates an unexpected dependence of synergistic interactions on the intensity of physical or chemical agents. This dependence is typical for various biological objects, interacting agents and the studied effect and indicates the existence of optimal dose rates of ionizing radiation, intensities of UV light, microwave radiation or concentrations of chemical compounds that provide the greatest synergistic interaction.

Our extensive data with diploid yeast cells subjected to simultaneous exposure to hyperthermia with 44 kHz ultrasound (Golovach et al., 1998; Petin et al., 1999) or cyclophosphamide (Omelchenko et al., 2014) were used in this section to represent the dependence of the synergistic enhancement ratio on the operating temperature. Figures 5.7 A, B summarize the results of the simultaneous action of ultrasound and high temperature on the inactivation of diploid cells of *Saccharomyces cerevisiae* (strain XS800). Figures 5.7 C, D exhibit similar results for the same cells that are simultaneously exposed to cyclophosphamide and high temperatures.

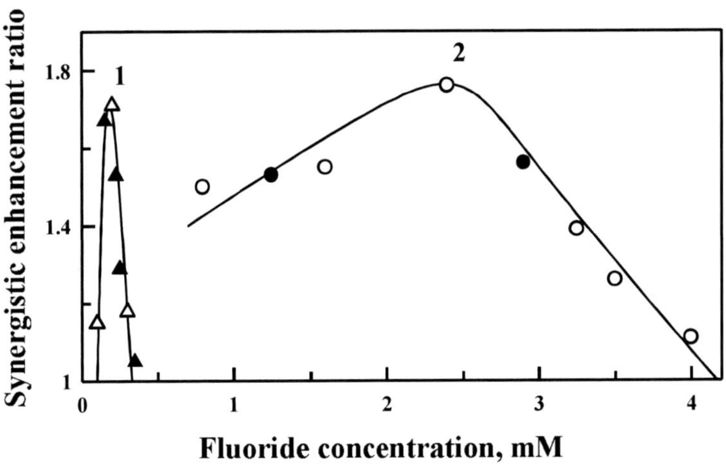

Figure 5.6. The dependence of the synergistic enhancement ratio on the fluoride concentration after the simultaneous action of fluoride and xylitol to inactivation of *Streptococcus sobrinus* 6715 at pH 5.5 and pH 7.0 (curves 1 and 2, respectively).

The relationship between the acting temperature and the intensity of ultrasound or the concentration of cyclophosphamide, providing the greatest synergistic interaction, is shown in Figures 5.7 B and 5.7 D, respectively. It can be concluded that the temperature at which cells are exposed to ultrasound and cyclophosphamide should be reduced in order to provide the greatest synergistic effect with a decrease in the intensity of the agents used. Extrapolating these data to lower intensities (dashed lines), it can be assumed that synergism can be observed even at low intensities of certain factors existing in the biosphere.

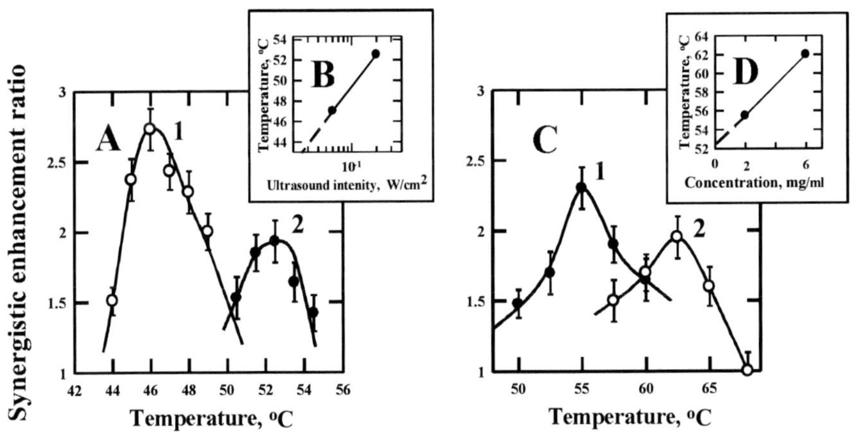

Figure 5.7. The dependence of the synergistic enhancement ratio on the acting temperature for the combined action of hyperthermia with ultrasound (A: curve 1 – 0.053 and curve 2 – 0.195 W/cm^2) and cyclophosphamide (C: curve 1 – 2 mg/ml, curve 2 – 6 mg/ml) on diploid yeast cells *Saccharomyces cerevisiae* (strain XS800); the relationship between the applied temperature and the intensity of ultrasound (B) or the concentration of cyclophosphamide (D), providing the greatest synergistic effect.

To confirm these data and thereby demonstrate the potential significance of synergistic interactions at low intensities of other agents and various biological objects, we presented the relationship of acting temperature with the dose rate of ionizing radiation (Figures 5.8 A, B, C, D), as well as with the intensity of microwaves and UV light (Figures 5.9 A, B) or with the concentration of chemicals (Figures 5.9 C, D), providing the greatest or other value of the synergistic enhancement ratio.

The data presented illustrate the relationship between the applied temperature and the intensity of the physical factors or the concentration of chemical compounds that provide the maximum or other level of synergistic effect. Sometimes we used not the maximum values of the synergistic effect, since in some cases the greatest synergism was not always achieved in experiments with mammalian cells. To obtain some of the results shown in Figure 5.8, the initial data were taken from a number of publications devoted to the simultaneous thermoradiation effect on bacterial spores (Reynolds, Garst, 1970; Reynolds, Brannen, 1973), bacteriophage (Trujillo, Dugan, 1972), mammalian cells (Johnson, Pavelec, 1973; Ben-Hur, Elkind, 1974; Ben-Hur, 1976; Urano et al., 1990) and heating rabbits (Petin et al., 2011).

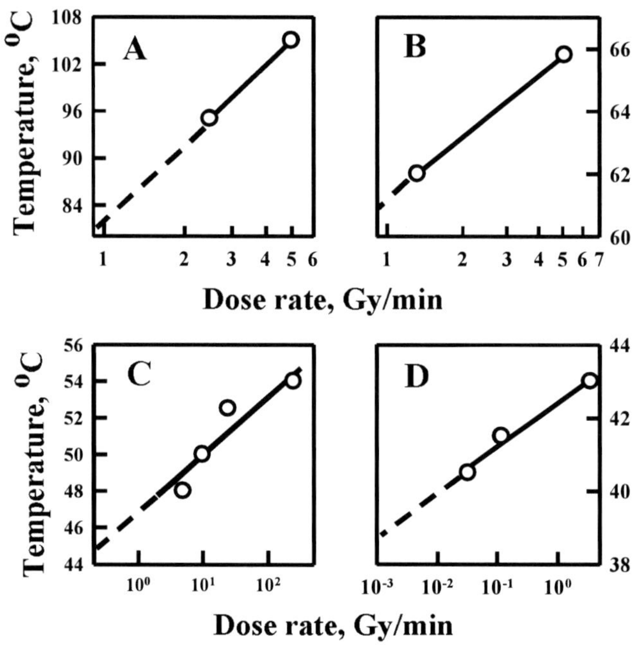

Figure 5.8. The dependence between the acting temperature and the intensity of physical factors that provide the greatest or other value of synergistic interaction. Simultaneous action of ionizing radiation and high temperature on inactivation of bacterial spores of *Bacillus subtilis* (A), bacteriophage T4 (B), diploid yeast cells *Saccharomyces cerevisiae* (strain XS800) (C), cultured mammalian cells CHO (D).

A linear relationship between the acting temperature and the intensity of the agents used has been demonstrated for various biological objects and tests (Figures 5.8 and 5.9) – at a lower intensity of a physical factor or concentration of chemicals, the acting temperature should be reduced to maintain the highest or some other selected synergy values. Extrapolation of these data to the region of lower intensities (dashed lines) indicates the potential significance of synergistic interactions at low intensities of unfavorable factors found in the biosphere.

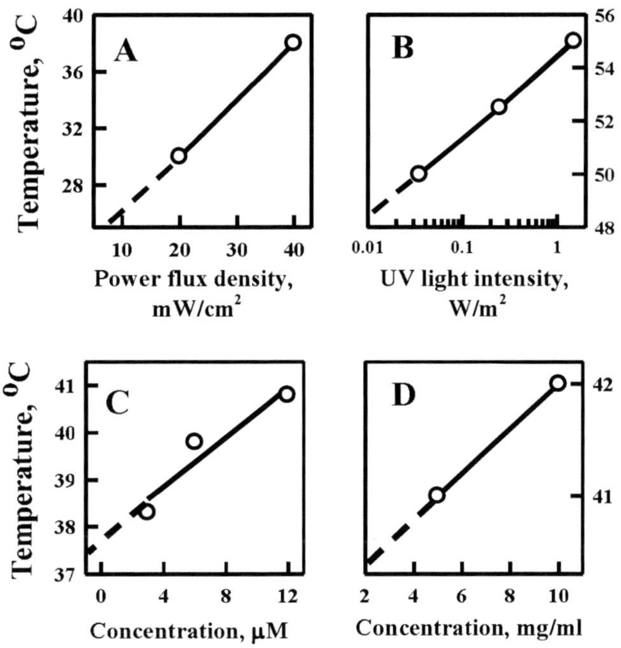

Figure 5.9. The dependence between the acting temperature and the intensity of physical factors or the concentration of chemicals that provide the greatest or other value of synergistic interaction: A – rabbit heating with microwaves (7 GHz) at an ambient temperature of 30°C; B – simultaneous action of high temperature and 254 nm UV light on inactivation of yeast cells; C – simultaneous action of high temperature and cis-diamminylchloroplatinum on inactivation of mammalian cells CHO; D – simultaneous action of high temperature and tris (1-aziridinyl)-phosphine sulfide on the inactivation of mammalian cells CHO.

These data allow us to conclude that with a decrease in the intensity of one of the agents used in combination, the intensity of another factor should also be reduced to ensure the greatest synergy. This follows a cardinal conclusion about the possibility of synergistic interaction of harmful environmental factors at their low intensities, which can occur in the daily life of a person and in the biosphere.

Although many of the described experimental data were obtained at temperatures outside the physiological range, it is possible that there are optimal intensities of harmful agents that can synergistically interact with each other or with the physiological temperature of homeothermic animals or humans. Consequently, the assessment of health risks or environmental quality under the action of numerous natural and man-made agents at the level of intensity or concentrations found in environmental and occupational conditions should take into account their synergistic interaction.

The main general patterns of synergistic interaction that follow from the presented results can be summarized as follows. The acting temperature, the intensity of the physical factors or the concentration of chemical agents are critical for the synergistic effect display regardless of biological tests, objects and the nature of the physical or chemical agents used in combination. An unusual bell-shaped dependence of the synergistic effect on the intensity of the agents used was discovered – for all analyzed cases, the synergistic enhancement ratio first increases, reaches the highest value, and then decreases with an increase in the intensity of physical agents or the concentration of chemical compounds (Figures 5.5, 5.6 and 5.7). These data indicate the existence of the intensity of physical factors or the concentration of chemical agents that maximize the synergistic effect. Any deviation of these values from the optimum leads to a decrease in synergy. This effect was obtained for various biological objects, tests and interacting agents. It can be concluded that this feature of the manifestation of the synergistic effect is universal and can be observed for other biological objects, tests and active agents.

The next most important results are represented by the linear relationships between the operating temperature and the intensities of the applied agents, providing the greatest or any arbitrary amount of synergy. It has been shown that with a decrease in the intensity of one agent, the intensity of another factor should also be decreased in order to ensure the greatest or other level of synergy (Figure 5.8 and 5.9). This effect has also been obtained for various biological objects, tests, and physical or chemical agents used in combination. Extrapolation of the obtained relations to the region of small values (dashed lines in Figure 5.8 and 5.9) also allows us to formulate an important conclusion. With long exposure times, which are important for radiation protection and ecology problems, harmful environmental factors at their low intensity can synergistically interact with each other, with the ambient temperature or even with the physiological temperatures of homeothermic animals and humans. This conclusion is important for assessing health risks under the combined effects of low intensities of influencing factors found in the environment, at workplaces and in the daily life of a person. These patterns of synergy might be useful for optimizing the combined use of various agents.

5.3. Biophysical Interpretation of Synergy Dependence on the Intensity of the Used Agents

In medical radiology for the treatment of tumors, along with the use of ionizing radiation, the use of hyperthermia has become widespread, inhibiting the process of recovery from radiation damage and synergistically interacting with many agents. In the first approximation, the duration of hyperthermia action may be considered as thermal "dose", while the analogue of the intensity will be the value of the current temperature. Considering this supposition, let us analyze the dependence of the synergism of the simultaneous action of hyperthermia with ionizing radiation and other agents on the applied temperature, i.e., on the intensity of thermal exposure. Figure 5.10 A showed the dependence of the synergistic enhancement ratio on the acting temperature after the simultaneous application of hyperthermia with ionizing radiation for bacterial spores of *Bacillus subtilis*.

It has been demonstrated above (Figure 5.5) that there is an optimal dose rate at which the greatest synergistic effect is recorded. For the same spores Figure 5.10 A shows the dependence of the synergistic enhancement ratio on the operating temperature. The bell-shaped dependence of this coefficient on the current temperature indicates the presence of an optimal temperature at which the greatest synergy is observed at a fixed dose rate (^{60}Co, 1.35 Gy/min). It is of important to analyze whether these patterns are preserved for eukaryotic cells. Figure 5.10 B shows the change in the synergistic enhancement ratio after the simultaneous action of hyperthermia with ionizing radiation (^{60}Co, 10 Gy/min) on the survival of diploid yeast cells *Endomyces magnusii* (strain K7, curve 1) and *Saccharomyces ellipsoideus* (strain Megri 139-B, curve 2) on the temperature at which the irradiation took place. It can be seen that for each of these yeast strains there is a certain temperature range that synergistically increases the effect of ionizing radiation. Within this range, there is a temperature, at which the synergistic effect is maximal. It is also seen that the temperature range, synergistically increasing the effect of ionizing radiation, is shifted to the region of higher temperatures for more thermally resistant *Saccharomyces ellipsoideus* strain.

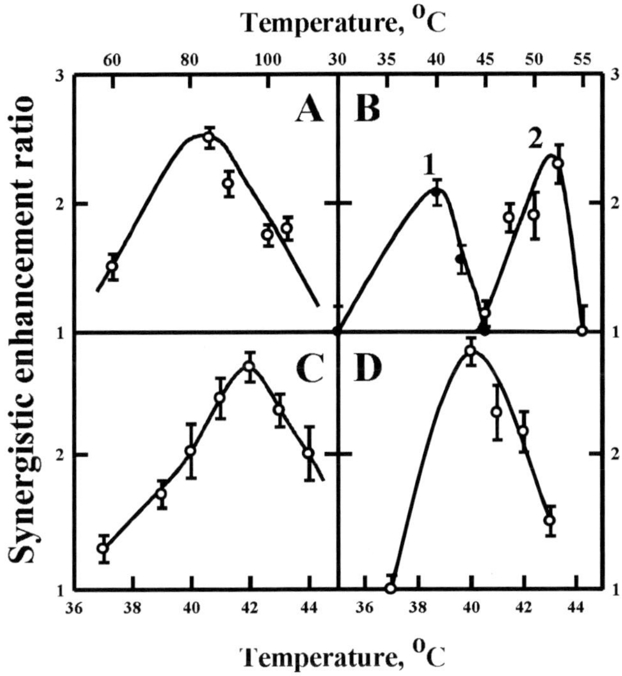

Figure 5.10. The dependence of the synergistic enhancement ratio on the acting temperature after the simultaneous application of hyperthermia with ionizing radiation (A, B) or chemical agents (C, D) on cell survival: A – bacterial spores of *Bacillus subtilis*, irradiated with ^{60}Co γ-rays (1.35 Gy/min), B – diploid yeast cells *Endomyces magnusii*, strain K7 (curve 1) and *Saccharomyces ellipsoideus*, strain Megri 139-B (curve 2), irradiated with ^{60}Co γ-rays (10 Gy/min); C – simultaneous action of hyperthermia with tri (1-aziridinyl) phosphine sulfide (10 μg/ml) on Chinese hamster cells (line V), D – simultaneous action of hyperthermia with cisplatin (1 μg/ml) on fibroblasts of Chinese hamster cells B14F28.

Using the data published in (Johnson, Pavelec, 1973) for Chinese hamster cells (line V) exposed to the simultaneous action of the cytostatic tris(1-aziridinyl)phosphine sulfide (TAFS) and hyperthermia, we calculated the dependence of the synergistic enhancement ratio on the operating temperature (Figure 5.10 B). Similar dependences were calculated for the simultaneous action of cisplatin and elevated temperature on fibroblasts of Chinese hamster cells B14F28 (Figure 5.10 D) based on the data published by other authors (Eichholz, Hietel, 1990). In these cases, there is also an optimal temperature at which the greatest synergistic interaction is observed. Note that the authors of these works (Johnson, Pavelec, 1973; Eichholz, Hietel, 1990) did not perform such calculations.

The set of data shown in Figure 5.10 and in the previous paragraph, shows that in all cases there is a bell-shaped dependence of the synergistic enhancement ratio on the applied temperature for fixed intensities of physical agents or drug concentrations. Bell-shaped dependences of the synergistic enhancement ratio on the intensity of the physical agents or the concentrations of chemical compounds acting at a fixed temperature or the intensity of another agent have been demonstrated for various biological objects and tests.

The presented examples of the synergism dependence of on the intensity of agents used in combination are original. What is their originality? It is well known that most radiobiological effects are unidirectionally dependent on the intensity of the damaging factors. As a rule, the effectiveness of physical agents decreases with a decrease in their intensity, for example, the dose rate of ionizing radiation. A poorly studied and rarely found the inverse dose rate effect is known. But biological reactions have never been observed for which the existence of an optimal intensity would be noted, when the biological effectiveness first increased, reached its greatest value, and then decreased with an increase in the intensity of the applied factor. It is these reactions that are obtained for synergistic effects in the above data observed for various biological objects and acting agents that indicate the universality of the revealed pattern.

It is of interest to analyze the possible mechanisms of this unusual phenomenon from the point of view of mathematical model of synergy described above. According to this model, the synergism is due to the formation of additional effective damage because of the interaction of sub-damages produced by each agent. The additional number of effective damages is determined by the minimum value of the sub-damage generated by each agent (equation (5.6)). For example, consider the simultaneous action of a fixed dose rate of ionizing radiation and hyperthermia. At low values of operating temperatures and long-term exposures providing a given value of the total effect (for example, set for the concreteness, 10% survival), the number of heat damage is much less than the number of radiation sub-damage and thermal damage are responsible for small value of the synergistic effect. The data shown in Figure 5.11 demonstrate the dependence of the yield of radiation (curve 1) and thermal (curve 2) sub-damage on the operating temperature.

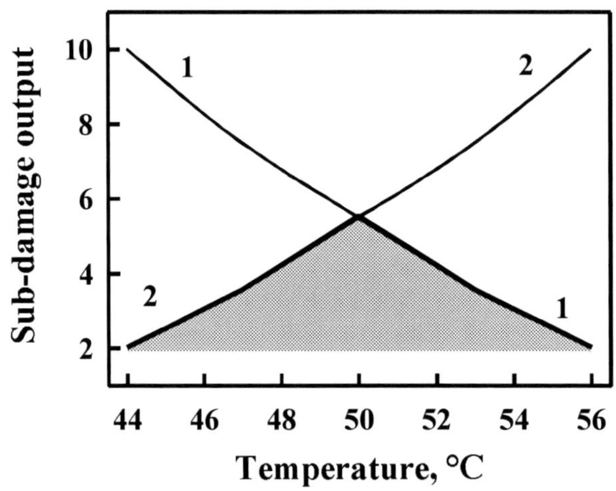

Figure 5.11. The dependence of the yield of radiation (curve 1) and thermal (curve 2) sub-damage on the operating temperature after the simultaneous action of ionizing radiation and hyperthermia; the shaded area qualitatively reflects the dependence of the synergistic enhancement ratio on the acting temperature.

Small values of the synergistic enhancement ratio or complete absence of synergism (k = 1) in the low-temperature region means a small number or complete absence of thermal damages and the corresponding sub-damage that determine the effectiveness of synergism

(Figure 5.11). An increase in the operating temperature leads to an increase in thermal damage and sub-damage and, accordingly, the effectiveness of synergistic interaction. This increase continues until the maximum value of the synergistic enhancement ratio is reached at equal sub-damage from each agent (equation (5.11)). This condition is violated with a further increase in the temperature above the optimal value. This leads to a decrease in the synergistic effect, but already due to a decrease in the number of sub-damages produced by hyperthermia due to the short duration of exposure. This decrease is conditioned by to a greater contribution to the lethal effect of elevated temperature and a corresponding reduction in the duration of the simultaneous action of hyperthermia with ionizing radiation. Then the area shaded in Figure 5.11 reflects the dependence of the synergistic enhancement ratio on the operating temperature, determined by the smallest number of sub-damages produced.

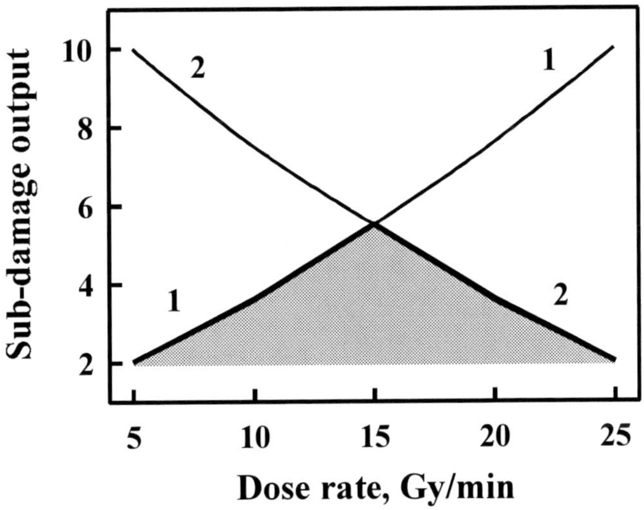

Figure 5.12. The dependence of the yield of radiation (curve 1) and thermal (curve32) sub-damage on the dose rate of ionizing radiation after the simultaneous application of ionizing radiation and hyperthermia; the shaded area qualitatively reflects the dependence of the synergistic enhancement ratio on the dose rate of ionizing radiation.

From the same standpoint, it is possible to interpret the bell-shaped dependence of the synergistic effect on the dose rate of ionizing radiation or the intensity of other physical factors, as well as the concentration of chemical agents used simultaneously with a fixed elevated temperature. In these cases, the optimal values of the intensities of physical factors are also observed, at which the condition for achieving the greatest synergy is realized (equation (5.11)), although the interpretation of the observed effects will slightly differ from the above. For example, Figure 5.12 demonstrate the production of sub-damage after the simultaneous action of a fixed temperature and changing dose rate of ionizing radiation.

To achieve a set value of the total effect (in our example 10% survival) at low dose rates and long-term exposure, few radiations and many thermal sub-damages should be produced. A further increase in the dose rate leads to a reduction in the duration of action and the corresponding decrease in thermal and increase in radiation sub-damage. This change continues until the optimal dose rate is reached, when the equality of sub-damages formed by each agent is registered, and the greatest value of the synergistic enhancement ratio is reached. A further increase in the dose rate leads to a decrease in the efficiency of synergistic interaction due to the continuing decrease in the number of thermal sub-damage due to a decrease in the total duration of exposure at a high dose rate of ionizing radiation.

So, our studies have demonstrated and qualitatively explained the unexpected dependence of synergism on the intensity of various agents used in combined actions. In the literature, there are other examples of the presence of an optimal intensity that provides the greatest synergy. For instance, in the work (Dobrovolsky, 1991), the "maximum principle" was noted in the case of chronic combined effects of radionuclides and hyperthermia on laboratory animals. The optimal dose rate of ionizing radiation was also revealed under the combined action of lead and chronic irradiation of *Arabidopsis thaliana* seeds growing for five years after the radiation accident at the Chernobyl nuclear power plant in places with different levels of radiation pollution within a 30-km zone (Dineva et al., 1993). However, the authors of these works did not try to explain these data.

5.4. General Biological and Clinical Significance of Synergy Ideas

Evaluation of the potential significance of the synergistic interaction of ionizing radiation with hyperthermia and chemical inhibitors in medical radiology and oncology remains an urgent problem. It is known that the maximum effect of the combined action of two agents is achieved after their simultaneous action (Dethlefsen, Dewey, 1982; Moding et al., 2013; Sucher, 2014; Evstratova et al., 2018). As was mentioned above the synergistic effect of two damaging agents is observed if it exceeds the independent sum of the effects of each agent. We emphasized the importance of the key word in this definition – an independent sum, when the probability of the total biological effect is determined by the product of the probabilities of the effects induced by each agent separately (Gmurman, 2014; Vasina et al., 2018). This means that under the independent action of two agents, the total number of damages is determined by the sum of damage induced by each agent (equations (5.2)). For a clearer understanding of the synergistic interaction in Figure 5.1 was presented an example of quantifying the synergistic enhancement ratio for exponential survival curves of yeast cells. Figure 5.13 shows a similar example for sigmoid survival curves of cultured mammalian HeLa cells.

These curves were constructed based on averaged data obtained by different authors (Miyakoshi et al., 1984; Sugahara, 1984; Saha, 2006; Hall, Giaccia, 2018). The theoretical curve expected for independent addition of the effects of ionizing radiation and hyperthermia is calculated using equation (5.1). To assess the effectiveness of interaction, we used the synergistic enhancement ratio k (Petin, Berdnikova, 1979; Belkina, Petin, 2014), determined by equation (5.5).

Since an increase in the time interval between the application of agents leads to a decrease or even the absence of a synergistic effect, this section of the work is devoted to the analysis of the possibilities of the main ideas of synergy in oncology after the simultaneous combined action of various physical and chemical agents. It is known that elevated temperatures synergistically enhance the lethal effect of ionizing and UV radiation, as well as many chemical compounds used in the treatment of cancer (Petin et al., 2012, 2020; Petin, Kim, 2016). On the basis of the revealed new patterns, these works describe the original concept of synergistic interaction and the mathematical model for optimizing and predicting the effects of synergy.

The main provisions of this model have been tested on numerous experimental results describing synergistic effects after the action of various agents on biological objects of increasing complexity – DNA, viruses, spores, bacterial and yeast cells, cultured mammalian cells of normal and tumor origin, plants, animals, and humans. In partular, synergistic effects are recorded for plants (Cebulska-Wasilewska et al., 1981; Dineva et al., 1993; Evseeva, Geraskin, 2001), whole organisms like animals (Petin et al., 2011; Grigoriev, Grigoriev, 2016; Zhavoronkov, Petin, 2018) and humans (Kim et al., 2007 a, b; Lantz et al., 2013; Melloni, 2014; Torres-Durán, 2014). General patterns of the observed synergy effevts for various objects were also noted in publications (Trujillo, Dugan, 1972; Chadwick, Leenhouts, 1981; Petin, Zhurakovskaya, 1995, 2014; Pronkevich et al., 2017; Evstratova et al., 2018; Helena et al., 2019; Evstratova, Petin, 2021). Below we summarize the main points of these observations.

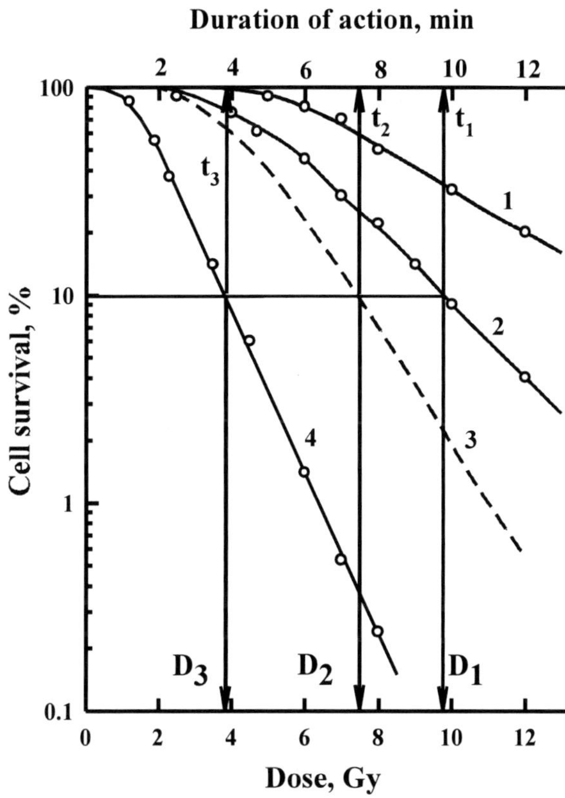

Figure 5.13. An example of a quantitative assessment of the synergistic enhancement ratio for cultured mammalian cells HeLa: curve 1 shows the action of hyperthermia alone (43°C), curve 2 is the survival curve under the action of X-ray radiation only (250 kV, 1 Gy/min), curve 3 is the theoretical curve, expected for independent summation of the effects of hyperthermia and ionizing radiation, curve 4 exhibits the survival curve of the same cell line with simultaneous thermoradiation exposure.

In accordance with the conceptual foundations of synergy (Petin, Komarov, 1997; Petin, Zhurakovskaya, 2009, 2014, 2015; Petin, Kim, 2016), at a certain ratio of influencing agents, the maximum efficiency of synergistic interaction should be recorded. This pattern was observed and reproduced well for pro- and eukaryotic cells for the combined action of hyperthermia with ionizing radiation (Petin, Zhurakovskaya, 1995; Petin, Kim, 2004), ultraviolet light (Petin, Anokhin, 2014), ultrasound (Petin et al., 1999), microwaves (Petin et al., 2011; Zhavoronkov, Petin, 2018) and various chemical agents (Evstratova et al., 2016; Filimonova et al., 2019). On this basis, it was concluded that with an increase in the operating temperature, the synergistic effect first increased, reached its greatest value, and then decreased. A similar conclusion was established for animal (Dobrovolsky, 1991), as well as in a study of heating rabbits under the simultaneous action of microwaves (7 GHz) and elevated ambient temperatures (Petin et al., 2011; Zhavoronkov, Petin, 2018). In these works the synergistic enhancement ratio reached a very high value (k = 12), showing that to achieve the same effect, the duration of the simultaneous action of elevated temperature and microwave radiation decreased 12 times in comparison with this parameter expected for an independent addition of the effects from each agent. High values of the synergistic enhancement ratio (k = 5–7) were also demonstrated for humans as a result of the simultaneous action of smoking and radioactive decay of radon and its daughter products (Lantz et al., 2013; Torres-Durán et al., 2014). It is important that these data show that synergistic effects are significantly more pronounced for animals and humans compared to the synergism observed for bacteria, yeast cells and cultured mammalian cells *in vitro*.

It is known that radon and its short-lived α-decay products are considered important sources of natural radioactivity. The synergistic interaction between tobacco exposure to smoking man and radon is a real problem for humans (Kim et al., 2007 a, b; Meenakshi, Mohankumar, 2013; Melloni, 2014). Therefore, of interest are our theoretical models (Petin, Komarov, 1997; Petin et al., 2012; Zhurakovskaya, Petin, 2015; Petin, Kim, 2016), which provide a mathematical description and prediction of lethal and carcinogenic effects after the combined action of hyperthermia with ionizing radiation, UV light, ultrasound and microwave radiation. The mathematical model used in these works is briefly described above. The model predicted the greatest efficiency of the synergistic interaction. The postulates of the model are based on the assumption that the mechanism of synergistic interaction should be expected due to the formation of additional effective damage arising from the interaction of sub-damages caused by each agent under consideration. The results of the model were verified by quantitatively comparing the predictions of the model with experimental data (Kim et al., 2007 a, b; Petin et al., 2000, 2011; Petin, Kim, 2004, 2016; Petin et al., 2012, 2020; Petin, Morozov, 2015; Evstratova et al., 2018; Zhavoronkov, Petin, 2018). In these works, some general patterns of synergistic interactions were revealed that do not depend on the agents used, biological objects and tests.

The most significant of them can be formulated as follows. The synergism is observed as a result of the combined effects of the two factors and can be observed only at a certain ratio of the effects induced by each agent. The most interesting pattern of synergistic effects display lies in the experimental observation that there are criteria that ensure the achievement of the greatest synergistic effect (Petin, Komarov, 1997; Petin, Zhurakovskaya, 2009, 2014; Zhurakovskaya, Petin, 2015). Interestingly, the effectiveness of the simultaneous combined action depends on the intensity of the agents used (Petin, Zhurakovskaya, 1995, 2015; Evstratova et al., 2018). For example, to ensure the optimal ratio of simultaneous thermoradiation exposure with decreasing the dose rate (or the intensity of other agents), it is necessary to reduce the operating temperature and *vice versa*. This pattern is presented in this work (Figures 5.7, 5.8, and 5.9). Throwing a retrospective look at the identified patterns of synergistic interactions, it is necessary to pay attention to their independence from the biological objects and tests used, as well as from the influencing factors, thereby emphasizing their general biological significance and the prospects for the development of new methods of combined action of agents in therapy (Martin, 2001; Steel, 2002; Chou, 2006; Saha, 2006; Woodcock et al., 2011; Pankratov et al., 2014; Kaprin et al., 2017). This data can be served as the basis for assessing the prospects for the practical implementation of the ideas of synergy in medical radiology and oncology (Evstratova, Petin, 2020). Before proceeding to a discussion of this problem, it is of interest to consider the mechanisms the synergistic effect appearance.

Some researchers believed that inhibition of recovery processes after combined actions can be considered as a reason for the mutual reinforcement of two active agents (Zaider, Rossi, 1980; Kurpeshev, Konoplyannikov, 1987; Kuzin, 1983; Kogl, 1986; Joiner, van der Kogel, 2009). In accordance with this approach, when high temperatures are combined with ionizing radiation or UV light, hyperthermia disrupts or suppresses the processes of cell recovery from radiation damage; the cells become more sensitive to radiation exposure, as a result of which the synergism of thermoradiation irradiation is observed. We have already noted that the recovery process is described by two parameters: the irreversible component of radiation damage and the recovery constant, which characterizes the probability of cell to recover per unit time. A quantitative assessment of these parameters after various combined impacts made it possible to conclude that the recovery constant does not depend on the conditions of combined impacts (Komarova, Petin, 2007; Evstratova, Petin, 2013, 2015; Evstratova et al., 2016). In these works, it was shown that a decrease in both the rate and the volume of recovery under conditions of combined actions is due to the formation of irreversible damage, and not by a violation of the recovery process itself. This means that the mechanism of synergistic interaction is not related with the recovery process damage. Then the inhibition of recovery observed by many authors after of combined actions is not the cause of synergy, but rather an expected consequence of the formation of irreversible damage from which the cell is unable to recover. This data can also be added to the already cited universal patterns of synergy display.

The existence of universal patterns of synergy indicated the need to develop a new paradigm and theoretical model of synergy, which should take into account the identified data. As was mentioned above, biophysical concept of synergistic interaction was proposed– synergism is due to the formation of additional effective damage arising from the interaction of sub-damage that are formed by each of the agents, but these sub-damages are not effective when they are applied separately (Petin, Komarov, 1989, 1997; Petin et al., 2012, 2020; Petin, Kim, 2016). This model was described in more detail above. For a better understanding of the general biological and clinical significance of synergy ideas, some comparative experimental data for yeast cells and cultured mammalian cells will be presented below.

Extensive data have been published on the inactivation of Chinese hamster cells (CHO) after the separate application of hyperthermia, ionizing radiation and cisplatin, as well as after the simultaneous action of hyperthermia with ionizing radiation or cisplatin (Miyakoshi et al., 1984; Kumar et al., 1985; Hall, Giaccia, 2018). Cisplatin is currently widely used in oncology as an antineoplastic agent (Helena et al., 2019; Kaprin, Mardynsky, 2019). Data on the synergistic interaction of hyperthermia with ionizing radiation or cisplatin were also obtained for diploid yeast cells (Evstratova, Petin, 2013, 2015; Evstratova et al., 2016). Using the averaged results presented in these works, we calculated the dependence of the synergistic enhancement ratio after the simultaneous action of hyperthermia with ionizing radiation or with cisplatin for Chinese hamster cells and diploid yeast cells. The results are shown in Table 5.1.

It can be seen that for mammalian cells, after the simultaneous action of ionizing radiation and elevated temperatures (37–41°C) from 1.0 to 3.7 and then decreased to 1.4 with an increase in the operating temperature to 42°C. The authors of the works (Petin, Kim, 2016; Hall, Giaccia, 2018) note that at higher temperatures, cell death is mainly due to the operating temperature, and not to ionizing radiation. Therefore, the synergistic interaction of these agents was practically absent at high temperatures. The greatest effect of synergistic interaction was revealed after the simultaneous action of cisplatin and hyperthermia on Chinese hamster cells (Table 5.1).

Table 5.1. The dependence of the synergistic enhancement ratio after simultaneous use of hyperthermia (T, °C) with ionizing radiation ($\boldsymbol{k_1}$) **or** cissplatin ($\boldsymbol{k_2}$). For mammalian cells, X-ray radiation (250 kV; 0.033 Gy/min) was used, cisplatin concentration was 3 μM. For yeast cells, ^{60}Co γ-rays (10 Gy/min) were used, cisplatin concentration was 0.25 mg/ml

CHO mammalian cells						
T, °C	**37**	**38**	**39**	**40**	**41**	**42**
k_1	1.0±0.1	1.1±0.2	1.3±0.2	1.5±0.3	3.7±0.5	3.5±0.4
k_2	1.5±0.2	2.5±0.2	3.6±0.3	3.1±0.3	2.0±0.2	1.4±0.1
Diploid yeast cells *Saccharomyces cerevisiae*, strain XS800						
T, °C	**42.5**	**45**	**47.5**	**50**	**52.5**	**55**
k_1	1.0±0.1	1.6±0.1	1.8±0.2	2.0±0.2	1.8±0.3	1.0±0.1
k_2	1.0±0.1	1.4±0.1	1.6±0.2	1.8±0.2	2.0±0.2	1.4±0.1

It can be seen that for mammalian cells, after the simultaneous action of ionizing radiation and elevated temperatures (37–41°C), the synergistic enhancement ratio
from 1.0 to 3.7 and then decreased to 1.4 with an increase in the operating temperature to 42°C. The authors of the works (Petin, Kim, 2016; Hall, Giaccia, 2018) note that at higher temperatures, cell death is mainly due to the operating temperature, and not to ionizing radiation. Therefore, the synergistic interaction of these agents was practically absent at high temperatures. The greatest effect of synergistic interaction was revealed after the simultaneous action of cisplatin and hyperthermia on Chinese hamster cells (Table 5.1).

To demonstrate the commonality of the revealed regularities, we used diploid yeast cells *Saccharomyces cerevisiae* (strain XS800), which in the stationary stage of growth were subjected to simultaneous action of hyperthermia with ionizing radiation (^{60}Co γ-rays, 10 Gy/min) or cisplatin of various concentrations. Table 5.1 shows the values of the synergistic enhancement ratio after the simultaneous action of hyperthermia with ionizing radiation (k_1) or cisplatin (k_2) on diploid yeast cells of *Saccharomyces cerevisiae.* The data given in Table 5.1 demonstrate the existence of an optimal temperature that provides the most effective synergistic interaction: $k_1 = 2$ at a temperature of 50°C, $k_2 = 1.8$ at a temperature of 52°C. The existence of an optimal temperature under combined actions of hyperthermia and ionizing radiations has also been demonstrated above for other cellular systems and influencing agents. This pattern emphasizes the universality of this pattern.

Another universal pattern for synergy manifestation is its dependence on the intensity of the agents used. For yeast cells, it has been shown that at a fixed operating temperature, there is an optimal dose rate of ionizing radiation or intensity of UV light that provides the maximum synergistic interaction (Petin, Zhurakovskaya, 2015). It has been proven theoretically and experimentally that with a decrease in the intensity of one of the agents used, the intensity of another factor should also be reduced to ensure the greatest synergistic interaction for simultaneous action of interacting agent. Using the averaged data for mammalian cells published in many works (Miyakoshi et al., 1984; Kumar et al., 1985; Hall, Giaccia, 2018), the dependence of the synergistic enhancement ratio on the temperature at which the irradiation took place were obtained (Petin et al., 2012; Evstratova et al., 2016; Petin, Kim, 2016). It turned out that these curves shift towards lower temperatures with a decrease in the dose rate of ionizing radiation. For example, for Chinese hamster cells, a twofold increase in the effect ($k = 2.0$) of the simultaneous action of hyperthermia and ionizing radiation was achieved for dose rates of ionizing radiation 3.6, 0.12, and 0.033 Gy/min at temperatures of 43, 41, and 40.5°C, respectively. This means that in order to achieve the same synergistic effect with decreasing dose rate, it is necessary to decrease the temperature at which irradiation occurs to provide the same synergistic effect. A similar pattern is confirmed in experiments with diploid yeast cells – with a decrease in the dose rate of ionizing radiation from 250 to 25, 10, and 5 Gy/min, in order to ensure the maximum synergistic effect, it was necessary to reduce the operating temperatures, at which the irradiation took place, from 54 to 52.5, 50 and 48°C, respectively.

It was of interest to analyze whether the revealed pattern will persist for the synergistic interaction of medicinal compounds and hyperthermia. As an example, we have chosen the simultaneous action of cisplatin and hyperthermia. As was mentioned above, under the action of chemical agents, the analogue of dose can be the duration of action, while the analogue of the intensity would be the concentration of chemical preparations. Then it can be expected that a decrease in the concentration of cisplatin should be accompanied by a corresponding decrease in the effective temperature in order to maintain any fixed value of the synergistic effect. We calculated this dependence for cultured Chinese hamster cells based on the data published by other authors (Miyakoshi et al., 1984; Kumar et al., 1985; Hall, Giaccia, 2018). It has been shown that with a decrease in the concentration of cisplatin from 12 to 6 and 3 μM, to ensure the synergistic enhancement ratio $k = 3.0$, it is necessary to decrease the operating temperature from 40.8 to 39.6 and 38.0°C, respectively. Similar results were obtained for diploid yeast cells – with a decrease in the concentration of cisplatin from 835 to 167 μmol/L, in order to ensure the maximum synergistic effect, it was necessary to decrease the operating temperature, at which the drug was applied, from 52.5 to 50.5°C, respectively. These data show that the lower the cisplatin concentration, the lower the applied temperature will provide the desired synergy value.

The results presented in this chapter confirm the existence of an optimal ratio of acting agents to achieve the greatest synergistic interaction when used simultaneously. The set of our new results with previously published data (Petin et al., 2012, 2020; Petin, Morozov, 2015; Petin, Kim, 2016) demonstrate the existence of optimal temperatures at which the greatest synergy is achieved under the simultaneous application of hyperthermia with various inactivating agents on cells of different origins. This pattern is universal, since its manifestation did not depend on the agents used, biological objects and tests. The second universal regularity of the manifestation of synergy lies in its dependence on the intensity of physical factors or the concentration of chemical agents at a fixed operating temperature. We used basic concepts of the mathematical model of synergy to interpret these universal regularities using the example of the simultaneous action of hyperthermia with ionizing radiation (Section 5.3).

Again, taking a retrospective look at the total results presented in this chapter and previously published data (Petin et al., 2012, 2020; Petin, Morozov, 2015; Petin, Kim, 2016; Zhavoronkov, Petin, 2018), a large complex of universal patterns of synergistic interaction of various factors was obtained, which does not depend on the agents applied, the analyzed biological effect and the objects used. Briefly, these patterns and the consequences arising from them can be summarized as follows.

1. For a constant dose rate of ionizing radiation or the concentration of a chemical agent, the synergism of simultaneously application of these agents with hyperthermia might be observed only within a certain temperature range. Within this range, there is a specific temperature that ensures the greatest synergistic interaction.
2. An increase in the temperature, simultaneously acting with physical or chemical factors, above the optimal value leads to a decrease in the efficiency of synergic interaction. In this case, the final effect is to a greater extent induced by hyperthermia.
3. A decrease in the temperature below the optimal value also leads to a decrease in the efficiency of synergistic interaction, while the observed effect is mainly due to the action of ionizing radiation or another agent used in combination with hyperthermia.
4. The range of temperatures synergistically interacting with physical or chemical agents shifts to the region of lower temperatures for thermosensitive objects or with a decrease in the intensity of another agent used that indicates the fundamental importance of synergistic effects at low intensities of factors that can actually occur in in everyday human practice and in the biosphere.
5. For a fixed temperature applied simultaneously with a varying dose rate of ionizing radiation or a concentration of a chemical agent that determines its intensity, synergy can be recorded only within a certain range of intensities of the applied factors.
6. Within this range, there is an optimal intensity, at which the the greatest synergistic interaction is obtained, while any deviation of the intensity of the acting agent from the optimal value leads to a decrease in the effectiveness of synergistic interaction.
7. At a fixed operating temperature, an increase in the dose rate of ionizing radiation or the concentration of chemical agents above the optimal value leads to a decrease in the efficiency of synergistic interaction due to a decrease in the duration of irradiation and a corresponding decrease in thermal damage.
8. A decrease in the dose rate of ionizing radiation or in the concentration of chemical agents below the optimal value at a fixed acting temperature also leads to a decrease in the efficiency of synergistic interaction. For this case, the observed effect is mainly due to a large yield of thermal damage because of the prolonged exposure at low dose rates of ionizing radiation or concentrations of chemical agents.
9. A linear relationship between the operating temperature and the intensity of the agents used has been demonstrated for various biological objects and tests – at a lower intensity of a physical factor or concentration of a chemical one, the operating temperature should be reduced to maintain the highest or some selected value of the synergistic effect.

The results obtained are not only of fundamental importance from the point of view of understanding the mechanisms of synergistic interaction and ways to optimize the influencing agents to achieve the greatest synergy, but are also useful for a number of practical applications in the field of biomedical technologies and environmental research. Comparing the new results obtained in this work with numerous data published by other authors, it can be stated that our results bring order among the chaos of mutually exclusive facts described in the literature. In general, the experimental and theoretical materials of this chapter give a fairly complete picture of the features, patterns, mechanisms of action and ways of further development of various manifestations of synergistic effects at the combined actions of harmful environmental factors on biological objects in the modern world.

It should be noted that the ideas of synergism described here are not used in oncological practice. It can be assumed that knowledge and consideration of the general patterns of synergy can be useful for specialists who use the simultaneous action of various agents to optimize combined treatment methods in modern medicine in general and in oncology in particular. The main heuristic value of the above results lies in the necessity and fundamental possibility of developing practical methods for simultaneous thermoradiation exposure to realize the greatest synergistic effect, which makes it possible to reduce the dose of ionizing radiation or the concentration of chemical agents during radiation therapy. Synergistic ideas can also be used to optimize the combined use of drugs in the treatment of various diseases.

Chapter 6

Peculiarities of an Osmotic Homeostasis Display

Abstract

Exchange with the external environment of cells is carried out using the system of osmotic homeostasis, including the membrane apparatus. For the effective exchange, it is necessary to maintain a certain pressure balance in the cell-environment system. Violations of this balance cause changes that are incompatible with the normal cell functions. For example, ionizing radiation or hyperthermia can lead to such violations, especially in conditions of a sharp change in the absolute values of the temperature of cell environment surroundings. The dependence of cell osmotic properties on the dose of ionizing radiation and temperature parameters is presented. The osmotolerance and cell ability to reproduce in environments with various osmotic pressures at different temperatures were described. New data related with the influence of tonicity and chloramphenicol on hyperthermic cytotoxicity and cell permeability are provided. Original data on cyclic heating and cooling on cell survival are also shown. The action of osmolytes on destructive effect of ionizing radiation, hyperthermia, microwave radiation and ultrasound are compared. Peculiarities of biological effects of electromagnetic non-ionizing radiation at cellular and molecular levels are discussed. It is concluded that osmotic homeostasis of cells, along with DNA, is an important target responsible for cell viability under the damaging action of various agents.

6.1. Introduction to the Problem of Osmotic Homeostasis

The problem of identifying and studying the critical cell targets that determine their resistance to the damaging effect induced by various radiation and other environmental factors is still far from its final understanding. It is of interest to analyze the development of ideas about the mechanisms of ionizing radiation action in the simplest cells. Lea (1946), Timofeev-Ressovsky, Zimmer (1947), and Alper (1970, 1979) systematized the factual material obtained in the study of the bactericidal effect induced by ionizing radiation and came to the conclusion that: a) survival curves of bacteria are always exponential; b) the mean lethal dose does not depend on the intensity of ionizing radiation and the temperature, at which the irradiation occurs; c) when vegetative forms of bacteria are irradiated, the mean lethal doses of radiation depend on its LET. Since these conclusions were in good agreement with the results obtained when studying the inactivation of viruses and mutations in Drosophila, it was suggested that the death of bacterial cells after exposure to UV light or ionizing radiation is due to lethal mutations. Thus, it was postulated that the main structure damaged by irradiation is the nuclear apparatus of cells or in modern terms DNA. In the same year (1946), a mutant of *Escherichia coli* strain B/r, resistant to UV light and ionizing radiation, was isolated from the *E. coli* strain B (Witkin, 1946). The study of this strain was a major event in radiobiology and served as a stimulus for the development of cellular radiobiology, since it became possible to study the nature of radiosensitivity at the genetic level. Subsequently, the study of the reactions of bacteria to irradiation began to be carried out in parallel on wild-type strains and radiosensitive mutants. It was shown that the survival curves of bacteria cells *E. coli* B/r, in contrast to the survival curves of *E. coli* B, always have a sigmoid shape both after exposure to UV light and ionizing radiation. The shape of the survival curve depended on the culture conditions up to irradiation. Subsequently, it was shown (Alper, 1970, 1979) that the shape of the dose-effect curves of *E. coli* B/r could also be modified after irradiation with UV light.

These new data have significantly complicated the interpretation of the mechanisms of lethal action of radiation. A number of authors in the same years considered it even fundamentally impossible to explain cell death by the occurrence of lethal mutations, since in order to interpret the change in the shape of the dose-effect curves depending on the cultivation conditions, it is necessary to assume that the number of targets can be arbitrarily changed by these conditions. Alper and Gillies (1960) made a significant step in explaining the nature of the different sensitivity of *E. coli* B and *E. coli* B/r strains, who showed that the sensitivity of these strains is influenced by factors such as the growth phase of the culture, the composition of nutrient media, pH and others. In this case, the same environmental factors in different ways influenced the sensitivity of the two strains. For example, if the cultivation of irradiated *E. coli* B cells on a minimal medium increased cell survival, then the same cultivation conditions reduced the survival of the irradiated *E. coli* B/r cells, and the survival curves of these strains coincided. The experimental data obtained allowed the authors to suggest that the difference in the sensitivity of these strains is associated with the peculiarities of their metabolism and that these differences are due to a mutation in one gene.

The authors of the cited works tried to answer the main question of radiobiology, which structures in the cell are the target, the damage of which is responsible for its death. Thus, in the monograph Alper (1979) the assumption was formulated according to which radiation causes two types of damage in the cell – "N" and "O." "N" is damage such as lethal mutations associated with a breakdown in the structure of DNA. These lesions are not modified by cultivation conditions. Type "O" – damage to other structures of the cell. According to Alper, the most likely candidate would be cell membranes. Damage type "O" can be modified by cultivation conditions.

Thus, for several decades, the central issue that has been discussed in the literature has been the question concerning the main target responsible for cell death. In those years, the argumentation in favor of one or another hypothesis about a radiosensitive target was based on indirect data (analysis of dose-effect curves, peculiarities of the reactions of bacterial cells under certain environmental conditions, etc.). In the early 60s, mutants of *E. coli*, highly sensitive to the action of various environmental factors, were isolated. Radiosensitivity loci were marked on genetic maps; as well as physicochemical methods were also developed for analyzing the effect of radiation on cells. From that moment on, the experimental solution to the problem entered a fundamentally new phase of its development.

The osmotic cell homeostasis system is one of the most important systems responsible for numerous cellular functions. Thanks to this system, cells successfully exchange with the external environment, absorbing and releasing water and physiologically significant compounds dissolved in it. Destabilization of the osmotic homeostasis system by various environmental factors can lead to some changes incompatible with normal cell activity. It is known that exposure of cells to such different in nature agents as penetrating ionizing radiation, ultraviolet light, ultrasound, microwaves and hyperthermia causes the cells to lose various intracellular compounds. Such a reaction of cells in response to exposure to cytotoxic agents is based on various damage into the same targets. These targets are primarily the cytoplasmic membranes of cells. Loss of various compounds to the external environment by damaged cells can also be due to other reasons, for example, an increase in intracellular osmotic pressure due to the processes of radiation catabolism (degradation), ultrasonic depolymerization and thermogenic dissociation of weak polyelectrolytes of cells.

The transition to the external environment of intracellular compounds is often accompanied by an increase in the size of cells and their transformation from rod- and disc-shaped forms to spherical due to the absorption of water from the external environment even under isotonic conditions, i.e., in the absence of an osmotic gradient. This feature of cell reaction after thermoradiation and ultrasound exposure indicates the destabilization of their osmotic homeostasis. The optimal ratio of osmotic pressures of the cell contents and the extracellular environment is one of the most important conditions ensuring the normal structural and functional status of the cytoplasmic membranes and cellular exchange with the external environment. To maintain this ratio in experiments with cells, various physiological fluids with a certain osmotic pressure are used, for example, physiological NaCl solution. In microbiological practice, when calculating the tonicity of suspension media of microorganisms, the temperature factor, as a rule, is not taken into account. In addition, if they are taken into account, then within the framework of generally accepted concepts, according to which physiological fluids for cells of homoeothermic (warm-blooded) organisms should contain higher concentrations of salts than for cells of poikilothermic (cold-blooded) organisms (Pirt, 1975). At the same time, there is reason to believe that the osmotic properties of these groups of cell organisms can vary significantly depending on the ambient temperature. This is indicated by numerous literature data on the dependence of salt requirements and osmotolerance of cells of pro- and eukaryotes (Novitsky, Kushner, 1975, 1976) on the temperature of the cell suspension. The fact that physiological fluids for cells of warm-blooded animals contain higher concentrations of salts than saline solutions for cells of cold-blooded organisms may also indicate the osmotic nature of this phenomenon and the important role of the temperature factor in cell responses.

At present, it is generally accepted that damage to lipid and/or protein components of cell membranes underlies the disruption of cell permeability as a result of high temperatures, ionizing radiation, and ultrasound (Mille et al., 2002; Laroche, Gervais, 2003). On this basis, it can be assumed that the change in cell permeability under the action of hyperthermia and other factors is also due to a violation of osmotic homeostasis of cells (Morozov et al., 1990, 1996, Rep et al., 2000; Tamas et al., 2000; Mager, Siderius, 2002). This assumption is based on the following known factors maintaining a certain osmotic balance in the cell – environment system.

First, at physiologically normal temperatures, viable cells, even under isotonic conditions, already exhibit a certain excess of osmotic pressure in comparison with the suspension medium due to the osmotic pressure of proteins and other polyelectrolytes. This excess pressure can lead to the loss of biologically important compounds into the external environment and thereby reduce the cell viability due to the membrane damage under the action of various factors.

Secondly, the composition of cell contents includes, along with others, high-polymer compounds, in contrast to suspension media containing such simple compounds as inorganic salts, which provide normal osmotic homeostasis of cells. Being strong electrolytes, the metabolism of these substances is due to a mutation in one gene are characterized by a high degree of electrolytic dissociation, close to unity. Consequently, they are completely dissociated into ions regardless of the temperature of the solution. In this case, the osmotic pressure of the solution π will be expressed (Nelson, Kox, 2021) by the formula

$$\pi = iCRT, \tag{6.1}$$

where i is the isotonic coefficient of the solution; C is the molar concentration of the solution; R is the universal gas constant; T is the absolute temperature.

As can be seen from this formula, the osmotic pressure of strong electrolytes increases with an increase in temperature due to an increase in the mobility of osmotically active elements – ions, and not due to an increase in their number at a given concentration of a substance. Another dependence will be recorded when heating solutions of weak polyelectrolytes, for example, nucleic acids, proteins, and amino acids entering the cell contents. In this case, the osmotic pressure of cells will increase with increasing temperature not only due to an increase in mobility, but also in the number of osmotically active particles due to intra- and intermolecular dissociation of ionizing groups and molecular complexes at a given concentration of the substance as a result of an increase in the degree of electrical dissociation. With regard to monomeric fragments of molecules of high polymer compounds, this process is described by the following formula (Nelson, Kox, 2021)

$$\pi = k\left(1 + \sqrt{\frac{\exp(-Q/RT)+B}{C}}\right)CRT, \tag{6.2}$$

where k is the osmotic coefficient for imperfect solutions; Q is the thermal effect of the reaction; B is a constant coefficient independent of temperature.

As follows from formula (6.2), the osmotic pressure for weak electrolytes depends, among other things, on the expression under the square root sign, the value of which increases with increasing temperature. This expression is a measure of the ionization of electrolyte molecules, which increases with an increase in the temperature of the solution, and reflects the degree of electrolytic dissociation. In other words, with an increase in the temperature of the cell suspension, the osmotic pressure of the cells will increase faster than the pressure in the suspension liquid, which is, for example, isotonic sodium chloride solution. This can lead to a violation of osmotic homeostasis of cells, because of which they are subjected to the damaging action of osmotic stress. The consequences of this stress will be all the more detrimental to cells, the higher the rate of change, the magnitude of the difference and the value of the current temperature. In this case, excessive osmotic pressure of cells can more likely destabilize membranes weakened by hyperthermia, thereby increasing their permeability. As a result, a stream of solvent-water will rush into the cell, and various dissolved compounds will be washed out of the cell. Then cell viability will decrease.

Published experimental data, the results of which are summarized in work (Petin, Morozov, 2015) together with the findings that will be described in this chapter, confirm the rationale for the high significance of the state of the osmotic homeostasis system in the fate of cells exposed to the damaging action of various environmental factors.

6.2. Influence of Tonicity and Chloramphenicol on Hyperthermic Cytotoxicity and Cell Permeability

Understanding hyperthermic cytotoxic mechanisms is important to the biological basis for the use of thermotherapy in oncology (Streffer et al., 1990). Hernan et al. (1981) have reported that an increase in the time taken to raise the temperature of eukaryotic cells *in vitro* led to decreased hyperthermic lethality. Later, similar data were obtained by Wu and Lin (1985) in Chinese hamster V79 cells. Both groups postulated that differences in thermal damage to the cell membrane associated with heating rate explain differences in the thermal response of cells. In addition, there is compelling evidence that heat shock proteins (HSP) are involved in protection of cells from heat damage (Li, 1985; Alexandrov, Kislyuk, 1994). Several lines of research, however, have shown that tolerance to heat can be induced in the absence of additional levels of HSP in cells (Hall, 1983; Anderson et al., 1986). Burns et al. (1986) failed to detect the synthesis of HSP 70 in leukemia cells, which were preheated over 3 h from 37 to 42°C and then exposed to 42°C for 1 h, while in rapidly heated cells, the heat inducible HSP 70 comprised 4% of the total protein synthesis after a 1 h heat treatment at 42°C. Also, Anderson et al. (1988) reported that thermotolerance in Chinese hamster cells may occur under some conditions in the absence of elevated levels of HSP 70. These authors as well as Tomasovich et al. (1983) concluded that other factors within the cell could confer thermotolerance, in addition to the contribution made by HSP, especially during mild heat challenges such as occur during slow heating rates.

A working hypothesis was proposed by Morozov et al. (1986, 1990) according to which a higher cell sensitivity to rapid heating rates is the result of a more profound destabilization of osmotic homeostasis caused by rapid versus slow heating which leads in a larger increase in intracellular osmotic pressure in rapidly heated cells. This hypothesis is strongly supported by several reports illustrating the dependence of osmotic features in pro- and eukaryotic cells exposed to different temperatures in cell suspension (Jozwiak, Paleczka, 1988; Merinov, Moiseev, 1991; Grishchenko et al., 1995). Novitsky and Kushner (1975, 1976) and Kushner (1978) reported that growing bacterial cells required an increase in extracellular tonicity in order to maintain viability when incubation temperatures were increased while the reverse situation is true for the tonicity requirements of cell growing under reduced incubation temperature. Further support for the osmotic hypothesis was our observation that protection of bacterial cells from thermal damage occurs when extracellular tonicity was increased as incubation temperature increased (Morozov, 1990, 1992; Morozov et al., 1990). Herman et al. (1981) reported an increase in cell volumes of rapidly heated CHO cells as opposed to slowly heated cells. These authors also proposed that the change in cell volumes observed in rapidly heated cells reflected damage in the plasma membrane with loss of ability to control the intracellular osmotic environment and cell volume.

Assuming this hypothesis to be valid it would be expected that hypotonic culture medium should increase the differential cytotoxic effects of various heating rates and, on the contrary, hypertonic medium should decrease these effects. Thus, the aim of the present study was investigating the interaction between tonicity of the culture media, hyperthermic cell killing and permeability changes. In addition, the influence of the protein synthesis inhibitor chloramphenicol on the development of thermotolerance at different heating rates was studied.

The repair-proficient strain of *E. coli* B/r, the repair-deficient mutant *E. coli* B_{s-1} and haploid yeast strain of *Zygosaccharomyces bailii* were used in the present investigation. Before heating, the strains were grown in YEP broth (yeast extract – 1%, glucose – 1%, peptone – 1%, NaCl – 1%) at 37 and 30°C during 18 and 48 h of incubation for bacterial and yeast cells, respectively. Cells were harvested 2x by centrifugation and washed in saline (0.85 NaCl, pH 7.0), then were diluted in saline. Final suspensions prepared in this manner contained approximately 10^8 cells per ml for bacterial and 10^6 cells per ml for yeast cells. Colony-forming ability was assessed by plating of cells on nutrient agar (YEP broth + 2% agar-agar) and counting macrocolonies (Morozov, 1983) after 24 and 48 h incubation of bacterial and yeast cells, respectively. Cellular permeability was estimated by measurement of optical activity ($\lambda = 260$ nm) of the supernatant of cell samples. Prior to heat treatment, cells from plateau phase were diluted in 0.01 M phosphate buffer containing 0.9 or 6% of NaCl, or YEP broth with or without chloramphenicol (10 mg/ml). This concentration of chloramphenicol is known to inhibit protein synthesis (Gillies, Alper, 1959; Yamamori, Yura, 1982; Kagramanova, 1995). Transfer of cell samples to solution with chloramphenicol was made at least 30 min before thermal treatment. The cells were then cooled at 0°C for 20 min prior to heating in order to increase heating rate and inhibit cell metabolism. This procedure was checked in special experiments and does not affect cell killing in subsequently heated cells. After the treatment, the cell samples were heated at different heating rates (mean heating rates were 0.015, 0.25 and 1.5°C per s) from 0 to 52°C over time period of 1 h, 3.5 min and 35 s. After heating reached the peak temperature (52°C), the cell survival was close to 100% independently of cell species and heating rate. Then the cell samples were heated to 52°C for different times. Temperature was measuring using an electrical thermometer probe (mode MT-54) with 0.1°C accuracy. The following methods of heat treatment were used: (i) 5 ml plastic tubes with the cooled at 0°C cell suspension placed into a water circulator (model TA-150) with water cooled at 0°C and then the temperature of water circulator and cell suspension was increased up to 52°C in 60 min (mean heating rate was about of 0.015°C/s); (ii) 5 ml plastic tubes with cell suspension cooled at 0°C placed into a water circulator previously heated to 52°C and then the cell samples were heated from 0 to 52°C over 3.5 min (mean heating rate was about of 0.25°C/s); (iii) 0.05 ml of cell suspension at 0°C was put into the tubes containing 5 ml suspension liquid preheated to 52°C, in this case, the cells were heated from 0 to 52°C in 35 s (mean heating rate was about of 1.5°C/s). After having reached and maintained the peak temperature (52°C) the cell samples were heated at 52°C for different times. Immediately after heating the samples were removed and diluted into saline at room temperature to ensure rapid cooling before plating and/or estimating the cellular permeability. All experimental series were repeated 3–5x. Figure 6.1 exhibits the effect of different rates of heating on *E. coli* B/r cells reflecting by heat survival curves.

In this and all other figures, the error bars show the standard deviation of the mean from different experiments (Armitage, 1971). It is apparent that hyperthermia action in isotonic media at different heating rates led to various levels of cytotoxicity. It is clear that the bacterial cells were much more sensitive to heat after heating at the highest heating rate (1.5°C/s).

To test the effect of HSP on the different cell responses to various heating rates, chloramphenicol (10 mg/ml) was added before and was present during hyperthermic treatment of cells in YEP broth and isotonic phosphate buffer. Figure 6.2 shows that slowly heated cells in nutrient broth reveal a higher level of thermotolerance than cells slowly heated in phosphate buffer. The additional thermotolerance was inhibited by chloramphenicol. However, even in this case, the majority of the induced thermotolerance was expressed even in the presence of chloramphenicol in the YEP broth media. In addition, it can be seen that the compound had no effect on cell thermotolerance after heating in phosphate buffer.

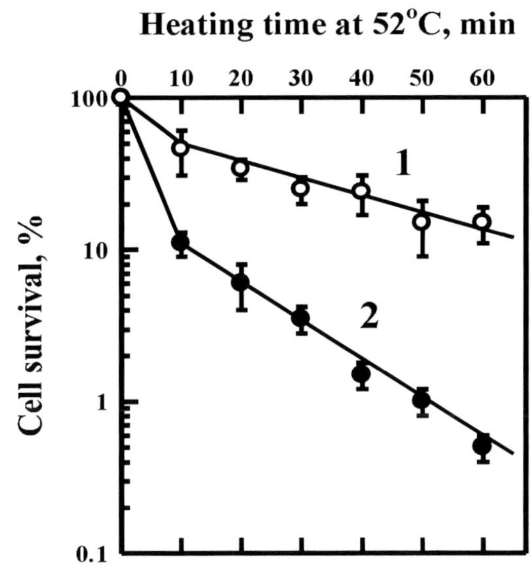

Figure 6.1. The effect of different heating rates (open circles – 0.015°C/s; closed circles – 1.5°C/s) on survival curves of *E. coli* B/r cells heated in isotonic 0.01 M phosphate buffer (pH 7.0 and 0.9% NaCl) at 52°C.

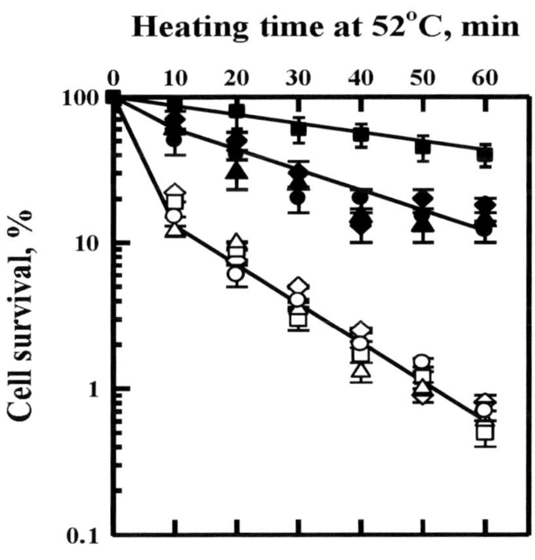

Figure 6.2. The influence of the heating time at 52°C on thermal cytotoxicity of *E. coli* B/r cells at different heating rates (open circles, triangles, squares, rhombuses – 1.5°C per s; closed symbols – 0.015°C per s) in isotonic 0.01 M phosphate buffer (with 0.9% NaCl) with (open and closed circles) and without (open and closed triangles) chloramphenicol (10 mg/ml) and in YEP broth buffer without (open and closed square) and with (open circles and closed rhombuses) chloramphenicol.

Figure 6.3 illustrates the dependence of *E. coli* B/r survival on the tonicity of heated cell suspension in culture exposed to different heating rates. As shown in this figure, the difference in cytotoxicity at various heating rates was more profound in hypertonic media (see Figure 6.1). The absence of heating rate effect in hypertonic condition seems to be due to thermoprotective actions of hypertonicity, which were more expressed in cells exposed to the highest heating rate. Comparable results also were observed for the repair-deficient mutant *E. coli* B_{s-1} (Figure 6.4) and yeast *Zygosaccharomyces bailii* (Figure 6.5).

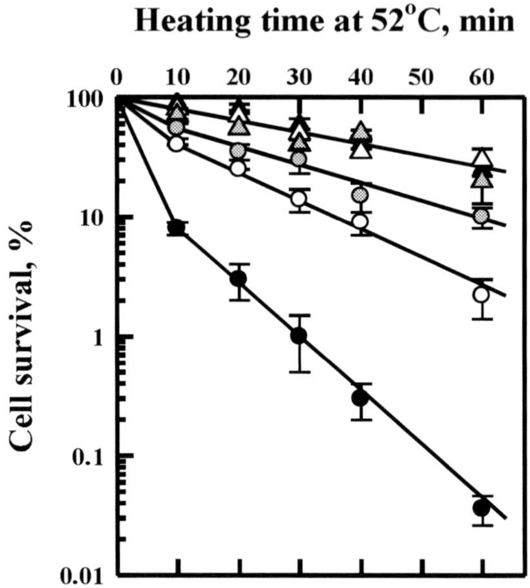

Figure 6.3. The influence of the heating time at 52°C on thermal cytotoxicity of *E. coli* B/r cells exposing at different heating rate: closed circles and triangles – 1.5°C/s; open circles and triangles 0.25°C/s, slightly darkened circles and triangles – 0.015°C/s in the heat media with various tonicity – closed, open and slightly darkened circles – hypotonic 0.01 M phosphate buffer pH 7.0; closed, open and slightly darkened triangles – hypotonic 0.01 M phosphate buffer pH 7.0 containing 6% sodium chloride.

The present results demonstrate the influence of heating rate on thermal cytotoxicity in bacterial and yeast cell systems. The influence of heating rate on thermal cytotoxicity for mammalian cells was also described (Herman et al., 1981; Wu, Lin, 1985). Taking our results and the data of these other authors as a whole, we can conclude that the effect of heating rate is expressed to both pro- and eukaryotic cells. We tested the possibility that the effect of heating rate on cytotoxicity might be caused by different induction of HSP. The assumption was presumed by other authors (Li, 1985; Anderson et al., 1988; Alexandrov, Kislyuk, 1994). To test the influence of HSP on the effects of heating rate, chloramphenicol was added before and was present during hyperthermic treatment of *E. coli* B/r cells to YEP broth and isotonic phosphate buffer. Chloramphenicol is known to be an inhibitor protein synthesis (Gillies, Alper, 1959) and bacterial cell thermotolerance (Yamomori, Yura, 1982; Kagramanova, 1985).

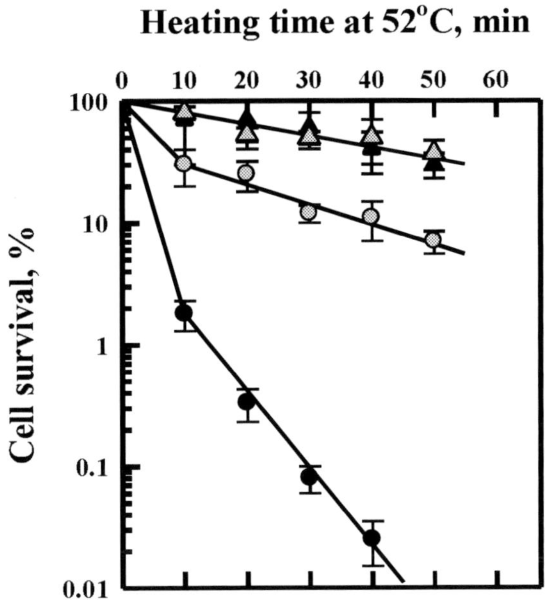

Figure 6.4. The influence of the heating time at 52°C on thermal cytotoxicity of *E. coli* B_{s-1} cells heated: at different heating rates: closed circles and triangles – 1.5°C/s; slightly darkened circles and triangles – 0.015°C/s) in the heat media with various tonicity (closed and slightly darkened circles – hypotonic 0.01 M phosphate buffer pH 7.0; closed and slightly darkened triangles – hypertonic 0.01 M phosphate buffer pH 7.0 containing 6% sodium chloride).

In our experiments, however, the presence of chloramphenicol did not affect the results when cells were heated in isotonic phosphate buffer and only partly prevented the development of thermotolerance in cell heated in nutrient medium. Cellular thermotolerance phenomena have mainly been investigated in nutrient media (Yamamori, Yura, 1982; Tsuchido et al., 1986; Anderson et al., 1988; Mizzen, Welch, 1988). Our results in bacterial cells demonstrated a minor additional degree of thermotoletance is induced during slow heating in nutrient medium compared with that observed in phosphate buffer and that this additional thermotolerance was prevented by chloramphenicol. These results suggest that no thermotolerance related to HSP levels is present in our experiments, although HSP levels were not directly measured. These results also suggest that other reason, besides the induction of HSP, may be responsible for differences in thermosensitivity of cells heated at various rates in non-nutrient media. Furthermore, it is apparent even in nutrient media the greater portion of thermotolerance was induced by slow heating even when chloramphenicol was present and, presumably, HSP induction was prevented.

The effect of heating rates and tonicity of medium on cell permeability was estimated by measurement of optical activity (λ = 260 nm) of the supernatant of cell samples. Cell lysis *per se* was not observed in these experiments since total cell count before and after heating was identical. Hence, the optical density changes of the supernatant were not related to cell lysis but only to loss of intracellular solution. As shown in Figure 6.6, cell membrane permeability was increased more profoundly when hyperthermia treatment of *E. coli* B/r cells were performed at the highest heating rate on hypotonic solution. The cells heated at hypertonic condition exhibited no such response. Membrane permeability of cells in hypertonic media was little altered and essentially the same when heating occurred slowly or rapidly and was far less affected even than cells heated slowly in hypotonic media.

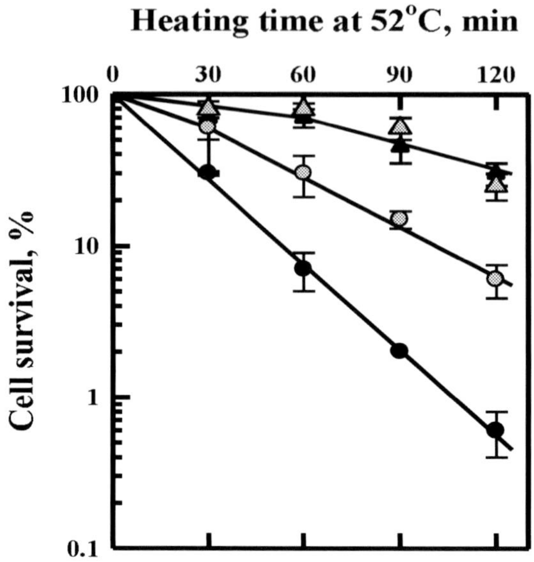

Figure 6.5. The influence of the heating time at 52°C on thermal cytotoxicity of *Zygosaccharomyces bailii* cells heated at different heating rates: closed circles and triangles – 1.5°C/s; slightly darkened circles and triangles – 0.015°C/s) in the heat media with various tonicity (closed and slightly darkened circles – hypotonic 0.01 M phosphate buffer pH 7.0; closed and slightly darkened triangles – hypertonic 0.01 M phosphate buffer pH 7.0, containing 10% sodium chloride).

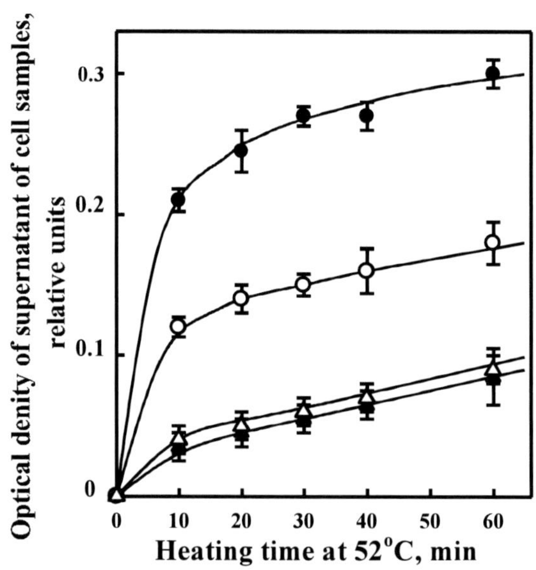

Figure 6.6. The effect of heating rates (open circles and triangles: 0.015°C/s; closed circles and triangles 1.5°C/s) and tonicity of heat media (open and closed circles: hypotonic 0,01 M phosphate buffer pH 7.0; closed and open triangles – hypertonic 0.01 M phosphate buffer pH 7.0, containing 6% sodium chloride) on optical activity (λ = 260 nm) of the supernatant of cell samples of *E. coli* B/r cells.

The data presented in this section provide support of the following conclusions: (i) effects of heating rate on thermal cytotoxic action are more expressed in hypotonic conditions; (ii) when similar experiments were carried out using hypertonic conditions no significant difference in cellular responses to heat treatment related to various heating rates were observed; (iii) suppression of the heating rate effects in hypertonic media was found to be due to the protection from thermal injury induced in the highest heating rate, but not to an alteration in the thermotolerance of slowly heated cells; (iv) the magnitude of cell killing and cell permeability were altered in the same direction and were directly proportional to heating rates; and (v) the major portion of the thermotolerance induced by slower heating rates is not due to the altered production of HSP in bacterial cells heated in nutrient media and thermotolerance appears to be unrelated to the induction of HSP when cells are heated in nonnutrient media.

Based on other experimental data, other authors (Tomasovich et al., 1983; Anderson et al., 1988) also concluded that other factors within the cell can confer differences in thermosensitivity of rapidly and slowly heated cells in addition to the contribution made by HSP. It seems logically to suppose the significant role of tonicity in the effects observed here. It might be explained by a more intensive increase in intracellular osmotic pressure after high heating rate in comparison with that which occurs after a low heating rate. The enlarged cell volume and increased cholesterol synthesis observed by Herman et al. (1981) may have resulted in high cell membrane rigidity during rapid heating as well as in increased cell permeability and cell killing (Herman et al., 1981; Wu, Lin, 1985; Morozov et al., 1990) and may represent inevitable cellular response to thermogenic increase in intracellular osmotic pressure induced by high heating rate. Therefore, these findings may well be applicable to heat response in mammalian cells and should be tested in such systems.

6.3. The Influence of Cyclic Heating and Cooling on *Escherichia coli* Survival

As was partly shown in previous section and has been previously found that when prokaryotic (Morozov, 1996 a; Morozov, 1990; Morozov et al., 1990, 1997) and eukaryotic (Herman et al., 1981; Wu, Lin, 1985) cells are heated at different rates to a lethal temperature and maintained at this temperature for a certain time period, they are damaged to a greater extent at a higher rate of heating. Furthermore, the damage to bacteria heated to a lethal temperature also increases with the rate of cooling of the cell suspension (Morozov et al., 1995, 2005). In both cases, the damaging effect depending on the rate of heating and cooling can be decreased or increased using suspension media with a certain osmotic pressure (Morozov, 1992; Morozov, Petin, 1988). Based on these data, it seemed to be worthwhile to investigate the influence of repeated heating and cooling on bacterial survival in media with different osmotic pressure in the range of lethal and physiologically normal temperatures. The urgency of such research was prompted not only by scientific interest (analysis of the mechanism of the lethal effect of hyperthermia on cells), but also by the practical importance of the influence of heating conditions on biological systems for the development of effective methods of application of the temperature factor in diverse fields of biology and medicine. The goal of the present part of our work was to determine the features and the nature of the damaging effect of repeated changing of heating and cooling conditions in lethal (2–52°C) and nonlethal (2–37°C) temperature ranges on *Escherichia coli* survival in liquid innutritious media with different osmotic pressure. With the same purpose and for comparison, the influence of cyclic changing of heating and cooling conditions on *E. coli* cells survival in the temperature range of 2–37°C in pasteurized whole milk was investigated.

The objects of investigation were collection strains *Escherichia coli* B/r and *E. coli* B_{s-1} kindly provided by V.D. Zhestyanikov (Institute of Cytology, Russian Academy of Sciences). Bacterial cells were incubated in nutrient broth (1% peptone, 1% yeast extract, and 0.9% NaCl) for 18 h at 37°C up to the stationary phase of growth. The cells were washed from the medium by twofold centrifugation of the cell suspension, and the cell precipitate was resuspended up to the concentration 10^8 cell/ml in 0.01 M phosphate buffer, pH 7.0 (hypotonic buffer), in 0.01 M phosphate buffer pH 7.0 containing 6% NaCl (hypertonic buffer), in buffers with different NaCl concentrations, or in pasteurized whole milk. To heat and cool the samples of cell suspensions at the average rate of 1.75 and 2.5°C per second, a medical syringe was used, which was connected by a clear cambric with a stainless-steel tube 1 mm in diameter and 0.05 ml in volume. Bacterial samples were collected into the tube up to the cambric level using the syringe, and the outlet of the tube was closed with molten wax. Then the tube was repeatedly immersed for 20 s (time of heating or cooling to reach the fixed temperature) into temperature-controlled vessels with water temperature 2, 37, or 52°C. To heat and cool the samples at the average rate of 0.42 and 0.10°C per second, stainless steel test tubes 6 and 10 mm in diameter were used, with 0.5 and 1.2 ml of the cell suspension, respectively. Afterwards, the tubes were repeatedly immersed into temperature-controlled vessels with water temperature 2 or 37°C for 1.5 and 2.0 min (time of heating or cooling to reach the fixed temperature). The average rate of heating was determined by means of a stopwatch and an electron thermometer equipped with a miniature temperature detector. The accuracy of temperature observation was ±0.1°C. Cell survival was determined according to number of visible by eyes macrocolonies after 18 h of incubation at 37°C of the transfers performed by the drop method (Morozov, 1983).

Figure 6.7 exhibits the survival curves of *E. coli* B/r suspended in hypotonic and hypertonic phosphate buffers after repeated heating and cooling in the temperature range of 2–52°C at the average rate of 2.5°C/s. The data presented on this figure demonstrate that this temperature treatment induced intense cell death in spite of the fact that even at the maximum number of heating and cooling cycles, the overall exposure dose of heating at 52°C was too small to cause a lethal effect of a commensurable magnitude. This can be seen from the comparison of the data in Figures 6.7 and 6.8. Figure 6.8 shows the survival curves of *E. coli* B/r heated at 52°C at the average rate of 1.75°C/s in the variants of bacterial suspension in hypertonic and hypotonic phosphate buffers.

It is of interest to compare the values of exposure thermal doses with equal biological effect during the heating of bacteria at 52°C in hypotonic medium (phosphate buffer), according to the data of Figures 6.7 and 6.8. For this purpose, at the first approximation of the dose corresponding to 25 cycles of heating and cooling was quantitatively calculated. Due to the immediate change of heating and cooling conditions, on reaching the 52°C temperature, the cells were kept at this temperature for only several seconds (never longer than 5 s) per cycle of thermal treatment.

Therefore, the maximum exposure thermal dose of bacteria at 52°C may be considered to be close to 2 min for 25 cycles of heating and cooling. Cell survival at this dose of heating corresponding to 25 cycles of temperature treatment in the hypotonic buffer was 2.5% (Figure 6.7, curve 1). A similar survival level of cells occurs in the case of constant heating for 20 min at 52°C in the hypotonic buffer (Figure 6.8, curve 2). Therefore, the values of equal-effect exposure doses of heating for continuous heating at 52°C in a hypotonic medium and for the heating and cooling treatment described above are hardly commensurable. Equal effect doses for continuous thermal treatment of cells at 52°C and under conditions of repeated heating and cooling within the 2–52°C temperature range are all the more incommensurable for bacteria suspended in the hypertonic medium (phosphate buffer) (compare Figure 6.7, curve 2 and Figure 6.8, curve 2). Comparison of the data presented in these figures with the calculated values of equal-effect exposure doses of heating at 52°C revealed that a 25 s exposure dose of heating over 5 cycles of heating and cooling in the temperature range of 2–52°C corresponded to a 60 min dose of continuous heating (Figure 6.7, curve 2); in both cases, death of 75% of cell population occurred.

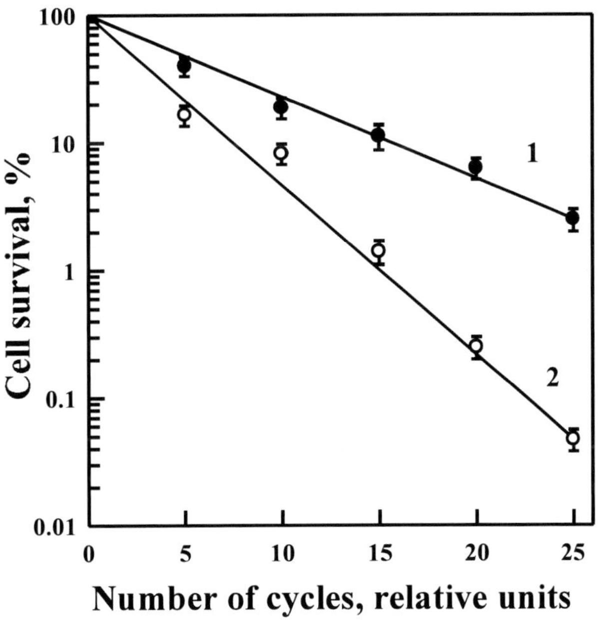

Figure 6.7. The influence of the number of heating and cooling cycles within the temperature range of 2–52°C at the average rate of 2.5°C/s on *E. coli* survival in hypotonic (curve 1) and hypertonic (curve 2) phosphate buffer.

Comparison of these results revealed that the killing effect under conditions of repeated heating and cooling within the temperature range of 2–52°C was substantially higher than under conditions of continuous heating of the cell suspension at 52°C. The nature of cell damage caused by repeated heating and cooling within the temperature range of 2–52°C is difficult to interpret in terms of the existing hypotheses of the mechanisms of thermal damage. Thermal death of cells is usually believed to be the result of thermogenic damage of cell lipids (Yatvin, 1977), proteins (Alexandrov, 1977; Alexandrov, Kislyuk, 1994), and membrane structures (Volkov, Polezhaev, 1983). This conclusion is supported by the data shown in Figure 6.9. This figure presents the survival curves of *E. coli* B/r after repeated heating and cooling within a nonlethal range of 2–37°C under different heating and cooling rates in media with different osmotic pressures. It can be seen that this treatment induces noticeable cell death even in the range of certainly nonlethal temperatures at moderate rates of heating and cooling, 0.4 and 1.75°C/s. In this case, interpretation of the reasons for cell death using the term "hyperthermia" make no sense as the upper level of the temperature used, 37°C, is the physiological optimum for *E. coli* growth, and the lower temperature level, 2°C, is the temperature for storing the transferring cultures for many weeks and even months (Meynell, Meynell, 1965; Schlegel, 1972; Morozov et al., 1986; Morozov, Petin, 1998). The similar results presented at Table 6.1 reveal that cells of the mutant *E. coli* B_{s-1} were also killed by repeated heating and cooling in the nonlethal temperature range of 2–37°C.

The data obtained indicate that the death of cells under repeated heating and cooling of cell suspensions was caused by the cyclic changing of conditions (heating and cooling), the rate of changing of temperature treatment, and the value of osmotic pressure of the medium, in which the cells were suspended, rather than by the thermal dose.

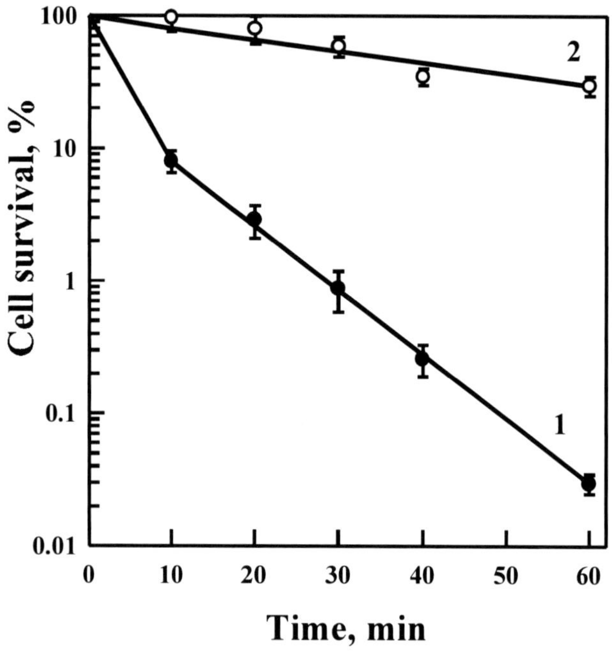

Figure 6.8. The influence of the duration of exposure at 52°C on *E. coli* B/r cells survival in hypotonic (curve 1) and hypertonic (curve 2) phosphate buffer.

The last conclusion is also confirmed by the data in Figures 6.1 and 6.3 illustrating an increase in the rate of cell death with increasing tonicity of the suspension media and of the rates of heating and cooling of bacterial suspensions during the cyclic changing of conditions of temperature treatment. Apparently, the nature of the damage causing cell death under repeated heating and cooling can be interpreted according to our conception that osmotic homeostasis is one of the critical systems in thermogenic cell damage (Morozov, 1990; Morozov et al., 1997). The stability of the osmotic homeostasis of cells is known to be one of the significant conditions for their normal functioning (Pirt, 1975). This stability can be changed within a certain range; beyond this range, cell survival decreases up to their death. Disturbance of cellular osmotic homeostasis can be caused by changing of the ratio of the concentrations of osmotically active elements of cell and of the environment. A similar phenomenon occurs in cell suspension upon temperature change because of initial differences of the electrolyte composition of the cells and of the suspension liquid.

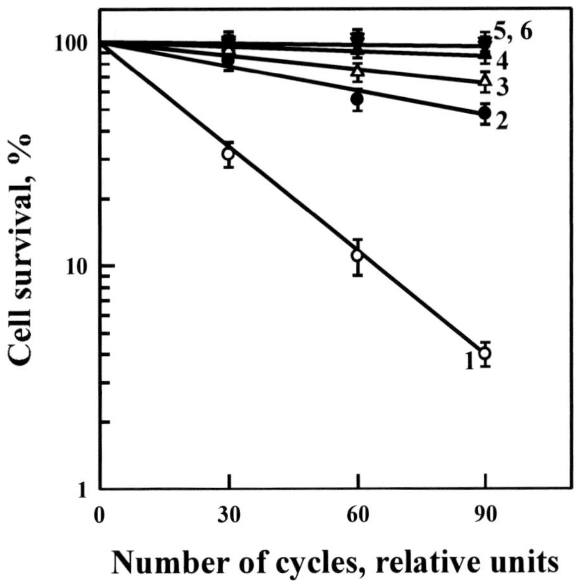

Figure 6.9. The influence of the number of heating and cooling cycles within the temperature range of 2–37°C at the average rate of 1.75 (curves 1, 2), 0.42 (curves 3, 4), and 0.10°C/s (curves 5, 6) on *E. coli* B/r survival in hypertonic (curves 1, 3, 5) and hypotonic (curves 2 4, 6) phosphate buffer.

Table 6.1. The influence of the number of heating and cooling cycles at the average rate of 1.75°C/s within the temperature range of 2–37°C on *E. coli* B_{s-1} survival in media with different tonicity

Number of heating and cooling cycles	Cell survival after the temperature treatment in hypotonic and hypertonic media	
	0.01 M phosphate buffer pH 7.0 with 6% NaCl	0.01 M phosphate buffer pH 7.0
1	100 ± 7.0	100 ± 7.0
20	23.5 ± 2.4	68.7 ± 6.3*
60	5.2 ± 0.6	54.9 ± 6.1*
90	0.9 ± 0.1	31.6 ± 3.9*

Notes: Each average value was obtained as a result of three determinations.
* The difference from the value obtained in the hypotonic medium is reliable ($p < 0.01$).

Unlike suspension media (for instance, Ringer's solution and other biological liquids), cells contain not only strong electrolytes with high levels of electrolytic dissociation weakly depending on the solution temperature, but also weak electrolytes such as proteins, nucleic acids and amino acids, and other compounds with a relatively low level of electrolytic dissociation, which, on the contrary, depends heavily on solution temperature (Morozov et al., 1990). These electrolytes dissociate upon temperature increase due to molecular heat motion and, consequently, they associate upon temperature decrease. Therefore, osmotic homeostasis is disturbed by temperature changes, even under conditions of the initial equality of osmotic pressures of the cell contents and the suspension medium; this process is examined in more detail in the works (Morozov et al., 1990; Morozov et al., 1997, 2005).

The osmotic pressure of the cells increases with heating and decreases with cooling. This conclusion is indirectly confirmed by the data of studies showing that hypertonic suspension media protect the cells from the damage upon heating (Morozov, Petin, 1998) and, conversely, increase the extent of damage upon cooling (Morozov et al., 2005). Increase (or decrease) of the intracellular osmotic pressure relative the osmotic pressure of the medium in the course of heating (or cooling) can induce disturbance in the cell membrane permeability and cause cell death. In this case, increased difference between the heating and cooling temperature and increased rate of temperature change will increase the depth of osmotic (hypertonic and hypotonic) shock, and thereby cause more intense cell death even in a nonlethal temperature range than can be caused by continuous thermal treatment; this result is demonstrated in the present work.

The conclusion that death of microorganisms upon repeated heating and cooling is caused by destabilization of the cellular osmotic homeostasis is also confirmed by the data of Table 6.2, representing the results of the influence of this treatment in a temperature range of 2–37°C on the survival of *E. coli* B/r suspended in pasteurized whole cow's milk.

Table 6.2. The influence of the number of heating and cooling cycles within the temperature range of 2–37°C at the average rate of 1.75°C/s on *E. coli* B/r survival in pasteurized cow milk

Number of heating and cooling cycles	Cells survival after the temperature treatment
0	100 ± 11.1
30	90 ± 9.9
60	102.7 ± 12.5
90	89.6 ± 9.1

Note: Each average value was obtained as a result of three determinations.

As can be seen from this table, repeated heating and cooling did not have a lethal effect on bacteria in this version of the experiment, although the average rate of heating and cooling was the same as in the case of similar temperature treatment in phosphate buffer with varying NaCl content. In the latter case, changing of the temperature conditions induced the death of *E. coli* B/r cells (Figure 6.3, curves 1, 2). The lack of lethal effect in the variants with cells suspended in milk can be explained by the stable balance between the intracellular osmotic pressure and that of the medium during the thermal treatment of the suspension, i.e., by the preservation of cellular osmotic homeostasis. The constancy of osmotic homeostasis of the cells under these conditions can be explained by the presence of both strong (e.g., NaCl) and weak electrolytes (proteins, amino acids, and other compounds with low levels of electrolytic dissociation) among the electrolytes contained in the cells and in milk. These compounds react in a similar way to temperature fluctuations by association and dissociation to osmotically active elements and have a similar effect on the osmotic pressure both of the cells and of the medium; the cells can therefore maintain osmotic homeostasis even under variable temperature conditions. Obviously, this is a possible reason for these drastic differences in bacterial response to the cyclic temperature changes in media with different electrolytic composition such as buffer with different NaCl content and cow's milk (see the data of Figure 6.3, curves 1–3, and Table 6.2).

The reasons for the increased cell damage under conditions of repeated heating and cooling at the average rate of 1.75 and 2.5°C/s in media with increased osmotic pressure can be explained in the following way. The cells suffer from hypotonic shock upon heating and from hypertonic shock upon cooling, with all the ensuing consequences, including cell death (Morozov, 1990). Obviously, these effects should increase during repeated heating and cooling, and they depend on the value of osmotic pressure in the suspension media. It seems likely that increased thermal death of microorganisms in hypertonic media under temperature changes is the result of this dependence of overall osmotically stabilizing, protective (upon heating), and osmotically destabilizing, damaging (upon cooling) effect of hypertonic conditions on cell physiology. Thus, repeated heating and cooling at a certain rate even within a nonlethal temperature range causes death of *E. coli*, particularly in suspension media with increased osmotic pressure. The assumption was made that processes of destabilization osmotic of cell homeostasis underlie lethal effect of this temperature treatment of microorganisms. The data obtained and their interpretation contribute significantly to the modern conception of the nature of cell thermal damage and can be used in various fields of biology and medicine.

6.4. The Action of Osmolytes on the Destructive Effect of Ionizing Radiation, Hyperthermia, Microwave Radiation and Ultrasound

Additive, synergetic or antagonistic interactions can be observed after combined actions of deleterious factors. As it was shown above, there are universal rules of synergism manifestation that does not depend on the nature of applied agents, biological objects and test systems. Synergetic interaction may be not observed at any random parameters of agents. For example, synergetic interaction resulting from simultaneous application of hyperthermia and various physical or chemical agents is registered only within a range of temperature. Inside this range, there is an optimal temperature for the greatest synergy. The result of the combined action depends on the intensity of the applied agents. A decrease in one factor applied should be accompanied with a decline in the other to retain the maximum synergistic value. It is not excluded that similar rules may be realized for antagonistic interaction of agents.

The radiosensitivity of Chinese hamster cells has been seen to depend on the sodium-chloride concentration (Raaphorst, Kruuv, 1976, 1977). Similar results with an extremum displacement depending on concentration have been reported for KCl, $CaCl_2$, NH_4OH, sucrose, and DMSO (Raaphorst et al., 1977; Maurice, Ferraris, 2008). Other physical factors produce a similar effect. Hypotonic solutions of NaCl, KCl, and glycerol augmented sensitivity of Chinese hamster cells V-79 to ultraviolet (UV) radiation (Romano et al., 1979) and hyperthermia (Lin et al., 1981) whereas hypertonic solutions decreased this effect. Impact of these solutions on the cell survival was verified for various microorganisms. A radioprotective property of 8.4% glycerol was demonstrated in *Escherichia coli* after UV, X-ray (Peak, Peak, 1980, 1984), and γ-irradiation (Bonura, Smith, 1976). Osmolytes have been seen to exert a radioprotective effect *in vivo* (Babicová et al., 2013). The range of variation in the glycerol concentration in these investigations was low. Radioprotective agents display a maximum efficacy only at optimal concentrations (Raaphorst, Kruuv, 1976, 1977; Romano et al., 1979; Rozhdestvensky, 1985; Bishayee et al., 2000; Ormsby et al., 2014). The efficiency of radioprotective action of anoxia also depends on the ionizing-radiation dose rate (Bychkovskaya, Ochinskaya, 1964; Ling et al., 1985). These and other fragmentary data show that manifestation of antagonistic interactions complies with the rules similar to those for synergetic interactions. However, the protective action of osmolytes (agents involved in the water transport and maintenance of water balance in cells) from ionizing radiation, hyperthermia, microwaves, and ultrasound has not been examined on *Escherichia coli* cells, which differ in their ability to repair radiation damage. The goal of this section is to fill the gap.

Bacterial cells *Escherichia coli* B/r, which are resistant to ionizing and UV radiation, and another strain B_{s-1} ($hcr^{-}exr^{-}fil^{+}$), which is hypersensitive to these agents due to impaired DNA repair, were used in experiments. Cells were subjected to different types of physical and chemical factors at the stationary growth phase. Their concentration was 10^7 cells/mL They were incubated in a nutrient broth with 1% peptone, 1% yeast extract (Chimmed, Russia), and 0.85% NaCl (Labtech, Russia) at 37°C for 18 h. Cells were irradiated with γ-ray-emitting ^{60}Co in doses of 90 and 450 Gy and at a dose rate of 20 Gy/min using "Isseledovatel" equipment. The ultrasound source was USDN equipment with a frequency of 44 kHz and intensity of 100 W/cm^2. A bacterial suspension (5 mL) was placed in the cavity determined by the equipment design. The source of electromagnetic radiation was microwave generator with a frequency of 7 GHz and continuous wave operation. Cells were irradiated in the region of the formed wave. The energy-flux density measured with PS-9 equipment was 100 mW/cm^2, and the specific absorption rate was 200 W/kg (Petin et al., 2000). Microwave heating is accompanied with gradual temperature increase to a certain stationary level. In our experiments, cell suspension was placed into tubes preliminarily heated to the required temperature (50 or 60°C) and immediately exposed to microwave radiation. Bacterial cell suspension in glass tubes with diameters of 10–12 mm was heated at various temperatures in a UT-15U4.2 water ultrathermostat. The temperature was registered with a TPE-1M electron thermometer (Russia). Sodium chloride and glycerol (Reakhim, Russia) were applied in doses of 0–20 and 0–60%, respectively. Each experiment was repeated from three to five times. Cell viability is presented as a mean ±SD.

The optimal ratio of intracellular and extracellular osmotic pressure is a key factor for structural and functional status of cytoplasmic membranes and cell exchange with the environment. This ratio was maintained in experiments by chemical solutions with a certain osmotic pressure. Figure 6.10 shows the effect of NaCl concentration in water solutions on survival of *E. coli* B/r exposed to irradiation with γ-ray-emitting ^{60}Co, hyperthermia, microwaves, and ultrasound.

Similar results were obtained with *E. coli* B_{s-1} strain (Figure 6.11). In these and other experiments, during exposure to these factors, the cell suspension contained osmolytes. A protective effect of these agents was evident only in case of simultaneous combined action of osmolytes and physical factors. Cell transfer in solutions of applied agents after heating resulted in cell death. Sodium chloride and glycerol in the doses applied did not affect the bacteria viability.

It is seen that the dependence of cell survival on NaCl concentration also have a bell-shaped appearance independent on the agents, i.e., cells have died at both high and low osmolyte concentrations. Hypertonic media in a certain concentration range protect both strain cells from damage produced by irradiation with ^{60}Co γ-rays, hyperthermia, microwave, and ultrasound. In all cases, there was optimal concentration of NaCl providing its maximum antagonistic interaction with physical agents. Uniform cellular response on the combined action of sodium chloride and exposure to different in nature physical factors means that there is a common mechanism for NaCl's protective effect. It seems hardly probable that impaired DNA repair may be a cell target responsible for cell sensitivity to various physical factors. DNA damage is not a key target for cell response after exposure to hyperthermia, microwave and especially ultrasound. The data shown in Figures 6.10 and 6.11 support this idea. It is seen that cell survival of thermosensitive *E. coli* B_{s-1}, with impaired DNA repair, in the dependence of NaCl concentration also have a bell-shaped appearance. This means that a maximum protective effect is reached at the optimal concentration independent on the nature of applied factors. Supposedly, the DNA repair system plays insignificant role in mechanisms of NaCl protective action.

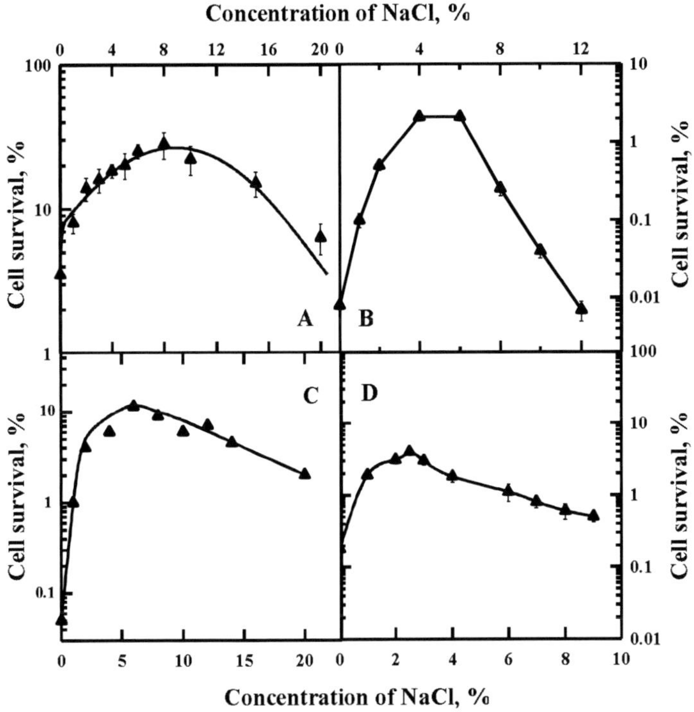

Figure 6.10. The effect of NaCl concentration in water solutions on viability of the bacteria strain *E. coli* B/r exposed to various physical factors. (A) γ-ray-emitting ^{60}Co, 450 Gy; (B) hyperthermia, 52°C, 15 min; (C) microwaves, 60°C 60 s; and (D) ultrasound, 44 kHz 3 min.

As was mentioned above, various agents can damage membrane permeability and cause a loss of intracellular compounds (Konev et al., 1970; Konev, Rudenok, 1973) due to increased intracellular osmotic pressure produced by radiation catabolism (Kuzin, 1973), ultrasound depolymerization (Elpiner, 1973), or thermogenic dissociation of weak cellular electrolytes (Voyutckij, 1960; Lehninger, 1972). The protection of sodium chloride from physical factors can be explained as being the result of the effect of osmolyte on cell properties. Media with increased osmotic pressure (to a certain level) may compensate osmotic pressure inside cells and, thereby, reduce the cytotoxic effect of ionizing radiation, hyperthermia, microwaves, and ultrasound. The other hand, media with reduced osmotic pressure facilitate the damaging effect of these factors. It was demonstrated in our experiments that hypertonic media could be protected only during interaction, but not after its termination. It was of interest to verify the dependence of antagonistic interaction of factors from their intensity as it was observed in the case of synergism (Petin, Kim, 2016). In the first approximation, treatment duration can serve as an analog of the dose for hyperthermia and chemical agents, whereas changes in temperature or chemical-agent concentration indicate their modified intensity. Really, an increased radiation dose enhances the amount of radiation damage, whereas the dose rate indicates its intensity, which affects the rate of damage formation. In the case of hyperthermia or chemical toxicants, the duration of treatment increases the amount of damage ("dose"), whereas changes in temperature or concentration of chemicals modifies the rate of damage formation ("dose rate").

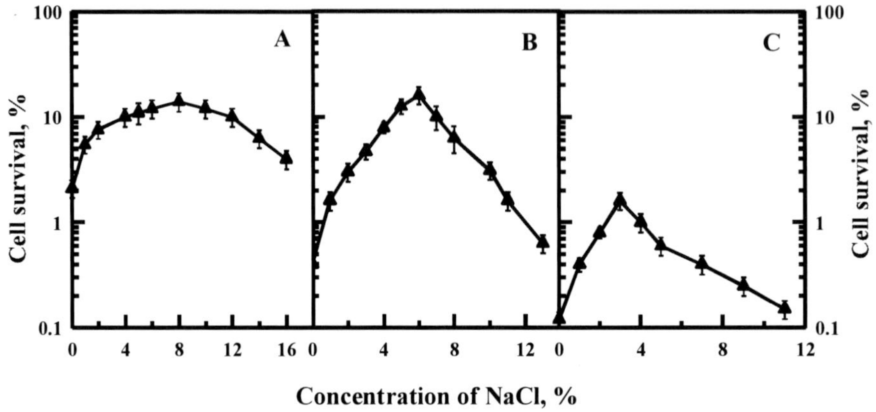

Figure 6.11. The effect of NaCl water solutions on viability of the bacteria cells *E. coli* B_{s-1} exposed to various physical factors: A – γ-rays of ^{60}Co, 90 Gy; B – hyperthermia, 52°C, 15 min; and C – ultrasound, 44 kHz, 2 min.

It was found (Figures 6.10 and 6.11) that the dependence of cell survival on NaCl concentration have bell-like pattern: cell survival first increases, then reaches a maximum, after which cell survival decreases with an increase in NaCl concentration, for all factors used. It might be expected that increased or decreased osmolyte concentration should be accompanied with altered temperature to retain their maximum protective effect.

The data presented in Figure 6.12 show the dependence of cell survival after the simultaneous action of temperature and osmolytes (NaCl or glycerol) on the intensity of these factors. In fact, these data reflect the patterns of manifestation of the antagonistic interaction of these factors with a change in the intensity of the agents used.

It was already noted above that an increase in cell permeability and death during heating can be largely due to, along with other damage to cells, destabilization of their osmotic homeostasis. A sharp temperature drop in the initial heating period can disrupt the adaptive capabilities of cells, leading them to an unstable state. As already mentioned, this violation is based on an imbalance in osmotic homeostasis of cells due to differences in the constants of electrolytic dissociation of molecules and intermolecular associates (components) of the contents of cells and the extracellular environment. The increased intracellular osmotic pressure because of heating, without meeting sufficient resistance from the suspension fluid, can deform the structural and/or functional status of cytoplasmic membranes, weakened by the direct effect of hyperthermia on their components.

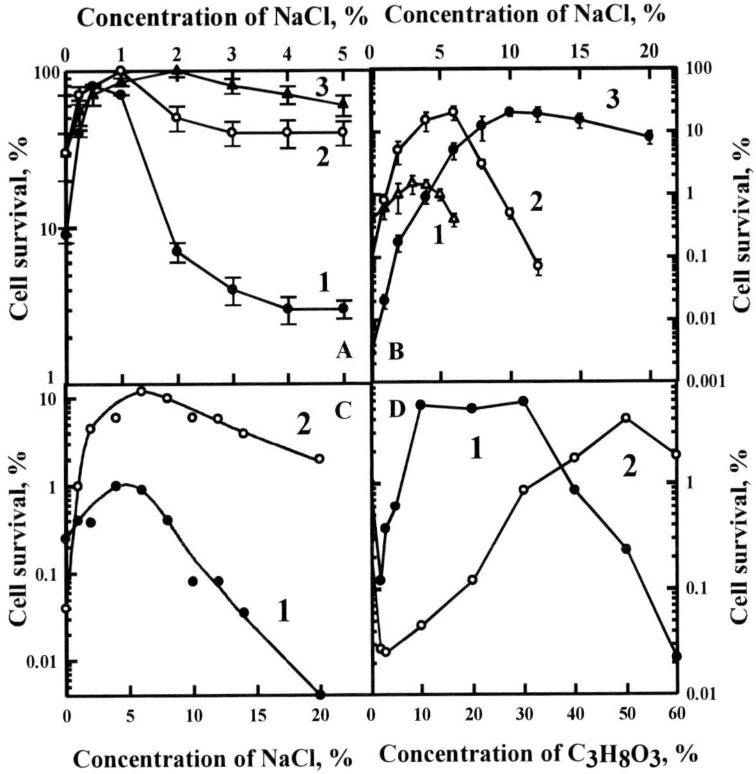

Figure 6.12. The viability of *E. coli* B/r bacteria exposed to low and high temperatures with different concentrations of NaCl and glycerol in the media. Panel A – NaCl + temperature 0 (curve 1), 20 (curve 2), and 37°C (curve 3), duration 1 h. Panel B – NaCl + temperature 50°C, 3 h (curve 1), 52°C, 15 min (curve 2) and 60°C, 90 s (curve 3). Panel C – NaCl + microwave at 50°C, 3 h (curve 1), 60°C, 60 s (curve 2). Panel D – glycerol + microwave at 50°C, 3 h (curve 1), 60°C, 60 s (curve 2).

In turn, the direct damaging effect of elevated temperature on the structure of membranes will increase under conditions of stress on their surface. In this case, there may be a violation of the structure and spatial-temporal coordination of the functions of membrane-bound enzymes, loss of biologically important compounds to the environment due to excessive pressure and increased membrane permeability and a decrease in cell viability.

It was already mentioned above that an increase in cell permeability and death during heating could be largely due to, along with other damage, cells destabilization because of their osmotic homeostasis. A sharp temperature drop in the initial heating period can disrupt the adaptive capabilities of cells, leading them to an unstable state. This violation is based on an imbalance in cell osmotic homeostasis due to differences in the constants of electrolytic dissociation of molecules and intermolecular associates of the contents of cells and the extracellular environment. The increased intracellular osmotic pressure, as a result of heating, without meeting sufficient resistance from the suspension fluid, can deform the structural and/or functional status of cytoplasmic membranes, weakened by the direct effect of hyperthermia on their components. In turn, the direct damaging effect of elevated temperature on the structure of membranes will increase under conditions of stress on their surface. In this case, there may be a violation of the structure and spatial-temporal coordination of the functions of membrane-bound enzymes, loss of biologically important compounds to the external environment due to excessive pressure and increased membrane permeability and, as a result, a decrease in cell viability.

If this assumption is correct, then suspension media with increased osmotic pressure should modify (increase or decrease) the magnitude of the bactericidal effects of heating at different operating temperature. To assess the effect of the tonic factor on the thermal stability of cells heated at different heating temperature, we studied the effect of a hypertonic (0.01 mol/L phosphate buffer pH 7.0, containing 8% NaCl) solution on the value of the bactericidal effect of heating under different heat exposure. The results are shown in Figure 6.12. Figures 6.12 B, C exhibit viability of *E coli* B/r bacteria exposed to low and high temperatures with different concentrations of NaCl (panels A, B, C), as well as glycerol (panel D) in the suspension media. The following temperatures were applied: panel A – 0 (curve 1), 20 (curve 2), and 37°C (curve 3) with duration 1 h; panel B – 50°C, 3 h (curve 1), 52°C, 15 min (curve 2) and 60°C, 90 s (curve 3). The results shown in Figures 6.12 C and 6.12 D were obtained under the action of microwaves (7 GHz) simultaneous with osmolytes and the following temperatures: panel C – 50°C, 3 h (curve 1), 60°C, 60 s (curve 2); panel D – 50°C, 3 h (curve 1), 60°C, 60 s (curve 2).

Let us highlight the main features of the manifestation of the protective effect of osmolytes applied with various concentration simultaneously with different temperatures. First of all, in all cases there was a pronounced effect of protecting cells from thermal damage. It is important that the maximum protective effect was manifested only at hypertensive concentrations. This fact indicates that the thermoprotective action of various salts, carbohydrates, glycerol might be based on a common mechanism associated with the system for maintaining cellular osmotic homeostasis.

It may also be noted that the efficiency of cell protection using NaCl and glycerol solutions against thermal damage induced at different heating temperatures was always bell-liked form. In other words, the existence of optimal concentrations of osmolytes, at which their protective effect is maximal, is existed. This pattern of antagonistic interactions of osmolytes with hyperthermia resembles a similar pattern of synergistic interaction of hyperthermia with various physical and chemical agents described in the first section of chapter 5. It recalled the dependence of synergistic interaction effectiveness on the intensity of the applied physical and chemical factors – the lower the intensity of one of the active agents applied, the lower intensity of another agent should be used to preserve the greatest synergistic effect and *vice versa*. A similar pattern is observed for antagonistic interactions. A decrease in the operating temperature from 60 to 52 and 50°C requires a decrease in the NaCl concentration from 10 to 5 and 3% to ensure to rich the maximum effectiveness of the protective action (Figure 6.12 B). A similar pattern is noted in Figure 6.12 C – a decrease in the operating temperature from 60 to 50°C requires a decrease in the NaCl concentration from 6 to 4% to ensure maximum antagonistic interaction. A very clear pattern of this peculiarity can be traced for the simultaneous action of microwaves and glycerin – at a decrease in the operating temperature from 60 to 50°C in order to maintain the maximum protective effect of glycerin, it was required to reduce its concentration from 50 to 10–30%.

We emphasize that the data in Figure 6.12 demonstrate a very important regularity – the maximum efficiency of the protective action of sodium chloride or glycerol shifts to the region of high concentrations of these compounds with more intensive heating (higher temperatures). These data indicate a more pronounced disruption of the balance of osmotic pressures in the cell-environment system during traditional or microwave heating, because of which the osmotic pressure of cells increases sharply in comparison with a lower heating intensity. Remember that a very similar correlation between the intensity of physical environmental factors and operating temperature was demonstrated in this chapter for synergistic interactions. Similar results were obtained for the thermosensitive strain of bacteria *E. coli* B_{s-1}. The survival rate of the cells of this strain after simultaneous exposure to temperatures of 0 and 37°C for 1 h in the presence of various concentrations of NaCl is presented in Table 6.3.

Table 6.3. Viability of *E. coli* B_{s-1} heated at 0 and 37°C for 1 h in the presence of various concentrations of NaCl

NaCl concentration, %	Cell survival, %	
	0°C	37°C
0	20,0 ± 1,8	17.3 ± 2.0
0.25	42.2 ± 3.0	52.5 ± 4.1
0.5	80.0 ± 5.0	73.1 ± 5.5
1.0	77.5 ± 4.1	88.6 ± 4.0
2.0	11.7 ± 1.9	98,9 ± 4.8
3.0	6.0 ± 1.7	87.1 ± 5.8
4.0	3.8 ± 0.8	72.9 ± 4.9
5.0	3.4 ± 0.6	66.0 ± 3.8

It is clear that in this case, a bell-shaped dependence of cell survival on the NaCl concentration and operating temperature is also recorded. For a temperature of 0°C, the maximum protective effect (antagonistic interaction of temperature and NaCl) is observed at 0.5% NaCl concentration. Any change in this concentration upward or downward reduced the viability of the cells. For an operating temperature of 37°C, the maximum protective effect of NaCl is observed at 2% NaCl concentration. Again, any deviation of this concentration from the optimal value led to a decrease in the effectiveness of its protective action. Another regularity common for synergism and antagonism is also fulfilled – the lower the intensity of one of the agents, the lower the intensity of another factor should be applied, to preserve the greatest protective effect. Indeed, the maximum protective effect of NaCl at 37°C is observed at its 2% concentration, and a decrease in the operating temperature to 0°C required a decrease in the concentration of NaCl to 0.5% to maintain the maximum protective effect.

The results presented in this section show that antagonistic interaction of osmolytes with hyperthermia is determined by both concentration and temperature. With increased temperature, the maximum protective effect of osmolytes is shifted toward their higher concentrations. As was shown above, analogous observations have been reported for synergetic interactions after combined treatment of various physical factors and chemical agents with hyperthermia. It has been demonstrated that the synergetic effect depends on the intensity of applied agents and that the maximum synergetic interaction with ionizing radiation, UV, microwave, ultrasound, and chemical agents is realized at optimal temperature. These common rules for synergetic and antagonistic interactions of environmental factors do not mean identical mechanisms underlying their manifestation. The mechanisms of synergetic interaction have been discussed earlier in the first section of this chapter related with the formation of additional effective damage due to the interaction of some sub-damages ineffective after their separate application. The nature of such sub-damage can be different in the interaction of agents of different nature. Antagonistic interactions can be explained as being due to cell osmotic properties that depend on temperature and concentration of osmolytes in the cell suspension. It has long been known that a temperature increases or decreases in a cell suspension results in a corresponding increase or decrease of intracellular osmotic pressure more than in the liquid surrounding the cells (Voyutckij, 1960). This is due to the fact that changes in the cell osmotic pressure with altered temperature in the cell suspension are produced by intramolecular and intermolecular association/dissociation, which is controlled by the temperature. Reduced temperature facilitates association processes. Increased temperature intensifies dissociation processes, due to the thermal motion of molecules and disturbed balance in osmotic pressure inside and outside of cells. High temperature enhances intracellular pressure, to compensate which increased concentrations of osmolytes are required. Our results show that membrane damage and misbalance of intra- and extracellular pressure, along with DNA damage, are possible mechanisms of cell injury produced by various environmental factors.

The totality of the data obtained and their comparative analysis with the literature data make it possible to formulate new point of view on the mechanism of thermal death of cells and tissues. According to this point, the biological effect of hyperthermia, including microwave hyperthermia, along with other disorders of the structures or functions of cells, is based on thermal destabilization of the membranes and osmotic homeostasis.

6.5. Peculiarities of Biological Effects of Electromagnetic Non-Ionizing Radiation at Cellular and Molecular Levels

There is one fundamental unresolved problem in the radiobiology of non-ionizing radiation – whether the biological effects induced by non-ionizing radiation are due solely to thermal damage, or some non-thermal damage also occurs. This problem exists for both unicellular and multicellular organisms. It is now believed that such a simple classification is not general and sufficient. Evaluation of published data is complicated by the variety of irradiation equipment; exposure conditions, biological objects used, and recorded effects. All researchers agree that heat plays a critical role in most of the observed effects. Temperature must be carefully controlled when conducting experiments related to biological effects induced by non-ionizing radiation.

It is known that non-ionizing electromagnetic radiation (EMR) of a certain range is characterized by noticeable physiological activity even at low, subthermal exposure intensities (Biological Effects, 1986; Belyaev et al., 1993; Lokhmatova, 1994; Grigoriev, Grigoriev, 2016), while EMR with other parameters does not cause pronounced deviations in the state of various biological objects until the appearance of ordinary thermal effects (Adair, 1983). Since the influence of EMR in some cases is determined by thermal and in other cases by a specific action, the question arises about their fundamental specific contribution to the total damaging effect of EMR, especially since in some cases the results of the thermogenic influence of this type of radiation is not possible to interpret within the framework of only thermal the concept of defeat (Pakhomov et al., 1993). In addition, microwaves, inducing in the cell substance the oscillatory processes of molecular dipole ensembles ordered in time and space, can more intensively than with ordinary heating, destabilize the state of the cytoplasmic membranes of cells, and, consequently, the system of osmotic homeostasis, which can manifest itself in the form of specific nonthermal and thermal effects of this type of radiation.

6.5.1. Thermal Effects

Cultures of various unicellular microorganisms have often been used to distinguish between conventional heat and microwave-induced effects. There is definitive evidence for the thermal effects of intense microwave radiation. A large number of publications indicate the absence of biological effects until microwaves cause a noticeable increase in the temperature of the irradiated cell suspension (Biological Effects, 1986). A huge number of articles show that cell death occurs only when lethal temperatures are reached (Sapareto et al., 1982; Millar et al., 1984 and many others). The results of these studies mean that the response of cells to microwave exposure is observed only with the induction of heat, and no biological effects are possible without an increase in temperature. Based on these data, it was concluded that, in a wide range of frequencies and intensities, any biological effects of EMF at the cellular and molecular levels are associated with an increase in temperature. By the way, the assumption that the only dangerous effect of microwave radiation is dielectric heating of tissue was the basis for the adoption of the permissible level of exposure to EMF on humans in the United States and many other countries (Klauenberg et al., 1995).

6.5.2. Specific Non-Thermal Effects

Although it is generally accepted that the main effects of microwaves are thermal in nature, the possibility of specific non-thermal EMF effects cannot be ruled out. Several authors have postulated that factors other than heat may be responsible for the effects of microwave irradiation observed at the cellular level or in unicellular organisms. Heller and Teixeira-Pinto (1959) observed chromosomal aberrations in growing onion roots exposed to 27 MHz EMF. They have eliminated heating effects due to the short pulse. Chen et al. (1974) irradiated mammalian cells *in vitro* with a 2.45 GHz field at intensities from 20 to 85 mW/cm^2 for a duration of 4 to 20 minutes. Chromosomal aberrations were induced by microwaves, but not by conventional thermal heating. Moreover, some data indicated that EMFs can induce some specific chromosomal aberrations (Baranski, 1971; Harte, 1975; George, 1978). Mutagenic effects of pulse-modulated fields (8.6–9.0 GHz) were observed in the work of Dutta et al. (1978), some effects at 10 mW/cm^2 were explained by a non-thermal mechanism. Many efforts have been directed towards demonstrating the existence of carcinogenic and mutagenic effects, as well as chromosome damage at low levels of microwave irradiation (Manikowska et al., 1979; Yao, 1982; Balcer-Kuiczek, Harrison, 1985; Spitz, Johnson, 1985). Blackman et al. (1975) performed carefully controlled experiments with a suspension of *E. coli* that was irradiated at 1.70 GHz in the near field or at 2.45 GHz in the far field. The only effect was an increase in the rate of cell growth in the irradiated cells as compared to the control. The temperature was controlled with an accuracy of $\pm$ 0.5°C. Szmigielski et al. (1975) found reversible functional changes in embryonic cells irradiated with 3 GHz EMF for 30 minutes in the far zone, when no temperature increase was observed. Chen and Lin (1978) irradiated Chinese hamster V-79 cells at 2450 MHz at 500 mW/cm^2 (SAR = 1059 W/kg). These authors reported a decrease in the growth rate and some morphological changes that were not due to temperature. Resonant-like non-thermal effects of millimeter waves have been obtained in yeast cells (Grundler, 1985). Resonant microwave absorption of DNA molecules was also observed (Swicord, Davies, 1982; Sagripanti, Swicord, 1986). Irradiation of purified plasmid

DNA with microwaves at non-thermal levels at the frequency range from 2.00 to 8.70 GHz led to the formation of single- and double-stranded DNA breaks (Sagripanti et al., 1987). The nonthermal effects of microwave radiation have also been reported in recent publications (Belyaev et al., 1993; Pakhomov et al., 1993, and others).

Despite numerous studies aimed at proving the existence of nonthermal effects, a critical analysis of many publications related to this problem raises doubts about the credibility of these data (Biological Effects, 1986; Klauenberg et al., 1995). Thus, it can be concluded that (i) there is no generally accepted evidence of a specific biological effect induced by EMF at the cellular and molecular level; (ii) the theory of nonthermal action of microwave radiation has not yet been developed; (iii) further experimental and theoretical research should be continued.

6.5.3. Specific Thermal Effects

The increasing practical use in the clinical therapy of tumors of local hyperthermia generated by microwaves raises the question of whether the same biological effect should be expected when traditional heat is replaced by microwaves. It is possible that the thermal effects of microwaves may differ from the effects of classical hyperthermic exposure. An aqueous suspension of algae (*Chlamidomonas segnis*) was irradiated with 2450 MHz EMF at the logarithmic stage of growth (Hamid, Badour, 1973). It was demonstrated that the stimulation of photosynthesis and growth was greater than when using traditional heat. The interdependence of heat and microwave effects was shown when studying the inactivation of *Bacillus subtilis* (Wayland et al., 1977) and bacteriophage T5 were more sensitive to 1.07 GHz EMF inactivation than heat alone (Moody et al., 1979). It was concluded that exposure to 2.45 MHz EMF at an energy flux density of 510 mW/cm^2 led to a significant increase in the mutation rate compared to the effects of traditional heat and that microwave irradiation disrupted the enzymatic activity of *Staphylococcus aureus*, which could not be explained by thermal effects alone (Biological Effects, 1986). Consequently, the results of the cited works show the existence of some biological responses to microwave irradiation that exceed the effect of traditional heating.

Thus, cellular studies related to biological effects produced by microwave radiation can be subdivided into the following groups: (i) demonstration of thermal biological effects induced by high intensity EMF; (ii) illustration of the specific effects of microwave radiation at low intensity levels; and (iii) evidence of specific thermal effects of microwave radiation that is superior to classical heating. Our own research below addresses this latter issue.

To distinguish the specific and thermal components of microwave exposure, a device has been developed that provides a strict identity of the heating kinetics in the experimental (irradiated with micronutrients) and control samples (heated from the experimental sample) of the cell suspension. The essence of this device consists in the coaxial connection of the vessels for the control and experimental samples of the cell suspension. In this case, the inner vessel intended for the control sample is made of radio-opaque and heat-conducting material. The experimental vessel (external) is made of radio-transparent material. The design features of the device allow for equivalent heating of the control sample by transferring heat from the experimental one. As found, in the overwhelming majority of cases, the differences in the temperature of the experimental and control samples did not exceed 0.1°C. Checking by the χ^2 criterion showed the insignificance of these differences. Consequently, the thermal "loads" in the control and in the experiment were identical. At the same time, it was found that in the initial period of exposure, cell death in the control and experimental samples was absent due to the low temperature. With an increase in the duration of irradiation, a synchronous increase in temperature was observed in both samples and cell death was noted. However, in all cases, cell damage in the experimental samples was more intense than in the control.

Prokaryotic (*Escherichia coli* B/r) and eukaryotic (*Saccharomyces cerevisiae*, diploid strain XS800) strains capable of recovery in the stationary growth stage were used in this study. The same cell suspensions of bacterial (10^8 cells/ml) and yeast (10^6 cells/ml) cells were placed in experimental and control vessels to ensure identical heating rates for microwave and heat exposure. After irradiation, the colony-forming ability of cells was assessed by seeding cells on nutrient agar and counting the formed macrocolonies after 24 h of incubation of bacterial cells at 37°C and after 3-5 days of incubation of yeast cells at 30°C.

The experimental results are shown in Figures 6.13 (bacteria) and 6.14 (yeast cells). These figures show the relationship between cell survival in control (conventional heating) and experimental (microwave heating) samples. The dotted line indicates the case if identical effects for microwave and conventional heating were recorded. The presented data mean that in all cases the cell damage in the experimental samples was more intense than in the control ones although the temperature in them has always been the same. These data indicate the existence of a specific thermal effect of microwaves. It is important to emphasize once more that irradiation always took place under conditions of identical heating kinetics of microorganism suspensions. It is seen that the damaging effect of microwave heat is greater than the usual thermal heating for both pro- and eukaryotic cells. In other words, the cells were damaged by microwaves more efficiently than traditional heat. Therefore, it can be concluded that there is a specific thermal effect of EMF, which differs from the usual thermal effect.

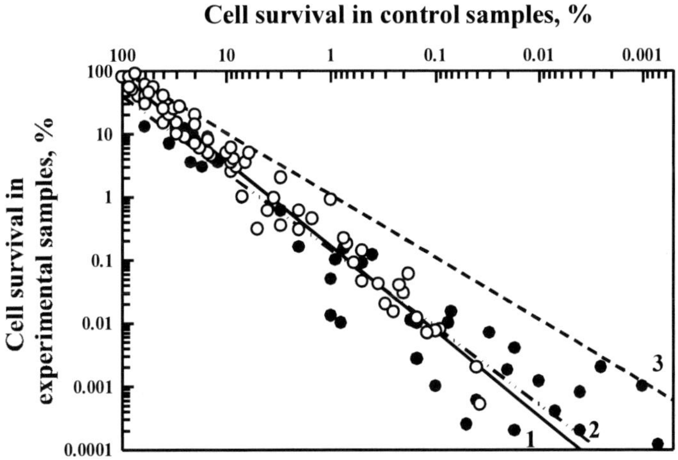

Figure 6.13. Relationship between the survival of *E. coli* B/r (open symbols, curve 1) and *E. coli* B_{s-1} (closed symbols, curve 2) under microwave (test samples) and heat (control samples) exposure under conditions of identical heating kinetics of microorganism suspensions, curve 3 - the expected relationship with the equal effectiveness of the effects of microwave radiation and convection heat.

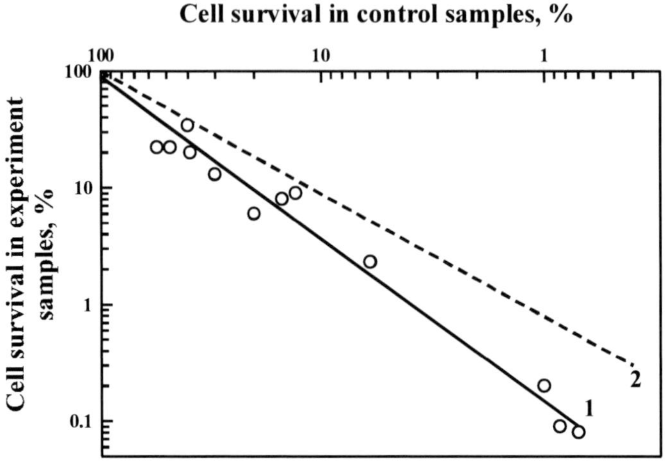

Figure 6.14. The relationship between the survival of the yeast *S. cerevisiae* under microwave (test samples) and heat (control samples) exposure under conditions of identical heating kinetics of microorganism suspensions (curve 1), curve 2 reflects the expected line after the equal effect in both samples.

In addition to the data on cell death, results were also obtained indicating the different effects of microwaves and convection heating on the cell permeability of pro- and eukaryotes. Figure 6.15 exhibits the relationship between the optical density of supernatants of cell samples of bacterial cells under microwave (test samples) and heat (control samples) exposure under conditions of identical heating kinetics of microorganism suspensions. As can be seen, heating by microwaves caused a more intense violation of cell permeability than the equivalent usual thermal heating. Hence, we may conclude again that the cells were damaged by microwaves more efficiently than traditional heat.

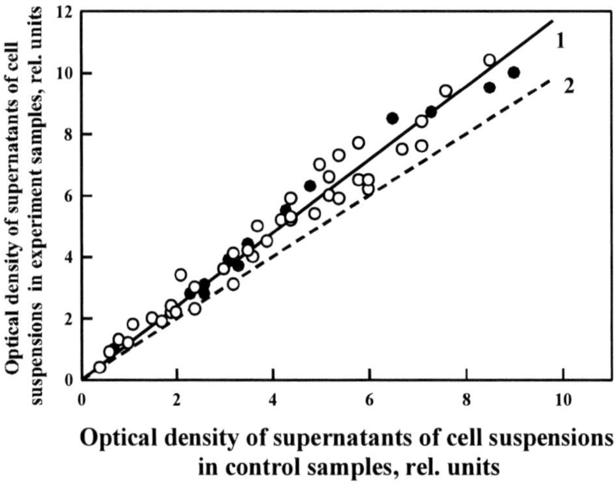

Figure 6.15. The relationship between the optical density of supernatants of cell samples of bacteria *E.coli* B/r (closed symbols) and *E. coli* B_{s-1} (open symbols) under microwave (test samples) and heat (control samples) exposure under conditions of identical heating kinetics of microorganism suspensions (curve 1), curve 2 is the expected curve for the equal efficiency of the effects of microwave radiation and convection heat.

Therefore, it can be concluded again that there is a specific thermal effect of EMF, which differs from usual heating. Similar data indicating an increased susceptibility to the functions of permeability of cell membranes in the case of equivalent microwave heating compared to convection heating were obtained on Ehrlich ascites carcinoma cells (Table 6.4).

Consequently, microwave heating caused additional damage to the functions of pro- and eukaryotic cell membranes compared to equivalent heat heating. This specificity of action is associated with thermal processes and can be manifested mainly only under heating conditions, since cooling of the cell suspension, as a rule, led to the disappearance of lethal and other damaging effects. At the same time, EMF of ascertain range, even at low (non-thermal) exposure intensities, have a pronounced physiological activity (Chiabrera et al., 1985; Belyaev et al., 1993; Lokhmatova, 1994; Grigoriev, Grigoriev, 2016). These data suggest that EMR of various ranges can induce physiological consequences in organisms, determined in some cases by the thermal, and in others – by the specific (informational) components of the absorbed dose.

Table 6.4. Influence of the duration of microwave heating (experiment) and thermal heating (control) in the temperature range 20–52°C on the intensity of the release of optically active material at 265 nm from Ehrlich ascites carcinoma cells into the environment

Duration of exposure, min	Optical density, relative. units	
	Control	Experiment
10	0.25	0.23
16	0.29	0.37
21	0.30	0.42
26	0.36	0.55

The peculiarities of the biological action of EMR of thermal intensity exhibit a dual nature of the damage they induce. At the same time, the specific contribution of thermal and specific components to the overall effect of damage is determined, on the one hand, by the radiation parameters, and on the other, by the individualities of the organization of living systems. The specific nature of EMR damage is most often observed in integral multicellular organisms (Akoev et al., 1986; Tigranyan, Shorokhov, 1991; Grigoriev, Grigoriev, 2016), or on certain cellular objects: cells of protozoa, nervous system, etc. (Biological Effects. 1986).

The nature of these effects may consist, first, in a more intense damage to cell membranes under the action of microwave-induced vibrational ensembles of molecules (dipoles) in comparison with chaotic vibrational processes of molecules heated by ordinary heat. Secondly, along with damage to membranes in a microwave field, thermal destabilization of the osmotic homeostasis system can also make a specific contribution to the specificity of microwave effects due to more intense microwave dissociation of weak cell polyelectrolytes, as evidenced by the data of studies on the increased cellular permeability of experimental samples of cell suspensions.

Figure 6.16 shows survival curves of *E. coli* B/r bacteria heated with equal thermal doses of convection heat or microwaves in isotonic NaCl solution and incubated in the post-radiation period in isotonic or hypertonic (10%) NaCl solutions. Curve 1 – thermal heating, curve 2 – microwave heating; curves 3, 4 – incubation in isotonic NaCl solution of bacteria heated once with heat (curve 3) or microwaves (curve 4); curves 5, 6 – incubation in a hypertonic NaCl solution of bacteria heated once with heat (curve 5) or microwaves (curve 6). It can be seen that the incubation of bacteria in media with different osmotic pressures led to the fact that cell survival either increased (under isotonic conditions) or decreased (under conditions of hypertension), regardless of the method of preliminary heating of microorganisms.

Figure 6.17 shows the survival curves of *E. coli* B/r bacteria exposed in a single isoeffective dose to hyperthermia induced by microwaves (curve 2) or convection heat (curve 1) in water with different contents of sodium chloride. It can be seen that the presence of this compound in a suspension medium in a certain concentration range increased the thermal stability of cells during both microwave and thermal heating.

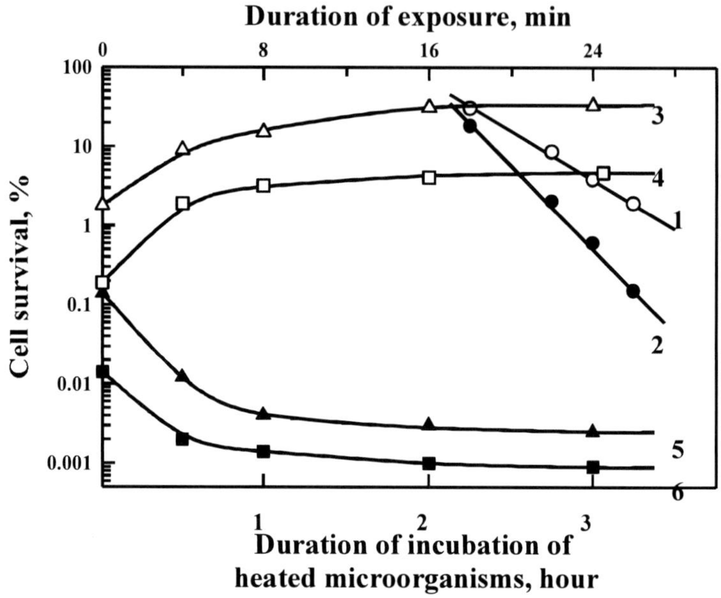

Figure 6.16. Survival curves of *E.coli* B/r bacteria heated with equal thermal doses of convection heat or microwaves in isotonic NaCl solution and incubated in the post-radiation period in isotonic or hypertonic (10%) NaCl solutions. Curve 1 – thermal heating, curve 2 – microwave heating; curves 3, 4 – incubation in isotonic NaCl solution of bacteria heated once with heat (curve 3) or microwaves (curve 4); curves 5, 6 – incubation in a hypertonic NaCl solution of bacteria heated once with heat (curve 5) or microwaves (curve 6).

It can be seen that the efficiency of thermal protection in the case of heating by microwaves was noticeably lower than in the case of heating with traditional heat (Figure 6.17). The difference in the volume of salt thermal protection of cells from damage induced by both agents is highly significant ($p<0.001$). Similar differences, less pronounced, were also obtained in a comparative study of the effect of aqueous solutions of common salt on the resistance of the *E. coli* B_{s-1} thermosensitive mutant to hyperthermia induced in a single dose by microwave irradiation or thermal heating (Table 6.5). In this case, salt also had the property to protect cells from thermal damage during heating by these agents. Note again that there is an optimum concentration of NaCl at which the protective effect is maximized. This pattern means that both synergistic and antagonistic interactions are characterized by the presence of the maximum effect at a certain ratio of the acting agents. Thermal damage caused by the influence of microwave and thermal heating was noticeably less pronounced. In this case, there was only a slight tendency towards a decrease in the efficiency of protecting cells of the *E. coli* B_{s-1} strain from thermal damage by microwaves compared to damage induced by thermal heating. The use of the difference method of processing the results (Armitage, 1971) showed that these differences were significant ($p<0.05$). It is apparent that there is an optimum concentration of NaCl, at which the protective effect is maximized. This pattern indicates that both synergistic and antagonistic interactions are characterized by the availability of the greatest effect at a certain ratio of the interacting agents.

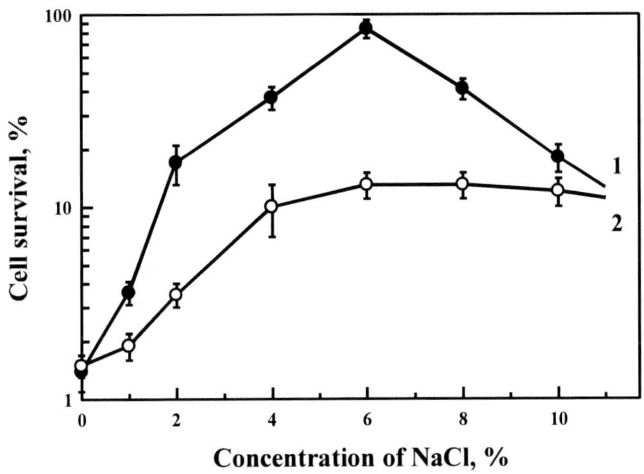

Figure 6.17. The influence of NaCl aqueous solutions of various concentrations on the survival of *E. coli* B/r bacteria during heating in a single isoeffective dose (from 20 to 54°C) with heat (curve 1) and microwaves (curve 2).

Table 6.5. The dependence of the survival of *E. coli* B_{S-1} bacteria on the concentration of sodium chloride in a suspension medium during exposed to a single dose of convection and microwaves heating

Concentration of sodium chloride in the environment of exposure, %	Survival, %	
	Convection heating	Microwave heating
0	2.2 ± 0.2	2.3 ± 0.3
1	5.7 ± 0.6	2.3 ± 0.3
2	7.9 ± 1.1	6.6 ± 0.7
4	17.0 ± 2.1	11.8 ± 1.3
6	27.9 ± 2.9	21.5 ± 1.0
8	24.1 ± 1.8	20.8 ± 1.5
10	16.8 ± 1.6	11.6 ± 0.9

Consequently, the volume of thermal damage modifications in cells exposed to convection and microwave heating, determined by the tests of recovery and additional damage, was approximately equal, while the ability of cells to thermal protection was less pronounced in the case of heating by microwaves. This indicates that thermal damage induced in bacterial cells by microwave heating is quantitatively or qualitatively different from similar damage caused by convection heating. It can be assumed that this specificity of action is due to the peculiarities of membrane damage under microwave irradiation, since the differences in the effectiveness of the protective effect of the suspension medium were associated with the osmotic pressure of this medium. This assumption is supported by the above results, which demonstrate a more intense damage to cell permeability under the action of microwaves compared to equivalent convection heating.

A number of reports have been published in the literature indicating a more intense damaging effect of heating by non-ionizing electromagnetic radiation (EMR) on cellular systems than equivalent convection heating (Ismailov, 1987; Morozov et al., 1995; Kudryashov et al., 2008). It has been shown that microwaves of thermal intensity induce more significant damage in bacteria, yeast and mammalian cells as well as in animals in comparison with conventional thermal exposure. It can be assumed that one of the leading reasons for this phenomenon is the difference in the mechanisms of energy supply to thermosensitive targets of cells by microwaves and convection heating (Kuznetsov, 1994). In addition, since hyperthermia destabilizes the osmotic homeostasis system, it is quite logical to assume that the specific thermal action of microwaves is based on the processes of thermogenic disturbance of the osmotic balance in the cell-environment system due to the peculiarities of the transformation of microwave energy to the molecular structures of cell. Considering the data demonstrating a more intense damaging effect of microwave heating on cellular systems than equivalent heating using the convection mechanism for supplying heat to cells, it was of interest to compare the features of heating by microwaves and convection heat (using a dry-air thermostat) of cell microorganisms and solutions of some basic compounds that make up the cells. Along with this, the dynamics of cooling of samples of cell suspension and solutions of various compounds, as well as the role of the state of the osmotic homeostasis system in the manifestation of these differences, were investigated.

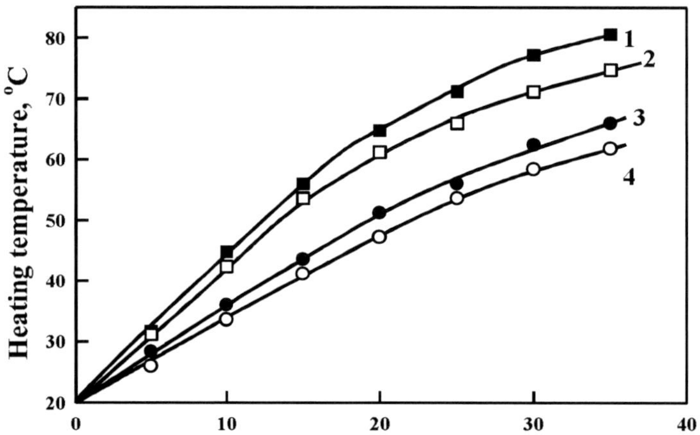

Figure 6.18. Results of microwave heating of suspensions of intact cells (curve 1), suspensions of inactivated cells (curve 2), a solution of the main molecular components of the cell (curve 3), and saline solution (curve 4).

In the experiments described below, an installation was used that emitted microwaves of the centimeter range (7 GHz) of continuous generation with an energy flux density equal to 200 W/cm^2. It was found that cellular and model samples were characterized by different heating rates during microwave irradiation (Figure 6.18). The intensity of heating increased in the series saline, model of cell composition, suspension of inactivated cells, and suspension of intact cells. Differences in the dynamics of heating were also noted upon irradiation with microwaves of solutions of individual compounds and cell suspensions (Figure 6.19). It can be seen from these figures that solutions of protein, DNA and sucrose were more heated than saline, and vegetable oil. As in the first case (Figure 6.18), cell suspensions were heated most intensively. It can be assumed that the differences in the heating rate of the samples are caused either by the features of microwave heating or by the unequal cooling rate. However, it was found that all samples, with the exception of vegetable oil, cooled at approximately the same rate (Petin, Morozov, 2015).

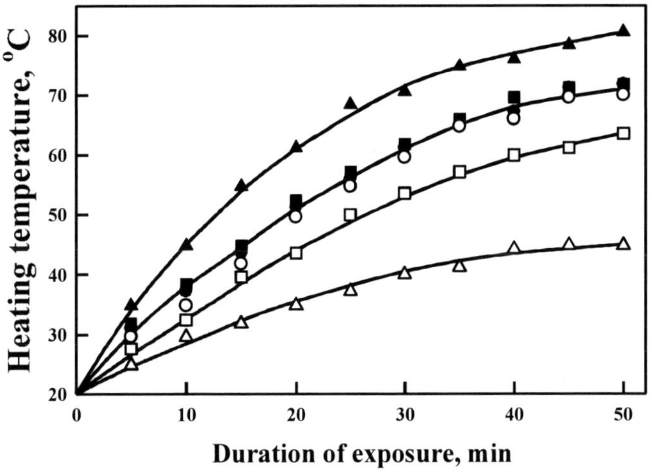

Figure 6.19. Results of microwave heating of suspensions of intact cells (closed triangles), peptone solution (closed squares), DNA and sucrose solution (open circles), saline (open squares) and vegetable oil (open triangles).

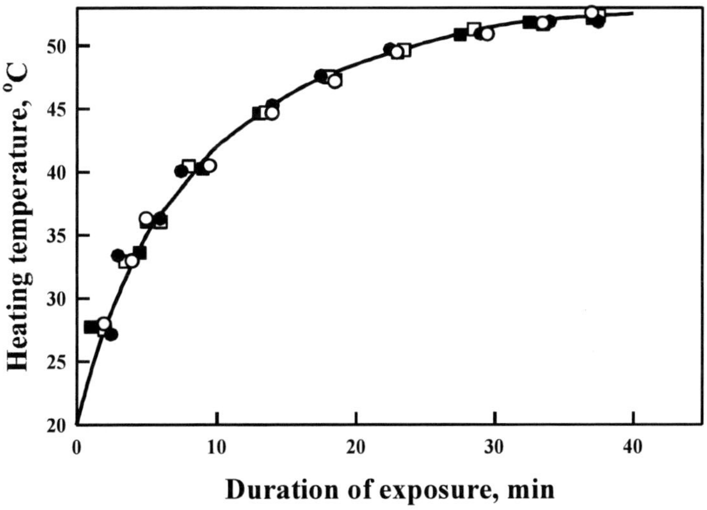

Figure 6.20. Results of convection heating of suspensions of intact cells (closed squares), suspensions of inactivated cells (open squares), a solution of the main molecular components of the cell (closed circles) and saline solution (open circles).

The data presented here indicates that the features of heating samples under the action of microwave radiation are due to the specificity of energy losses by microwaves in the substance of these samples. A completely different result was observed when the samples were heated by convection thermal energy (Figures 6.20 and 6.21). In this case, there were practically no differences in the rate and magnitude of heating of the cell suspension and solutions of compounds, with the exception of vegetable oil, which was heated more intensively.

The data obtained indicate significant differences in the dynamics of heating model and cell samples, as well as the solutions of molecular components of cells under heating conditions by microwaves and by convection heat. These differences, apparently, are due to the fundamental difference in the mechanisms of energy supply to the sample substance by both ways. One of the features of the energy loss of microwave radiation in the decimeter and centimeter ranges in the substance of the irradiated object is the simultaneous absorption of this energy in the entire irradiated volume of the object in comparison with the convection method of heat supply (Petin, Morozov, 2015). As a result, the heat transfer rate by microwaves to the object's substance is higher than with the convection mechanism of its supply at equivalent heating doses (Kuznetsov, 1994; Kudryashov et al., 2008).

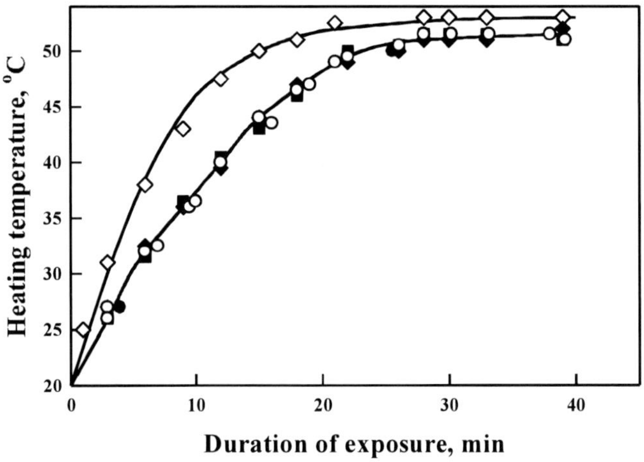

Figure 6.21. Results of convection heating of vegetable oil (open diamonds), saline solution (open squares), sucrose solution (open circles), DNA solution (closed circles), peptone solution (closed squares), and suspensions of intact cells (closed diamonds).

Another feature of microwave interaction with irradiated objects is the microgeometric inhomogeneity of the microwave energy absorption by the molecular components of the object due to differences in the dielectric constant of these components. It is this feature of microwave energy losses is primarily responsible for the differences in the heating dynamics of all the samples studied. It is quite clear that this property of microwave energy absorption in comparison with the convection method of heat can cause biological consequences that differ from the effects of convection heating in equivalent dose. In addition, it is likely that the differences in heating of cell samples by microwaves and heat might be due not only to the microgeometric heterogeneity of the molecular components of cells, but also to their increased dissociation into osmotically active components under the action of microwaves, which more intensively destabilize the system of osmotic homeostasis of cells (Morozov et al., 1995, 1997). Consequently, the specificity of the organization of cellular systems largely determines the features of the thermogenic action of microwave heating. The biological consequences of its influence are fundamentally different from traditional thermal heating due to fundamental differences associated with the inadequate microheterogeneity of the absorption of microwave energy by molecular components of cells.

Thus, the features of the biological effect of high-intensity EMF and the usual thermal heating of cell suspensions and some of the main components of cells show that microwave irradiation can cause a specific thermal effect that differs from the effect of ordinary heating.

6.6. Dependence of Cell Suspension Heating in the Microwave Field on Cell Concentration

The effect of microwave radiation on the cells and the whole organism is the subject of thorough investigation (Biological Effects, 1986; Ismailov, 1987; Schwan et al., 1988; Schwan, 1988; Klauenberg et al., 1995). The use of the microwave electromagnetic field energy in the medical, microbiological, and food industries is determined by the specific properties of the microwave heating of irradiated samples: the possibility to heat the entire volume, to control the rate of energy absorption and the temperature of the sample. Heat treatment is necessary for sterilization of biological liquids, with maximal preservation of their consumer properties. The development of optimal conditions for microwave heating is based on the knowledge of the biophysical patterns of the microwave radiation interactions with cells, and on the data on the mechanism of microwave energy absorption by cell suspensions (Petin, Berdnikova, 1977; Ignatov et al., 1978). Raising the microorganism concentration was shown to result in increased cell mortality under microwave irradiation (Dergacheva et al., 1998). The nature of this phenomenon is unknown. It may be assumed that the increased cell death in more concentrated cell suspensions exposed to a microwave radiation is associated with selective absorption of microwave energy by the cells, with a change in the viscosity of the irradiated suspension, or with progressive growth of the interface between the cells and the suspended liquid as well as the peculiar biophysical properties of microwave radiation absorption at the interface.

The goal of the present part of our work was to study the kinetics of microwave heating of cell suspensions at different concentrations under varied thermal insulation of the irradiated samples. Bacterial cells *Escherichia coli* and yeast cells *Saccharomyces cerevisiae* were used. The suspension medium was a physiological solution in Tris buffer (0.01 M) pH 7.0. Cell concentrations were studied in the range of 10^8–10^{12} cells/ml for bacterial cell suspensions and in the range of 10^6–10^{10} cells/ml for yeast cell suspensions. The bacterial mass was grown on a rich nutrient medium based on fishbone agar for 18 h at 37°C. The yeast biomass was grown on wort agar for 5 days at 30°C. The concentration of the bacterial suspension was evaluated by counting the visible colonies formed on a solid nutrient medium, with the drop method employed. The concentration of yeast cells was determined with a Goryaev cell.

Cell suspensions were irradiated in radioparent polyethylene 5 ml tubes that were placed in a closed phenoplast support (for better thermal insulation) or in an open support of organic glass. The source of microwave electromagnetic field was a 7 GHz generator operating in the continuous mode. The energy flux density was 100 mW/cm^2, the calculated power of the absorbed dose (specific absorption rate) was 200 W/kg. The tubes containing cell suspensions were placed in the microwave electromagnetic field parallel to the E vector. Irradiation was carried out in an anechoic chamber (5x10x4 m) at 25°C at 1.5 m distance from the funnel-shaped antenna. The samples were irradiated for various time intervals. At 5-min intervals of irradiation, the generator was switched off and the temperature of the irradiated suspension was registered with an electronic thermometer. The primary element of the thermometer was placed precisely in the center of the suspension volume, to avoid possible errors connected with the decrease in the temperature of the peripheral parts of the suspension as compared with the central part. During irradiation, the primary element of the thermometer (thermistor) was not kept in the irradiated suspension, to prevent errors in estimating the heating dynamics. The microwave electromagnetic field was shown to induce a more intense heating of cell suspensions than of the cell-free buffer solution, regardless of the microorganism type (Figures 6.22 and 6.23).

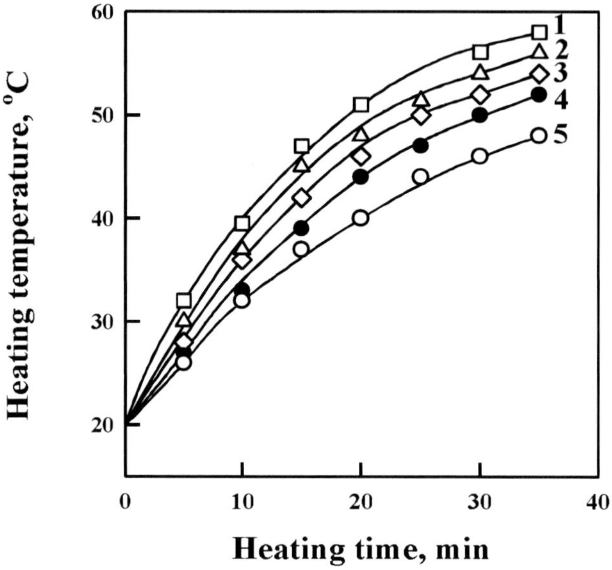

Figure 6.22. Temperature value of the sample of *Escherichia coli* cell suspensions of different concentrations as dependent on the duration of heating in microwave electromagnetic radiation under thermal insulation: curve 1 – 10^{12} cells/ml; curve 2 – $5 \cdot 10^{11}$ cells/ml; curve 3 – $1.25 \cdot 10^{11}$ cells/ml; curve 4 – $2 \cdot 10^{11}$ cells/ml; curve 5 – buffer solution.

In these and subsequent figures, the results of temperature measurements in a typical experiment are presented. The figures show that the rise of the cell concentration in irradiated samples of yeast and bacterial suspensions has led to a marked increase in microwave heating intensity. Under thermal insulation conditions, the greatest difference between the temperature of the buffer physiological solution and that of cell suspension at the highest concentration tested was observed after 15–20 min of exposure, and amounted to 12 and 14°C for the bacterial and yeast suspensions, respectively. The irradiation of samples using a support of organic glass (without thermal insulation) resulted in a smaller difference between the temperature values of the physiological solution and those of cell suspensions. Figure 6.24 presents the kinetic curves of microwave heating of bacterial suspensions at different concentrations as dependent on the time of exposure, under weak thermal insulation of the irradiated samples.

When Figures 6.22 and 6.23 are compared, the dependence of the microwave heating on the concentration of the cell suspension is seen to be less pronounced in the case of cell suspension irradiation using an organic glass support, without additional thermal insulation, under conditions of a higher intensity of heat dissipation. However, the observed dependence was maintained: suspensions of higher concentrations were heated to a higher temperature during the same time of exposure to the microwave electromagnetic field.

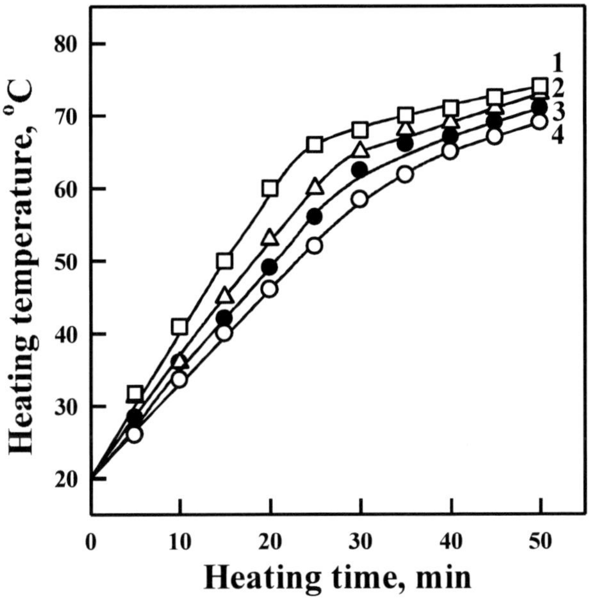

Figure 6.23. Temperature value of the sample of yeast *Saccharomyces cerevisiae* cell suspensions of different concentrations as dependent on the duration of heating in microwave electromagnetic field under thermal insulation: curve 1–$7 \cdot 10^9$ cells/ml; curve 2 – $3.5 \cdot 10^9$ cells/ml; curve 3 – $9 \cdot 10^8$ cells/ml; curve 4 – buffer solution.

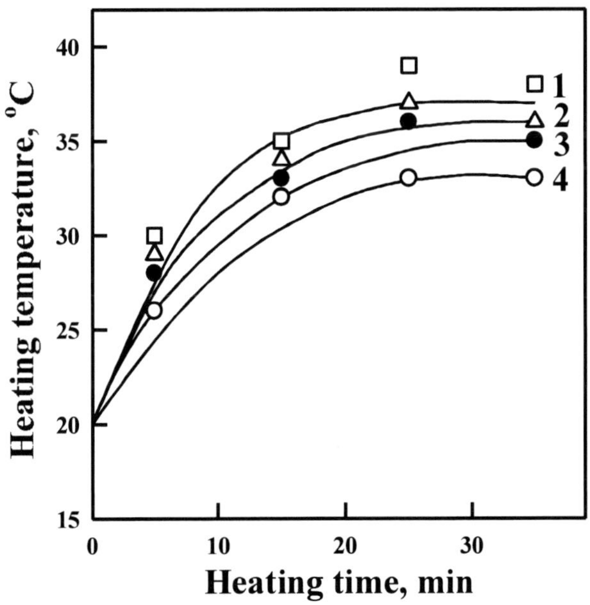

Figure 6.24. Temperature value of bacterial suspensions of different concentrations as dependent on the duration of heating in the absence of thermal insulation: curve 1 – 10^{12} cells/ml; curve 2 – $2.5 \cdot 10^{11}$ cells/ml; curve 3 – $4 \cdot 10^{10}$ cells/ml; curve 4 – buffer solution.

As it can be seen from Figures 6.22 and 6.23, the microwave heating of the bacterial and yeast cell suspensions proceeded in a similar fashion, though some difference was observed. In both processes the microwave field-induced increase in the rate of heating and, consequently, in the temperature increment was dependent on the concentration of cell suspension. The temperature growth dependence on the time of exposure was exponential for both bacterial and yeast cell suspensions. However, with further prolongation of the exposure the concentration dependence of microwave heating of the yeast cell suspensions became less pronounced, in contrast to the bacterial cell suspensions, where practically no change was registered. This difference may be accounted for by the structural properties of the yeast cells, in particular, by the larger size, as compared with the bacterial cells. The average volume and the weight of a yeast cell is known to be 100-fold to 200-fold greater than that of a bacterial cell (Schlegel, 1972). For this reason, during microwave irradiation the yeast cells sedimented and the supernatant of the yeast cell suspension with a lower cell concentration was heated less intensely; this resulted in the insignificant difference between the heating of suspensions with different cell concentrations. The precipitation of yeast cells with the increased microwave heating was promoted by cell death and a decrease in the cell surface charge leading to cell agglomeration.

In Figure 6.24, the dependence of the temperature increment observed in cell suspensions under microwave heating on the concentration of yeast and bacterial cells is presented. It is seen that the dependence of microwave heating on the cell concentration is exhibited for the yeast cell suspensions at considerably lower concentrations than for the bacterial cell suspensions.

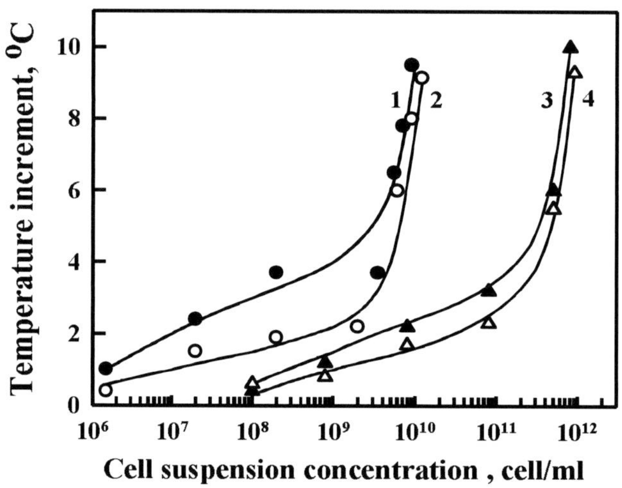

Figure 6.25. The dependence of temperature increment (relative to the buffer) of yeast (curves 1, 2) and bacterial (curves 3, 4) cell suspensions on the cell suspension concentration for to exposure to microwave electromagnetic field: 30 min (curves 2, 4) and 35 min (curves 1, 3).

Apparently, the larger size of yeast cells was the cause of the more intense heating in the microwave field as compared with the bacterial cell samples at the same concentrations of the cell suspensions. The isoeffective heating of the yeast samples was registered at the concentration approximately two orders of magnitude lower than that of the bacterial cell suspension (Figure 6.25); this corresponds to the difference between the volumes occupied by the yeast and bacterial cells in the suspensions.

This issue is illustrated in Figure 6.26 showing the changes in the temperature increment as a function of the volume occupied by the cells in the suspension. In this case, the data for yeast and bacterial cells are identical. The results obtained show the microwave electromagnetic field induces a more efficient heating in a microorganism cell suspension than in buffer solution, regardless of the type of microorganism, the extent of heating increasing with the cell concentration. This may account for the results of the study (Ignatov et al., 1978) where an increase in the death rate of microorganism cells with a rise in the cell concentration under microwave irradiation was reported.

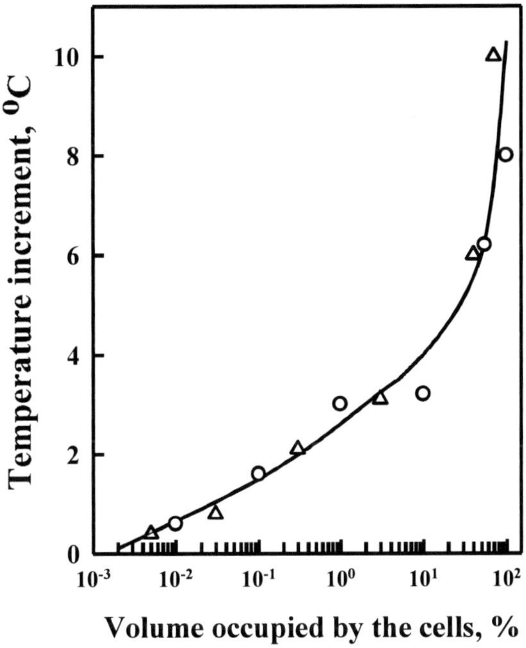

Figure 6.26. The dependence of temperature increment of yeast (open circles) and bacterial (open triangles) cell suspensions (relative to the buffer) on the volume occupied by the cells in suspension under 35-min microwave irradiation.

It is interesting to consider the nature of this phenomenon. The temperature of an object is known to be determined by the rate ratio of heating and cooling, and cooling depends on the medium viscosity, which rises with an increase of the suspension concentration. Thus, it may be assumed that in a concentrated suspension the withdrawal of the heat generated by microwave radiation is retarded because of the decreased rate of heat transfer from the center of the suspension to the peripheral parts, caused by the slowing down of convection processes. If this is the case, the temperature in the center of the sample registered after a certain time upon the removal of the microwave field must be higher at higher concentrations of the cell suspension.

In order to clarify the influence of this factor on the variations in the temperature of microwave-irradiated samples of physiological solution and cell suspensions, the dynamics of cooling of bacterial cell suspension with the highest concentration used (10^{12} cells/ml) and of the physiological solution was registered at room temperature. For this purpose, the suspension of bacterial cells and physiological solution were placed in identical tubes, heated to 56°C, then the temperature of the samples was measured repeatedly in the process of cooling. To achieve equal cooling conditions, each tube was placed in a glass vessel. The data of temperature measurements are presented in the Table 6.6. It is clear that the dynamics of cooling of the bacterial cell suspension and physiological solution is identical. Thus, the assumption that the variations in the microwave heating of cell suspensions of different concentrations were due to different rates of cooling was not true.

Table 6.6. Dynamics of cooling at room temperature of the samples of physiological solution and bacterial cell suspension at a concentration 10^{12} cells/ml preheated to 56°C

Time, min	Temperature, °C	
	Physiological solution	Cell suspension
0	56.0	56.0
2	50.6	50.6
4	45.5	45.4
6	41.2	41.0
8	38.0	37.6
10	35.1	34.9
12	32.7	32.4
16	29.7	29.3
20	27.1	26.6
24	25.5	25.2
32	23.9	23.9
40	22.5	22.8

The observed dependence of microwave heating on the concentration of cell suspension cannot be related to resonance phenomena, which are determined by the ratio of the size of the irradiated object to the wavelength in free space (condition more than $2\pi a/\lambda$, where a is the effective radius of the object, λ is the wavelength). This condition is proved by the results showing the dependence to be preserved when the cuvette for irradiation with an effective radius of 1 cm is replaced with that with an effective radius of 3 cm. In this case, the ratio $2\pi a/\lambda$ changed from 1.5 to 4.5.

According to literature data (Biological Effects, 1986; Ismailov, 1987; Schwan, 1988; Klauenberg et al., 1995), water makes the main contribution to the dielectrically loss in a biological object under microwave irradiation with a frequency higher than 1 GHz. The dielectrically loss is determined by the relaxation period of the dielectric, which is proportional to the viscosity of the medium (Dergacheva et al., 1998). Thus, it may be assumed that the increase in the temperature of microwave heating of yeast and bacterial cell suspensions with the increase of cell concentration is related to the rise in the suspension viscosity. The dependence of this effect on the fraction of the volume occupied by the cells in suspension for different types of cells (Figure 6.25) may be considered the evidence of the proposed mechanism. In favor of this assumption are our recent data on the more intense microwave heating of the concentrated solutions of high-molecular substances (sucrose, peptone, DNA) and of inactivated cell suspensions in comparison with the physiological solution (0.154 M NaCl) (Morozov et al., 1996).

Besides, in the analysis of the causes of a more intense microwave heating of cell suspensions, a possible effect of the heterogeneous structure of the aqueous phase should be taken into consideration. Along with the "free water" with the maximum absorption of microwave radiation at frequencies higher than 10 GHz, and the water within the ion hydration shell and in the vicinity of the cell surface absorbing at frequencies up to 2 GHz, there must be a fraction of water with intermediate lability and characteristic relaxation frequency in the range of 2–10 GHz. Since in our experiments the frequency of the acting electromagnetic field was 7 GHz, these partially immobilized water molecules may be assumed to contribute to the absorption of the electromagnetic radiation and its subsequent dissipation. If it is so, an increase of the cell concentration in the suspension or substance concentration in the solution must increase the fraction of the partially immobilized water, and thereby enhance the microwave heating, just as it was observed in our experiments.

Chapter 7

Radiation Hormesis after Low Doses of Ionizing Radiation

Abstract

This chapter deals with various approaches to the concept of low doses. A review of data on radiation hormesis, the positive effect of low doses of ionizing radiation, is presented. Many healthy effects induced by small doses of ionizing radiation on biological objects of various complexity are considered. Data on the favorable effect of the increased natural background of ionizing radiation are discussed. The manifestations of radiation hormesis were also revealed as a result of the Hiroshima and Nagasaki bombing, as well as various radiation accidents, professional and medical exposure. The fact that low doses of ionizing radiation are beneficial to humanity seems unlikely, but the health effects of chronic exposure to residents who have lived in ^{60}Co-contaminated apartments for over 25 years in Taipei (Taiwan). This unintentional human experiment proved that chronic exposure to humans at specific dose rates could effectively reduce the rate of human cancer. It is noted that cancer mortality among adults living in these buildings for 9–20 years was only 3% of the total cancer deaths of adult population.

7.1. Small Doses of Ionizing Radiation and Radiation Hormesis

The biological effect of acute doses affects highly radiosensitive human organs – the hematopoietic system, bone marrow cells, the gastrointestinal tract, and the central nervous system. Such effects were called *deterministic*. Irradiation at low dose rates can cause a variety of pathological changes, including radiation sickness and death, with a lower probability or stochasticity. These *stochastic* effects of radiation can induce cancers and some inherited diseases. Irradiation at low dose rates can cause a variety of pathological changes, including radiation sickness and death, with a lower probability or stochasticity. The stochastic effects of radiation can induce cancers and some inherited diseases.

For a long time, the following three paradigms have existed in radiobiology and radiation safety, which still exist to this day. These paradigms based on the data obtained from the analysis of the consequences of the atomic bombing: (a) any absorbed dose is harmful; (b) there is a linear relationship between the radiation dose and the recorded effect; (c) the effect of ionizing radiation is cumulative. Data on the positive effects of ionizing radiation include over 3000 reports (Luckey, 1991, 2007).

This concept is supported by the analysis of a large number of reports related with people living in conditions of naturally increased levels of radiation. The population of most countries receives a natural background radiation (NBR) of an average of 2.4 mSv/year. According to data published by International Committee on Radiation Protection (ICRP), many people around the world are exposed to higher doses of NBR, but they do not suffer from increased cancer mortality or reduced life expectancy compared with adequately matched controls. The

ICRP formulated the concept of a linear non-threshold dependence of the probability of stochastic effects on the dose, according to which ionizing radiation can harm to human health at an arbitrarily small dose. In principle, this concept was useful until the extensive series of appeared works devoted to the positive effect of low doses of ionizing radiation on a variety of objects, including humans (Luckey, 1980 a; Kuzin, 1991, 1995; Calabrese, Baldwin, 2000, 2002; Buldakov, Kalistratova, 2005; Luan et al., 2006; Koterov, 2010; Petin, Pronkevich, 2011). In 1994, after a twelve-year delay of prepared material, the UN Scientific Committee on the Effects of Atomic Radiation (UNSCEAR) published a review of the evidence on the positive effects of low doses of ionizing radiation (UNSCEAR, 1994). Analysis of a large amount of literature, presented in this chapter, has shown that there are levels of exposure to radiation, which have a beneficial effect.

The concept of "small doses" does not have a single definition. There are several approaches to determining low doses. Let's dwell on some of them. Many biological effects induced by ionizing radiation are described by the theory of the dual action of radiation (Kellerer, Rossi, 1972) or by the linear-quadratic dependence of some effect, for example, cell survival S, on the dose D of ionizing radiation

$$S(D) = \alpha D + \beta D^2\,, \tag{7.1}$$

where α and β are proportionality coefficients. Some radiobiologists propose to consider radiation doses as small if they are lower than the dose at which the number of damages induced by the one-impact (exponential) mechanism is equal to the number of damages arising from the two-impact (quadratic) mechanism (Tsyb, Mardynsky, 2010).

According to microdosimetric definition, at low doses, on average, one track of an ionizing particle passes through the nucleus (target) of the cell (Bond et al., 1988). The calculations carried out by these authors show that, upon irradiation with ^{60}Co γ-rays at a dose of 1 mGy, on average, one ionizing particle passes through the nuclei of mammalian cells (diameter 8 μm). At such a dose, one ionizing particle will pass through the nuclei about of 1/3 of the cell population, about 1/3 of the cells will not have a single track, and the rest will have two or more tracks crossing the cell nucleus (Booz, Feinendegen, 1988). This implies one important consequence: with a further decrease in the dose, there may exist a dose range within which, with a decrease in the dose, the number of targets through which the track of an ionizing particle has passed, and the dose per cell will not change with a change in the radiation dose. Already from this consequence, one can expect the features of the action of such small doses.

The results of many epidemiological studies show that with acute exposure to sparsely ionizing radiation at doses of 200 mGy and below, no cases of the development of deterministic consequences (radiation sickness) of radiation were observed. Moreover, it is not possible to detect an increase in the number of stochastic events (for example, malignant tumors). Therefore, UNSCEAR recommended that doses of 200 mGy and below may be attributed to low doses and dose rates of 0.1 mGy/min and below to low dose rates (UNSCEAR, 2008, 2010). The UNSCEAR report with the data on the positive effects of low doses of ionizing radiation (UNSCEAR, 1994) actually casts doubt on the general idea of the harmful effects of even small doses exceeding the natural background radiation (Yavorovski, 1997). Consequently, the concept of a linear non-threshold harmful effect of radiation in the region of low doses should be questioned. Many authors have cited arguments indicating the need to replace this paradigm (Cohen, 1995, 1997; Korogodin, Korogodina, 1997; Cai, 1999; Keirim-Markus, 2000; Rozhdestvensky, 1985, 2006).

The ideology of rationing safe levels of radiation was based on new information received: epidemiological data on the increase in malignant neoplasms among radiologists, as well as data on an increase in cancer among survivors of the atomic bombings in Hiroshima and Nagasaki. The likelihood of such "stochastic effects" occurring is proportional to the magnitude of the dose (Clarke, 2001). These data served as the basis for the formation of the concept of a linear non-threshold dependence of radiation carcinogenesis on the dose of ionizing radiation.

To denote the positive effect of small doses of influencing factors, the term "hormesis" was introduced. This concept is taken from pharmacology, in which it was known since the time of Paracelsus that strong poisons, taken in small concentrations, could be valuable medicines. The term radiation hormesis is used to denote the positive effect of small doses of ionizing radiation. For more than a century, the beneficial effects of radiation have been observed after low-dose exposure or chronic low-dose ionizing radiation exposure (Luckey, 1980 a, b 1991; Calabrese, Baldwin, 2000).

7.2. The Action of Small Doses of Ionizing Radiation on Biological Objects of Various Complexities

It is shown that a decrease in the background of natural ionizing radiation is accompanied by a lack of growth and development of biological objects. On the contrary, a slight increase in natural radiation is accompanied by an improvement and activation of metabolic processes, i.e., radiation hormesis (Luckey, 1980 a, b, 1991, 1996; Kuzin, 1991, 1995). A similar picture is observed with a lack of other natural physical and chemical factors, such as UV light flux, electromagnetic fields, vitamin A, vitamin B6, calcium, iron, selenium. The lack of some ingredients in food can also lead to pathology, such as a lack of ionizing and ultraviolet radiation, and if completely excluded, to dramatic events. The regulated addition of these agents compensates for the deficiency and leads to the restoration of metabolic processes to normal or higher levels. Note that the monograph by A.M. Kuzin (1991), describing the results obtained with a camera that reduces the natural radiation background by 10 times, provides data on a significant decrease in the body weight of developing rat cubs (8- and 21-day-old) kept for 10 days in a low-background chamber. This once again demonstrated the need for NBR, since its reduction slowed down the growth of young animals. In passing, we note that Planel's camera reduced the radiation background by 20 times due to the use of lead smelted in the pre-atomic era.

One of the indicative experiments on artificially changing the radiation background was the experiments of French scientists who irradiated three groups of *Paramecium tetraurelia* (Planel et al., 1987). The first group was irradiated with gamma radiation 1.75 mGy/year (natural background), the second and third – 0.3 mGy/year, while the last two groups were additionally covered with a lead screen with a thickness of 5 and 10 cm, respectively and the background dose rate was reduced 10 and 20 times, respectively. On the 8th day, the number of paramecium from the second group was 60% of the control, and the third group – 40% (Figure 7.1). This means that the lowered background radiation slowed down the multiplication of paramecia. These data indicate the need for a natural level of ionizing radiation for the functioning of the biosphere.

In experiments on the effect of reduced NBR on *Drosophila melanogaster* eggs (Planel et al., 1968), developmental inhibition was also observed: larvae appeared 24–48 hours later than control. When ^{60}Co was introduced into lead boxes in amounts corresponding to the irradiation power of 1.25 mGy/year, i.e., when the NBR was restored, the delayed effect was completely removed (Figure 7.2).

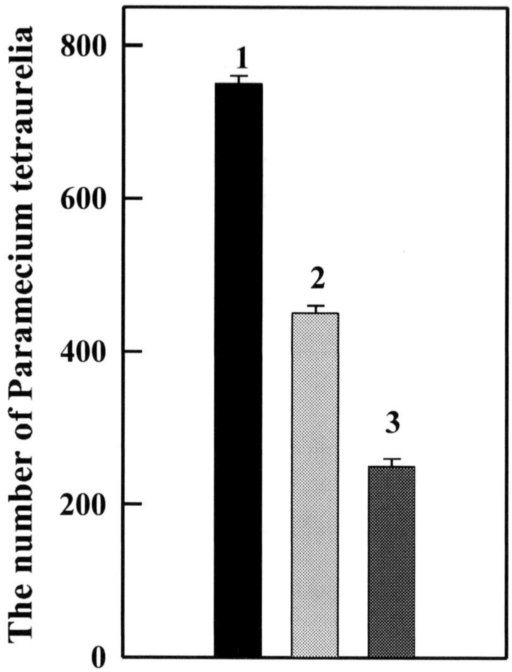

Figure 7.1. The number of *Paramecium tetraurelia* after reproduction for a fixed time at NBR (1) and in a chamber with a lowered background: 2 – decrease in NBR by 10 times, 3 – decrease in NBR by 20 times (Planel et al., 1987).

Analysis of fluctuations in the change in the body weight of the rat cubs showed a shift of the curves to the left with a decrease in the NBR compared to the control that means a decrease in the weight of the animals with an insufficient background of ionizing radiation (Figure 7.3). When uranium salts were introduced into the chamber, i.e., for compensating of the background, the effect of a decrease in body weight was removed; body weight gain was equated with control.

Since a large number of authors, according to many criteria and many biological objects, observed a decrease in various biological parameters with a decrease in the NBR, it can be assumed that a slight increase in the NBR may be beneficial. Indeed, it was shown on highly radioresistant cells of *Paramecium tetraurelia* after chronic γ-irradiation for 100 hours at a dose rate 10 times higher than that of the NBR led to an acceleration of fission (Croute et al., 1982). Chronic irradiation of chicken eggs (7.5 NBR, 1.5 μGy/h) showed not only the stimulation of the development, but also an increase in the embryo's resistance to unfavorable incubation conditions, which caused a decrease in their death by 14% in relation to the control (Kozlov et al., 1987).

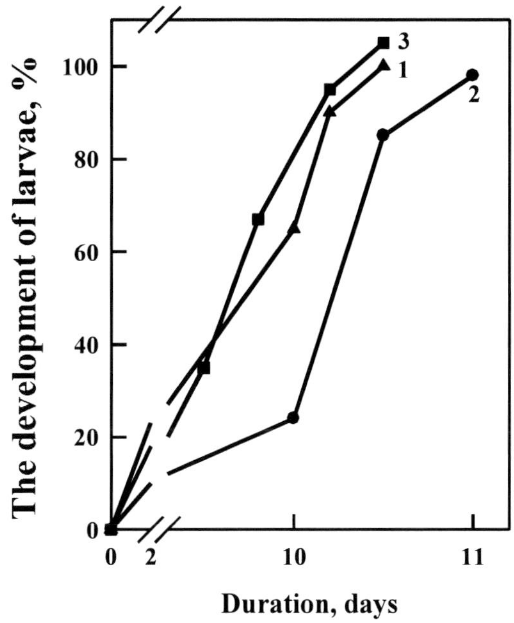

Figure 7.2. The development of larvae from Drosophila melanogaster eggs during NBR (curve 1) and in a low-background chamber: curve 2 – the reduced background, curve 3 – compensated background (Planel et al., 1968).

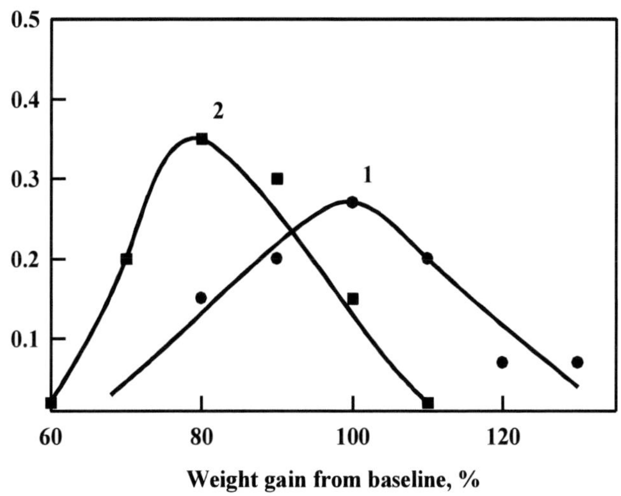

Figure 7.3. Fluctuations in the body weight of rat cubs with NBR (curve 1) and with a reduced background (curve 2) (Kuzin et al., 1991). The ordinate shows the ratio of the number of animals with a given weight gain to the total number of animals.

It has been shown that the development of chickens is accelerated as result of γ-irradiation occurring once at low doses (Kirshin et al., 1965). On cell cultures of the Chinese hamster and the cells of the Syrian hamster BHK-21, it was demonstrated that their single irradiation at the G1 stage at a dose of 1.6 cGy leads to some stimulation of cell division 48 hours after placing the irradiated cells on a nutrient medium (Mantsygin et al., 1981) (Figure 7.4).

The effect of gamma irradiation (60 cGy) from a ^{226}Ra source on the lifespan of *Drosophila melanogaster* was studied (Moskalev, 2007). It was shown that the average life expectancy increased in the irradiated group in comparison with the non-irradiated control.

For a long time ago, the results of lifelong X-ray irradiated mice (LAF1) were published (Lorenz, 1950; Lorenz et al., 1955). The dose was 1.1 mGy for 8 hours of irradiation, proceeded 6 times a week. There was an increase in the average lifespan of the mixed group of male and female mice by 8.2%, and for males by 14.4%. Experiments on irradiating DBA/2 mice at a dose of 1500 mGy for 5 and 12 days prior to infection with the leukemia-inducing virus showed that unirradiated, virus-infected mice died within 40 days, while the irradiated infected animals recovered (Shen et al., 1988). The exposure of mice for 30 minutes daily for a month in an atmosphere enriched with radon, led to the irradiation of the lungs and skin for this month, exceeding the NBR by 200 times. The authors showed (Volkova et al., 1961) that under the influence of such an increase in the radiation background, the phagocytic activity of leukocytes increases by 125–180% after exceeding the NBR by 200 times, indicating that animal immunity is increased.

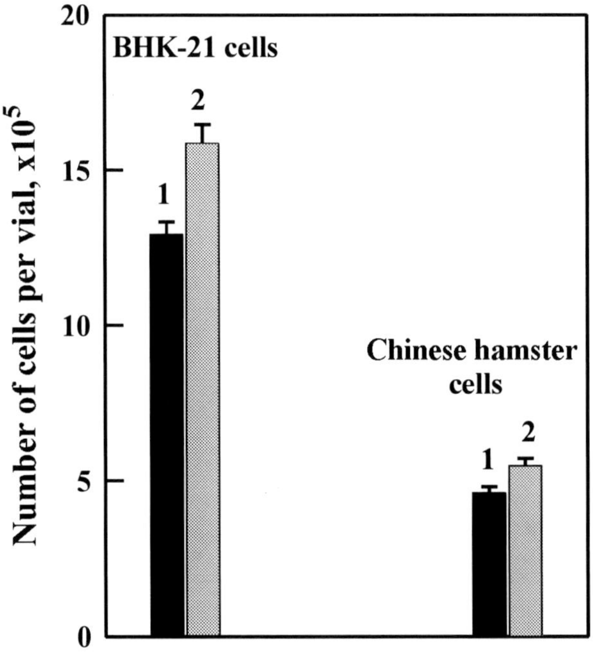

Figure 7.4. Stimulation of cell division – change in the number of cells 48 hours after a single irradiation with a dose of 1.6 cGy at the G1 stage (Mantsygin et al., 1981): 1 - control, 2 – experiment.

Japanese researchers (Muramatsu et al., 1964) studied the fertility of mice of three generations under chronic irradiation with a dose rate of 4.3 Gy/day. For first generation, the total dose was 344 mGy. It was shown that the fertility of chronically irradiated mice in all three generations was higher than in the control (Figure 7.5).

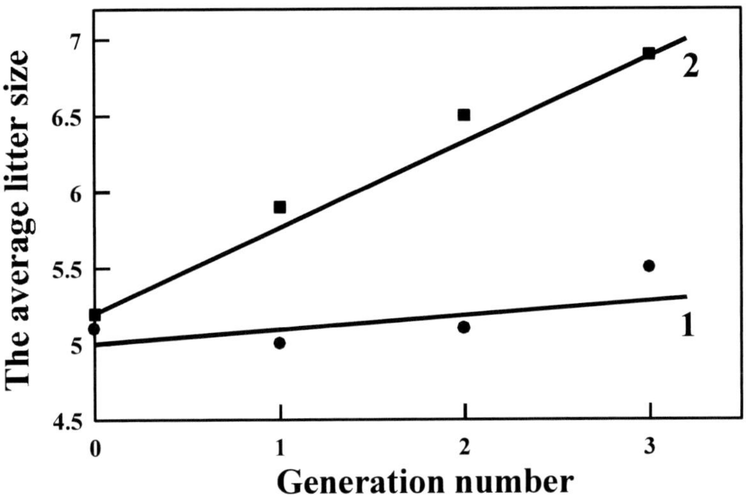

Figure 7.5. An increase in the average litter size of mice during their chronic irradiation with a dose rate of 4.3 mGy/day for three generations (Muramatsu et al., 1964): curve 1 – control, curve 2 – experiment.

An increase in the body weight of mice was shown under the action of small doses. Bustad et al. (1965) controlled the weight of developing mice (C57 BLX101), which from 6 to 56 weeks of age were irradiated with ^{60}Co γ-rays for 365 days for 8 hours per day at a dose rate of 0.017 and 0.033 mGy/min, after which the irradiation was stopped. In the group irradiated at a dose rate of 0.017 mGy/min, a significant increase in the weight of animals was observed by 6–8% compared to the control, and this increase persisted up to 22 months (total dose of 2900 mGy). A higher dose rate no longer led to this effect. The results of irradiation of mice (C57BL/6J) with five different dose rates in the range of 0.005–0.4 mGy/min have been published, the total cumulative doses were 200, 600, 1800, 5400, 11200 mGy (Spalding et al., 1981). Despite the diversity of the observed reactions, it is clearly seen that smaller dose groups of mice, irradiation of which began at 2 months of age, showed a longer life span (13–36%). Mice that were exposed to radiation at a later age did not show such a reaction. Many other facts of an increase in the life expectancy of animals after exposure to small doses of ionizing radiation are known. Caratero et al. (1998) found that low doses of γ-rays – 70 mGy/year and 140 mGy/year (25 and 50 times more NBR) increased the lifespan of 600 female C57BL/6 mice compared to control (300 animals) (Figure 7.6). The increase in life expectancy was reliably demonstrated in the irradiated groups as compared with the non-irradiated ones (at the LD50 level of 549 days in the control, 673 days in both groups of irradiated animals).

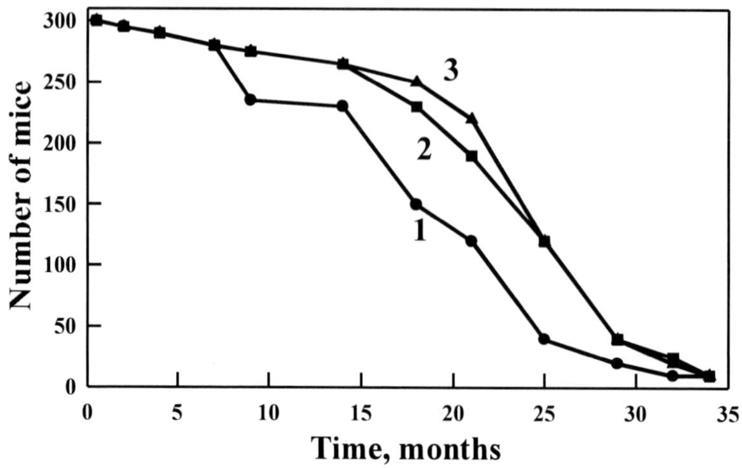

Figure 7.6. Survival curves of female C57BL/6 mice (Caratero et al., 1998): curve 1 – control, curve 2 – irradiation at 70 mGy/year, curve 3 – irradiation at 140 mGy/year.

The results of experiments on chronic irradiation with low doses of ^{60}Co γ-rats (0.04 Gy 3 times a week for 4 weeks) of female C3H/He mice (a line with an increased frequency of spontaneous breast tumors) were published (Luckey et al., 1978; Luckey, 1991). It has been shown that chronic irradiation with low doses was ineffective in slowing the growth of spontaneous breast cancers, while irradiation in combination with a calorie restriction diet resulted in a decrease in breast tumors. Figure 7.7 shows the dependence of the formation of L1 lung sarcoma cell colonies in mice on a single dose of X-ray irradiation (Nowosielska et al., 2008). These data show that experiments with doses of 0.1 and 0.2 Gy resulted in significant inhibition of colony development. On the contrary, no statistically significant change was observed in comparison with the control at a dose of 1 Gy.

In experiments with mice, the incidence of leukemia, solid cancers and sarcoma was lower in those irradiated with ^{137}Cs gamma radiation at doses of 2.5–20 mSv than in non-irradiated control animals. The number of all malignant tumors in irradiated animals was 30% less than in controls (Jaworowski, 1995). A lot of data on the positive effect of small doses of ionizing radiation have been obtained in experiments with rats. Carlson et al. (1957, 1959) showed that irradiation of male rats (Sprague-Dawley) throughout of their life at a dose rate of 0.006 mGy/min increased their lifespan to 131% of control. Carlson et al. (1959) showed that irradiation of rats for 1 year at dose rates of 0.002 and 0.004 mGy/min, exceeding the NBR by 500 and even 1000 times, gives a statistically significant increase in average life expectancy to 110 and 114%, respectively. Of particular interest in these studies are experiments carried out in unfavorable environmental conditions, achieved by increasing the living temperature to 35°C, which reduced the life span of animals to 60% of the control. Under these conditions, an increase in the NBR by 2.2 and 8.6 times increased the average life expectancy by 159 and 174%, respectively.

Figure 7.8 exhibits that with the injection of small amounts of ^{241}Am (0.37 and 3.7 kBq/kg), the lifespan of the rats was higher than the control (Moskalev, 1963). The work (Calabrese, Baldwin, 2002) summarizes the results of experimental data published by different authors on rats and mice in relation to lung tumors, lymph nodes, mammary gland tumors, bone sarcoma, etc. under the influence of general external gamma radiation, X-rays, neutrons. There was a decrease in the number of lung adenomas in RFMf/Un mice, irradiated with γ-rays in the range of 0.1–1 Gy in comparison with the control ones.

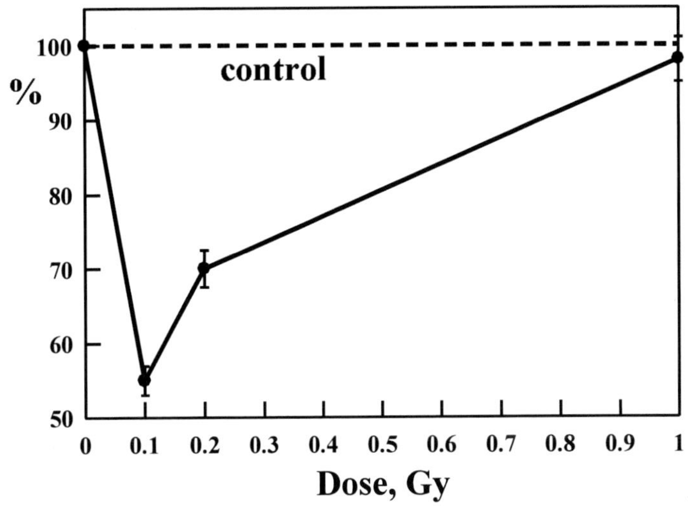

Figure 7.7. The number of L1 sarcoma cell colonies formed in the lungs (in % along the ordinate) grown in mice previously irradiated with different doses of X-ray radiation (Nowosielska et al., 2008).

Results of six-year chronic irradiation (^{60}Co) of 144 dogs (daily for 22 hours) at dose rates of 0.0004; 0.0011; 0.0024 mGy/min are described in the work of Grigoriev (1983), the total doses for 6 years are 1260, 3780, 7500 mGy, respectively. Despite the fact that the total doses were beyond the low doses recommended by the UNSCEAR, the author concludes that such chronic exposure did not lead to a violation of the viability of the dogs. The animals remained active; their overall morbidity did not exceed the control. The number of deaths of irradiated dogs was the same as in the control (about 10%).

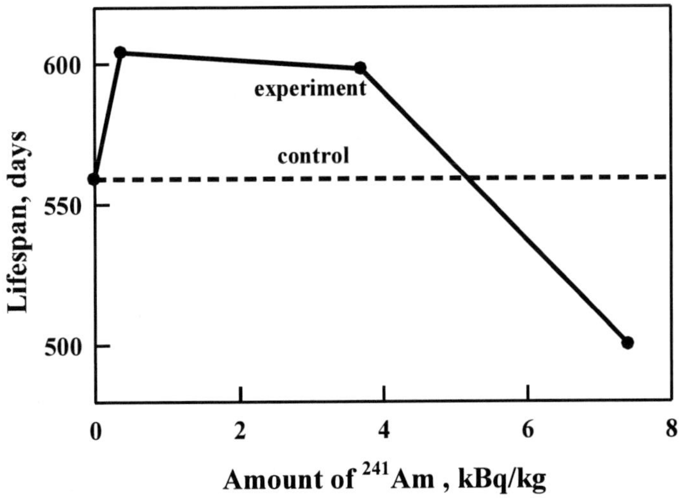

Figure 7.8. The dependence of the lifespan of rats on the amount of intraperitoneally injected ^{241}Am (Moskalev, 1983).

The work (Angleton et al., 1988) describes the results of gamma irradiation of ^{60}Co (doses of 160 and 830 mGy) of pregnant dogs. The number of offspring of dogs after irradiation with a dose of 160 mGy on the eighth day after mating increased to 108–110% compared with the control. Irradiation at a dose of 160 mGy at all stages of development did not reveal any negative effects on the health of the offspring of dogs, although at a high dose (830 mGy) the number of offspring in the litter decreased.

Thus, the totality of the described data on the effect of low doses of ionizing radiation, determined in accordance with the recommendations of the UNSCEAR (1994, 2008, 2010), on mice, rats, rabbits and dogs demonstrates a positive effect on the life expectancy of animals, the yield of oncological diseases, immunity, fertility and other indicators of the vital activity of animals. Moreover, in some of the works cited above, the positive effect of higher doses of radiation is noted.

7.3. Increased Natural Background of Ionizing Radiation

As was already mentioned, the NBR is chronically acting at a low dose rate on the biosphere in general and on human being in particular. The population in most countries lives with a natural background of ionizing radiation on average 2.4 mSv/year. According to the data published by the International Commission on Radiation Protection, many people around the world are exposed to higher NBR doses (5 mSv/year and above), but they do not suffer from increased cancer mortality or reduced life expectancy compared to adequately matched controls.

Mason, Miller (1974) analyzed mortality from all types of cancer in the population living in rural and urban areas of the United States located high above sea level (more than 3000 feet), in which the population was exposed to increased radiation, and in areas located at sea level. The analysis of the standardized mortality ratio (SMR) was carried out, which gives an estimate of mortality in the studied cohort compared with the background mortality rate of the population (taken as a unit) compared with the average mortality in the United States in 1950-1969. It was reliably shown ($p < 0.01$) that the ratio of the mortality rate of the white population living at high altitudes (4390 feet above sea level) with a high level of background radiation to the average mortality in the United States from all types of cancer is on average 0.81 (for both sexes). A similar ratio for residents living at sea level in San Francisco is 1.23 for men and 1.15 for women, respectively (Figure 7.9). The authors conclude that cosmic radiation at high altitudes may be responsible for the reduction in cancer deaths.

In 1950-1968, a study was conducted (Frigerio, Stowe, 1976) in 7 US states with an increased NBR (2.1 mGy/year) and in 14 with a low background (1.18 mGy/year). It was shown that the annual mortality from cancer per 100 thousand people averaged 147 in the United States, and 126 in areas with increased NBR. Yalow (1994) noted that in Colorado, with a high background radiation dose due to cosmic radiation, radiation from rocks containing an increased amount of uranium from buildings built from rocks, as well as due to radiation received during work in the nuclear industry, the incidence and mortality from all types of cancer are lower than in other regions with lower natural radiation levels.

Analysis of the dependence of cancer on the amount of natural radiation (Craig, Seidman, 1961), conducted in 163 regions of the United States, showed that the number of deaths from leukemia and lymphoma in 1949–1951 was 19% lower for males and 6% lower for females living at 2,000–5,000 feet above sea level than those living in areas below 500 feet. It is obvious that an increase in the dose from cosmic radiation not only does not increase mortality from these types of disease, but also reduces the risk of death.

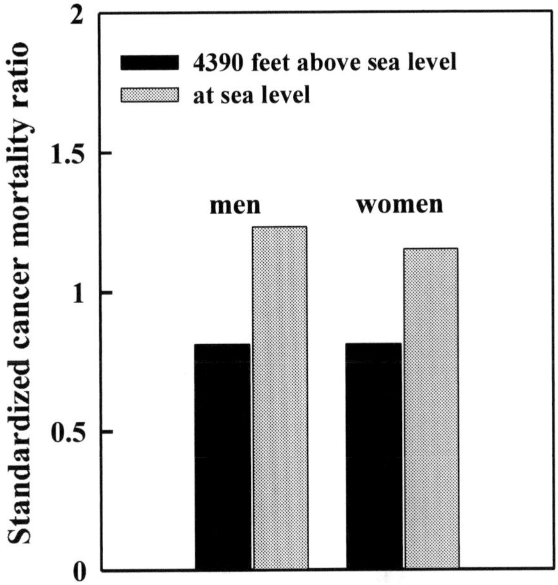

Figure 7.9. Cancer mortality rates in white populations in areas of the United States with different radiation backgrounds (Mason, Miller, 1974).

The same inverse relationship was observed between the increased background and mortality from leukemia (Eckhoff et al., 1974). It has been shown that the number of deaths from leukemia per 100 thousand people per year in the United States correlates with altitude: mortality from leukemia rises with an increase in altitude to 2000 feet above sea level, but then this figure significantly decreases at an altitude of 2000 feet and above. The work (Cohen, 1980) presents the results of a study of the relationship between the background of gamma radiation in 48 states of the United States and cancer mortality. It was demonstrated that this indicator slightly decreased (on average, from 18 to 14 deaths from cancer per 10,000 population per year) with an increase in the background of gamma radiation from 0.5 to 1.7 mGy/year. Analysis of cancer mortality in five French departments with relatively low NBR and in two departments with high NBR showed a negative correlation between the level of NBR and mortality of children from leukemia, women from breast cancer, and men and women from thyroid and lung cancers (Tirmarche et al., 1988).

Studies that also demonstrate the beneficial effects of increased background radiation were carried out in the United States from 1990 to 1994 (Jagger, 1998). Cancer mortality rates were compared in high-altitude areas with increased NBR in Colorado and New Mexico (with average annual dose 765 μSv/year) and in areas at sea level (Louisiana, Mississippi and Alabama) with average annual dose 195 μSv/year. Cancer mortality in areas with an increased background was 1.26 times lower than in controls (Figure 7.10).

The results of studies carried out in China serve as an indicator of the safety of increased natural exposure. In 1970–1978 the vast Yangjiang province with an increased background radiation (2.1–4.27 mGy/yr) from monasite, a thorium-containing mineral, was compared with the control province (0.77–1.65 mGy/yr) (Wei et al., 1990). The authors emphasize that the surveyed families have lived in these areas for 6 or more generations, are characterized by the same population structure, nationality, employment, food, and epidemiological status. It was shown that mortality after 1972 from all types of cancer was 14.6% lower in the province with an increased background compared with the control population. The work (Liu et al., 1987) provides data on increased immunological protection and reduced mortality from cancer in the inhabitants of this province. In the period from 1979 to 1995, examined 125,079 people from the same province (average annual effective dose 6.4 mSv). There was no increase in cancer risk associated with high levels of natural radiation (Tao et al., 2000).

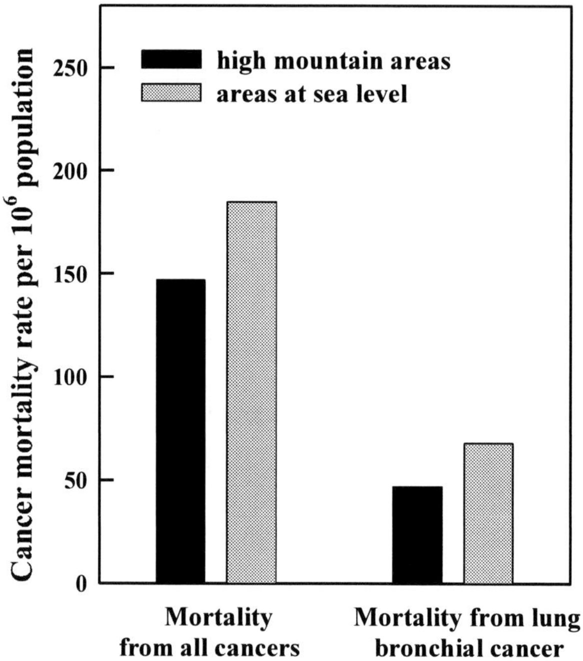

Figure 7.10. Cancer mortality among US residents living under various natural radiation backgrounds (Jagger, 1998).

For centuries, the inhabitants of Ramsar, a coastal region in northern Iran, have lived in houses with an increased NBR (population about 2,000). With an average effective dose in this region of 10 mSv/y, the range of dose rate variations was 1–131 mSv/year (Sohrabi, 1997). In some of the houses, the dose rate reached 260 mSv/year. The authors of works (Mortazavi, Karam, 2005; Monfared et al., 2006) report that the incidence of cancer and heart disease in the residents of Ramzar was less than in areas with a normal background radiation. It was shown (Nambi, Soman, 1987) a negative correlation between SMR from all cancers and NBR doses in various cities of India (Figure 7.11). The same correlation was noted for SIR (Standardized incidence ratio) – a standardized incidence ratio showing the ratio of the incidence rate of the studied population to the national level.

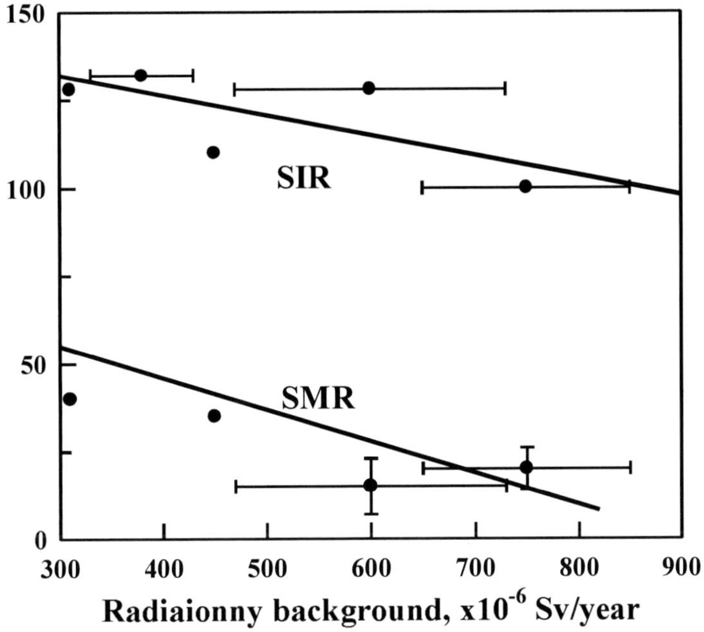

Figure 7.11. Cancer and cancer deaths per 10^5 population in cities of India with different NBRs (Nambi, Soman, 1987).

Since 1990, research has been carried out on the inhabitants (385,103 people) of the Karunagapilli Peninsula (Karunagappilly, Kerala, southwestern coast of India), known for a high level of background radiation (up to 35 mSv/year, and in some places up to 70 mSv/year) because of thorium contained in coastal sand (Nair et al., 2009). The analysis did not show a statistically significant dependence of the incidence of all malignant neoplasms on the background radiation level; cases of leukemia (30 cases) also did not significantly depend on the background radiation

A beneficial effect of chronic exposure to a threefold higher natural background (7–8 mGy/year) in China was found among women over 35 years old. These women were more fertile than women of the same age in the control area, $p < 0.05$. There is no difference in fertility in highly irradiated younger women compared to controls (Wei et al., 1990). Monitoring of two groups of newborns (26,150 from an area with an increased background radiation and 10,654 from an area with a normal background (dose load of about 1.5 mGy/year) showed no differences in any reproductive parameters (Jaikrishan et al., 1999). It has been established (Cohen, 1993, 1995) that in the United States, under the influence of different radiation levels, there is an inverse correlation between the concentration of radon in homes and the incidence of mortality from lung cancer Figure 7.12.

These data are valid for both men and women. The results of these studies included 1,700 different US areas and 272,000 individual measurements in typical homes, which are home to 90% of the US population. It has been shown that, within a certain range of radon concentrations in residential air, lung cancer mortality in the United States decreased in direct proportion with increasing radon levels. The following specific figures are given for the decrease in mortality from lung cancer (the number of events per 10,000 inhabitants per year) with an increase in radon concentration: 18.5 mBq/L – 7; 37 mBq/L – 6.2; 74 mBq/L – 5.3;

185 mBq/L – 4.4. To prove the correlation of mortality from cancer precisely with the concentration of radon, the absence of correlation of mortality from cancer was demonstrated for 50 other indicators (gender, age, smoking, etc.). Figure 7.12 provides more detailed data by the same author (Cohen, 1993, 1995) on the dependence of mortality in the United States on radon concentration in homes. Lung cancer mortality was considered. It can be seen that the concentration of radon from one to five pCu/L has a favorable effect – the SMR from lung cancer is less than one. Interestingly, lower and higher concentrations of radon begin to have a harmful effect. Consequently, there is an optimal concentration of radon for the manifestation of radiation hormesis.

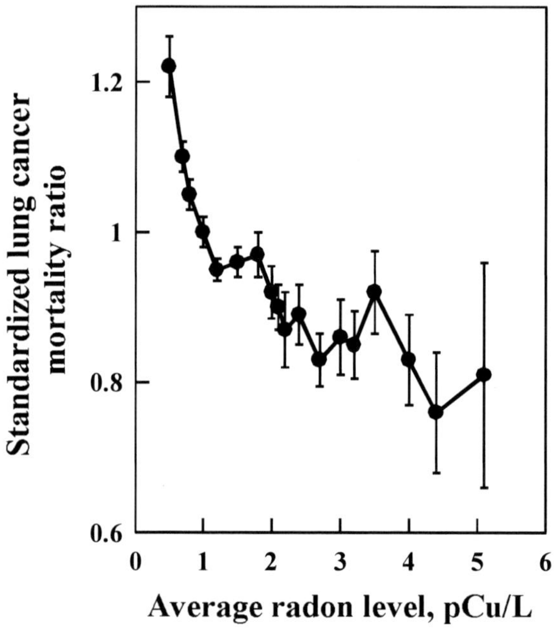

Figure 7.12. The dependence of lung cancer mortality on the radon concentration in US homes (Cohen, 1993, 1995).

Studies demonstrating the beneficial effect of increased radon background have also been carried out in southern Saxony, where the natural background of radon is increased (Becker, 2003). Similar results were obtained in Japan (Mifune et al., 1992; Kondo, 1993; Suzuki et al., 1994; Sobue et al., 2000). In the city of Misasa (Japan), an increased concentration of radon (26–57 mBq/L) was noted. Analysis of the standardized mortality ratio showed that for all cancer cases, the SMR was significantly lower (0.538 for women and 0.463 for men) in areas with increased radon concentration compared to the neighboring area without radon sources (external concentration of Rn – 11 mBq/L) – 0.850 for women and 0.770 for men (Figure 7.13).

Poisson's regression analysis, carried out in work (Mifune et al., 1992), showed that the relative risks among residents of this city were lower than in the control area for deaths from all cancers (0.67), from lung cancer (0.55) and stomach (0.59). It has been shown that the risk of lung cancer is not associated with increased radon concentration in Misasa (Sobue et al., 2000). These results are confirmed by the work (Suzuki et al., 1994), which provides data on the increased, compared with the control, cancer mortality of the population of the region in Japan with a reduced radon background.

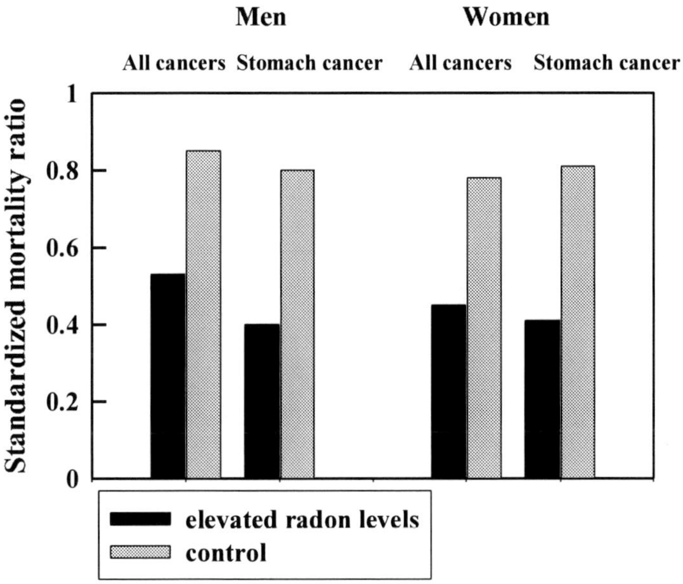

Figure 7.13. Comparison of SMRs from all cancers and gastric cancers in the Misas area and the control area (Mifune et al., 1992).

Similar results were obtained in Hungary (Lázár et al., 2003). Radon exposure was analyzed in 1,077 homes in two villages, and cancer deaths in the last 30 years were studied. Among 1,072 people, 60 cancers were found, only 4 of them were lung cancers. The minimum incidence of cancer was recorded in groups with a radon exposure of 110–185 mBq/L, where the relative risk of cancer formation was 4.1 times less than in the group with less radon exposure, and 5 times less than in the group with higher radon exposure. The results did not show an increase in lung cancer in the groups with radon levels up to 200 mBq/L, while the groups with radon exposure above 200 mBq/L had an increased risk of lung cancer.

These data demonstrate that radiation hormesis can be induced not only by sparsely ionizing radiation, but also by radiation with high LET (alpha particles of radon and its decay products). At the same time, an interesting regularity is revealed, previously described in works (Luckey, 1991; Kuzin, 1995), about the existence of some optimal value of the radiation intensity at which radiation hormesis might be observed.

7.4. Consequences of the Bombings of Hiroshima and Nagasaki

After the atomic bombing on August 6 and 9, 1945, respectively, of the cities of Hiroshima and Nagasaki in Japan, a register of hibakushi (residents exposed to radiation), who received external irradiation in a large dose range, has been functioning for more than 75 years, while the average radiation dose was 220 mGy. The register has data for 86,572 peoples (Table 7.1).

Table 7.1. The number of people in the dependence on the radiation dose received during the atomic bombing of Hiroshima and Nagasaki according to the hibakushi register

Radiation dose, Sv	Number of persons
Less 0.005	36459
0.005 – 0.02	16921
0.02 – 0.05	9390
0.05 – 0.1	6538
0.1 – 0.2	5467
0.2 – 0.5	6308
0.5 – 1	3202
More 1	2287

The main conclusions of the researchers of the Japanese registry are used as the basis for the recommendations of the UNSCEAR, ICRP, the World Health Organization (WHO), and the International Atomic Energy Agency (IAEA) for assessing the radiation risks of oncological and non-oncological diseases. Among the total survivors, leukemia cases were 21% higher than the Japanese average. For solid tumors, it was revealed that only 4.3% of oncological diseases could be caused by increased acute radiation. Some data from Japanese researchers indicated that acute exposure to low doses (below 100 mSv) could be not only harmless, but even beneficial to health (Kondo, 1993). In 1962, in an extensive study of mortality among survivors of the atomic bombing of Hiroshima and Nagasaki, the authors (Beebe et al., 1962) reported that the population living outside the cities at the time of the bombing and surveyed as "unirradiated control" was characterized by unexpectedly lower mortality in relation to the death rate of the entire population of Japan. In fact, this was the population exposed to low doses. The ratio of observed mortality to expected mortality is 0.77 with a high degree of reliability ($p < 0.01$).

An analysis of the total mortality from all causes in Hiroshima and Nagasaki was carried out (Jablon et al., 1965). For men who received low radiation doses (less than 0.31 and 0.31–0.61 Gy), the ratio of the observed values to the expected values was 0.79 and 0.76, respectively. For women, this ratio was 0.77 with a slightly higher radiation dose (0.62–1.24 Gy). At doses less than 0.31 Gy, it was below unity (0.84). These data have been confirmed in subsequent studies of mortality over the period 1950–1978 (Kato et al., 1982).

According to Kondo (1993), against the general increase in the incidence of cancer with increasing radiation dose, there are dose levels lower than 40 cGy, when the incidence of cancer either does not differ from the control level, or is significantly lower than in the control. This provision applies not only to solid neoplasms, but also to leukemia. Confirmation is the data summarized in Table 7.2 (Shimizu et al., 1990).

Table 7.2. Frequency of mortality from cancer for the period from 1950 to 1985 years among the population of Hiroshima and Nagasaki

Dose, cGy	Number of subjects	Leukemia		Other malignant neoplasms	
		Number	Frequency, %	Number	Frequency, %
Control	34272	58	0.17	2443	7.13
1–9	23321	38	0.16	1655	7.10
10–99	15346	53	0.34	1283	8.36
100–199	1946	23	1.20	221	11.4

As can be seen from this table, the induction of malignant neoplasms, including leukemia, during the first 40 years after the atomic bombardment does not increase in comparison with the control frequency at doses up to 9 cGy of instantaneous irradiation. Moreover, it can be even assumed that the mortality rate even slightly decreases, and at doses above 100 cGy, it significantly increases.

The data published in (Hattori, 1994 a, b) show that in atomic bomb survivors, mortality from leukemia only at a dose of 40 cGy was significantly higher than the control and amounted to 300 cases per 10^5 people over 35 years (Figure 7.14). At a dose of 10 cGy, the mortality rate was 40 cases per 10^5 people over 35 years. In the control, the mortality rate was about 180 cases per 10^5 people over 35 years.

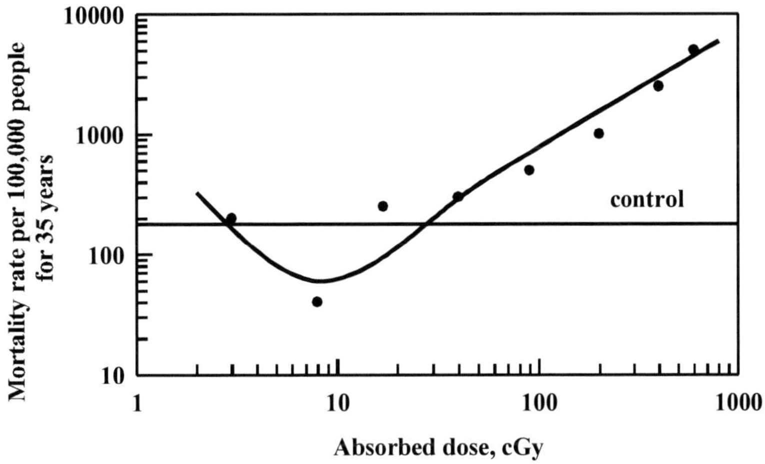

Figure 7.14. Dose dependence of mortality from leukemia in people bombed in Hiroshima and Nagasaki (Hattori, 1994 a, b).

Filyushkin and Petoyan (1988) compared the carcinogenic effects induced at different dose rates of ionizing radiation. It is based on a cohort of people who survived the atomic bombing in Japan. The authors conclude that the carcinogenic risk is weakened by protraction of the dose compared to the instantaneous implementation of the dose.

The US Committee on the Biological Effects of Ionizing Radiation reported no mortality from leukemia among 2,527 survivors of the Nagasaki bombing who received doses ranging from 31 to 69 cGy (BEIR, 1990). Any dose deviation outside this range led to an increase in the incidence of leukemia. The absence of cases of leukemia was confirmed in the work (Land, 1980). When analyzing mortality in both cities for the period from 1950 to 1978, it was shown (Kato et al., 1987) that the minimum mortality from all types of cancers was recorded at a dose of 7 cGy ($2.1 \cdot 10^3$ people/year), while the death rate in the control was $2.38 \cdot 10^3$ people/year. Very similar results were published in (Shimizu et al., 1992). These authors identified a dose range (0.5–20 cSv), within which mortality from leukemia was lower than in the control.

Despite the tragic consequences of the atomic bombings of Japanese cities, even under these circumstances, the above data indicate a possible radiation hormesis after exposure to small doses of ionizing radiation and thus demonstrate the possibility of the existence of a threshold for the harmful effects of radiation.

7.5. Radiation Accidents

In September 1957, a serious incident occurred at the "Mayak" enterprise (Eastern Urals), and a huge amount of radioactive waste got into the Techa River and the environment. As a result, thousands of people received high doses, but no direct deaths from high doses of radiation were recorded. The work (Kostyuchenko, Krestinina, 1994) provides data on 7,852 people who received average doses of 496, 120, and 40 mSv after the explosion of the high-level waste storage tank at the "Mayak" enterprise. It has been shown that at doses less than 300 mSv (Figure 7.15), the mortality rate of exposed residents was below the national level. It was also demonstrated that, over the entire studied dose range, mortality from cancer of exposed residents was lower than the non-irradiated population of the Chelyabinsk region (Figure 7.16).

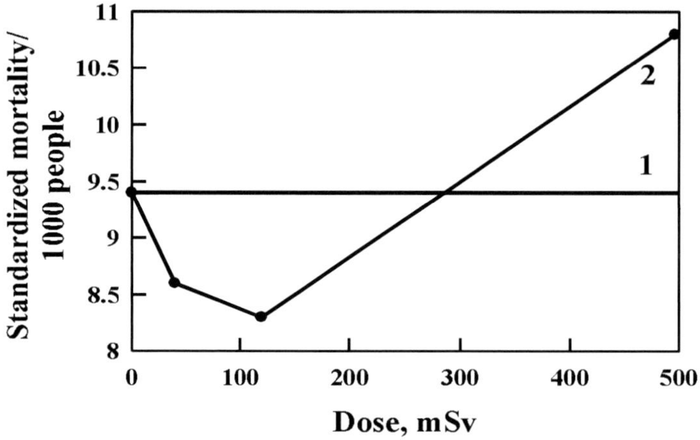

Figure 7.15. The dependence of the mortality rate of the population after the accident at the "Mayak" enterprise on the received radiation dose (Kostyuchenko, Krestinina, 1994): curve 1 – national mortality rate in the country, curve 2 – mortality of the exposed residents.

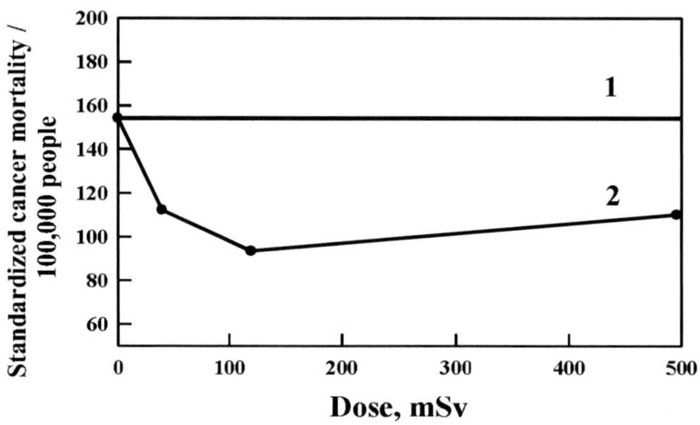

Figure 7.16. The dependence of cancer mortality on the dose received by the population after the accident at the "Mayak" enterprise (Kostyuchenko, Krestinina, 1994): curve 1 – unirradiated population of the Chelyabinsk region, curve 2 – exposed residents.

The most serious was the accident at the Chernobyl nuclear power plant on April 26, 1986. It was indeed a disaster and the saddest tragedy in the nuclear industry. To eliminate the consequences of the accident, about 200,000 liquidator workers were involved. In 1986 and 1987 they received a dose of about 100 mSv. The health status of these workers was studied by the International Chernobyl Project, supported by the IAEA and WHO. Scientists from Russian medical centers and agencies have also joined them. In the work (Luan et al., 2006), it is noted that according to the results of the implementation of the International Chernobyl Project on the basis of a linear non-threshold model, 150 additional deaths from leukemia were predicted among 200,000 liquidators within 10 years. Assuming that the liquidators had a similar cancer mortality as the Japanese survivors, an additional 100 deaths from leukemia were predicted within 10 years after the accident. However, in fact, only 48 cases of leukemia were verified within 13 years after the accident (Ivanov, Tsyb, 2002; Ivanov et al., 2001, 2004). The authors of the work (Luan et al., 2006) concluded that chronic exposure of the liquidators reduced their mortality from leukemia. They suggest that the overall cancer mortality rate for these workers should also be reduced.

The results of medical studies carried out on the basis of the Russian National Radiation and Epidemiological Register indicate that in the Bryansk region, which was most exposed to radiation in Russia after the Chernobyl accident, the incidence of leukemia in the population does not exceed the same value for the country as a whole; no increase and overall mortality from cancer was recorded (Ivanov, Tsyb, 2002; Ivanov et al., 2004). Another study of 65,905 liquidators 8 years after the accident also showed that mortality from cancer and other causes was somewhat reduced (Ivanov et al., 2001; Ivanov, 2007). The analysis of the standardized mortality ratio carried out in these works showed that this ratio for all causes of death of liquidators and for mortality from malignant neoplasms (1991–1998) was lower than 1, – lower background mortality in the country. An analysis of the mortality rate of 105 liquidators who lived in the Kaluga region for 10 years after the accident showed (Lushnikov, Lantsov, 1999) that 41.9% of them died from injuries and poisoning, 33.3% died from diseases of the circulatory system. The mortality structure corresponds to that for men of working age. This conclusion is confirmed by the results of work (Tsyb, Ivanov, 2006).

7.6. Professional and Medical Exposure

Two US organizations – the American Nuclear Society (ANS) and the Health Physics Society (HPS) – have stated that 50 or 100 mSv/year over a lifetime has no apparent deleterious effect on nuclear workers. Indeed, if 100 mSv of acute exposure may be beneficial (Kondo, 1993), then it is logical that 100 mSv of chronic exposure may also be beneficial.

Numerous studies of life expectancy and cancer incidence have been carried out on radiographers professionally irradiated by X-rays at work (USCEAR, 1982). The data obtained directly confirm radiation hormesis. Until 1940–50, when there was no adequate protection for workers related with X-rays and professionals received high doses of radiation, their high mortality and a significant increase in the incidence of cancer were noted (Yoshinaga et al., 2004). As the protection increased and, consequently, the individual radiation dose decreased, the differences in life expectancy and cancer incidence among radiologists and other medical doctors decreased. Aoyama et al. (1987) studied mortality among 2,028 radiological technicians in Japan over the period 1969–1982. It turned out to be 28% lower in comparison with the mortality rate of the entire population. A retrospective analysis of 12,195 male radiologists who received low doses of radiation for a long time showed (Yoshinaga et al., 1999) that for the period from 1963 to 1993 only 1,097 deaths were recorded, including 435 from cancer. Significant reductions in SMR have been shown for all cancers, including stomach and lung cancers, compared to cancer deaths in the general Japanese population. Similar results (Mohan et al., 2003) were obtained in a survey of 146,022 American radiology technicians (of which 73% were women). The mortality risk of radiologists was lower for all cancers (SMR = 0.82) compared to the US population.

In accordance with the prognosis based on a linear, threshold less model, 1.5–2% of additional cancers due to computed tomography were expected from all cancers in the US population (Scott et al., 2008). As a result of the analysis, the authors conclude that the use of computed tomography can reduce, rather than increase, the risk of cancer. Thorough medical examinations of nuclear power professionals are carried out in all countries.

The results of extensive epidemiological surveys have been published in many works (Abbatt et al., 1983; Matanoski et al., 1984; Beral et al., 1988; Gilbert et al., 1989 a, b; Wing et al., 1991; Kendall et al., 1992; Gribbin et al., 1993; Wiggs et al., 1994). Based on the data from UK National Register of Nuclear Workers, a cause-of-death analysis was carried out (Kendall et al., 1992) for workers in the UK's main nuclear industry (95,217 in total). It was shown that the risk of mortality from all malignant neoplasms in the study cohort was lower (SMR = 0.84) than in the country. For leukemia, the death rate was equal to that expected for the country. The manifestation of the beneficial effect of radiation in reducing the mortality rate was observed in the dose range from 10 to 30 cSv over 33 years of work. The authors emphasize (Kendall et al., 1992) that such a conclusion can be considered significant, because it was obtained in sufficiently large populations in which workers were equally healthy both upon entering the job and during the period of labor activity, i.e., the "healthy worker" effect did not appear. The concept of a "healthy worker" includes increased health requirements for workers at nuclear enterprises, their regular medical examinations and their greater material security. Further evidence that the mortality rate in the nuclear worker cohort is below the UK national level is provided by Darby et al. (1993) showing that the observed death rate for all cancers was below control (SMR = 0.86). In a survey of 22,552 workers at the UK Atomic Weapons Complex between 1951 and 1982 mortality from all causes was found to be 23% lower than the national level and from cancer by 18% lower (Beral et al., 1988).

Similar monitoring of the health and mortality of people associated with the nuclear industry has been and is being carried out in the United States. A study of 35,000 men and women workers at a nuclear plant in Hanford (USA) showed (Gilbert et al., 1989 b) that the mortality rate at the plant was significantly lower for all cancers (SMR = 0.85) than for the country as a whole. A study of workers at several US nuclear facilities (Gilbert et al., 1989 a) found no correlation between radiation exposure and mortality from all types of cancer, including leukemia. In a study of 9,000 workers at the Calvert Cliffs nuclear power plant (USA) from 1969 to 1988 346 deaths were identified, 101 of them from malignant neoplasms (Jablon, Boice, 1993). The average cumulative dose was low – 21 mSv, 12% of workers received a dose of more than 50 mSv. No relationship was found between the dose of ionizing radiation and possible death from neoplasms. Only two deaths from leukemia have been reported, with four nationally expected. A study (Wiggs et al., 1994) was conducted of a cohort of 15,727 workers at the Los Alamos National Laboratory (Nuclear Center). The results indicate that all-cause mortality among workers employed from 1943 to 1977 was significantly below control (SMR = 0.63). Cancer mortality was also significantly lower than the control (SMR = 0.64). A comparison was made of three groups of workers at a US nuclear shipyard – 27,872 people who received cumulative doses of more than 5 mGy; 10,348 workers who received doses less than 5 mGy; 32,510 unirradiated shipyard employees (Sponsler, Cameron, 2005). It was shown that mortality in the cohorts of exposed workers was 24% lower (SMR = 0.76) than in the non-irradiated control. In the irradiated cohorts, mortality from all malignant neoplasms was 17% lower than in controls.

In Finland, a survey of 15,619 operators of nuclear reactors was carried out (Auvinen et al., 2002). There was no increase in cancer incidence in general. All-cause and cancer mortality analysis conducted by French scientists from 1977 to 2004 on a cohort of 9,285 nuclear fuel cycle workers showed that these figures were lower than those in the country (Metz-Flamant et al., 2009). Results from a study of 2,776 Slovak nuclear workers showed that SMRs were 0.39 and 0.46 for men and women, respectively, for all deaths compared to the general population of Slovakia (Gulis, 2003). When analyzing mortality from cancer of nuclear workers in Japan, no statistically significant deviations were found compared with the Japanese population (SMR = 0.98), for cases of diseases not associated with tumors, the SMR was 0.86 (Iwasaki et al., 2003). Workers were surveyed at 5 Belgian nuclear facilities. Mortality from all causes and from cancer from 1969 to 1994 among men it was below the national level (Engels et al., 2005). 15% fewer cases of cancer have been reported among workers in the Australian nuclear complex than among ordinary people in the same region of Australia (Habib et al., 2006). 4,844 workers who had worked at least 10 years at nuclear facilities in Germany were examined. The results showed (Hammer et al., 2008) that mortality from all causes and from cancer in the study cohort was much lower than the national one (SMR = 0.54 and 0.66, respectively).

The body of research findings on the health of nuclear industry professionals from different countries includes a large number of workers and carefully selected non-irradiated controls. The control cohort included people from the same territories, taking into account gender, age, smoking, and socio-economic factor. In all of the above publications, there is a decrease in cancer mortality with increasing doses accumulated over the period of work experience (average 34 years). Luckey (2007) averaged the results of these groups of researchers (bold line in Figure 7.17) and found that 15 mSv/year can completely prevent cancer mortality before aging. Of course, no linear extrapolation is possible in biology, nevertheless these data are impressive.

Epidemiological results from studies of the health status of nuclear industry professionals from different countries were included in a joint work published by the International Agency for Research on Cancer (Cardis et al., 2007; Thierry-Chef et al., 2007; Vrijheid et al., 2007). The results of this large-scale analysis were obtained by averaging data from cohorts of workers in the nuclear industry from 15 countries of the world: Australia, Belgium, Canada, Finland, France, Hungary, Japan, Republic of Korea, Lithuania, Slovakia, Spain, Sweden, Switzerland, United Kingdom and the United States. These studies included 407,391 nuclear workers who worked for at least one year. The authors report (Cardis et al., 2007) that the estimates obtained are slightly higher, but statistically consistent with the estimates on which the current radiation protection recommendations are based.

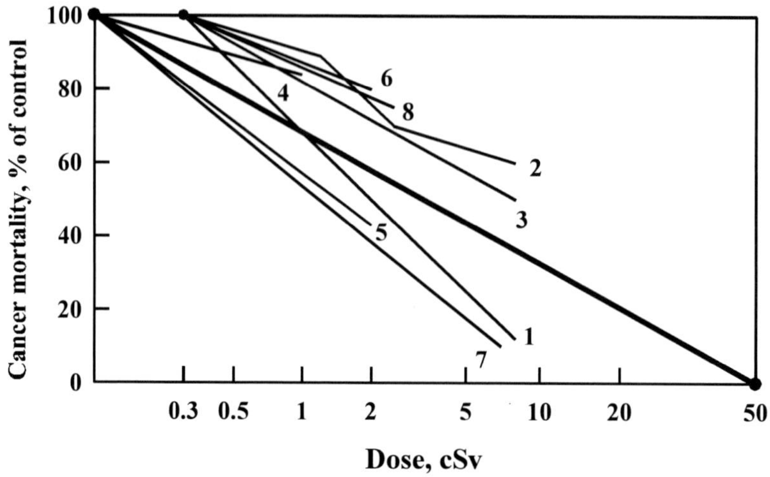

Figure 7.17. Dose dependence of cancer mortality among exposed workers in different countries (Luckey, 2007). Curve 1 – Kendall et al. (1992); curve 2 – Beral et al. (1988); curve 3 – Gilbert et al. (1989 b); curve 4 – Gribbin et al. (1993); curve 5 – Wing et al. (1991); curve 6 – Matanoski et al. (1984); curve 7 – Abbatt et al. (1983); curve 8 – Wiggs et al. (1994).

When analyzing these averaged results, it was shown (Fornalski, Dbrzyński, 2010) that mortality among workers in nuclear enterprises is independent of dose and that data on total mortality of professionals have significant scatter, on the basis of which it is impossible to draw definite conclusions about the risks. However, these authors suggest that protracted exposure to low doses of ionizing radiation is not associated with an increased risk of cancer and cancer mortality. Luckey (2007) compared the health status of 152,000 nuclear workers who received an average dose of 55 mSv and 149,000 unirradiated workers with similar working conditions. He showed that among the exposed workers, cancer mortality was lower than in the control group.

Similar results were obtained by Korean scientists who analyzed 11 epidemiological studies of professionals working at nuclear facilities around the world. There was a significant decrease (SMR = 0.75) in mortality from all cancers (Park et al., 2010).

7.7. Taiwan Incident

The fact that low doses of ionizing radiation are beneficial to humanity seems unlikely, but the health effects of chronic exposure to residents who have lived in ^{60}Co-contaminated apartments for over 30 years in Taipei (Taiwan) definitely indicate that the results other scientists about the usefulness of small doses of radiation, which were mentioned above, may be quite correct.

In 1982–1983 a ^{60}Co source (or several sources) accidentally ended up in a landfill for metal waste, which was melted down and used for the construction of houses. In 1983–1984 180 buildings were built, which housed about 1,700 apartments. The first contaminated apartment was discovered in 1992, and then, one after another, the others were also identified. About 10,000 people have lived and continue to live in these apartments, being exposed to chronic irradiation at a dose rate up to 1 mSv/h, and the estimated maximum annual dose reached 1 Sv. In the very first year of residence (1983), the dose rate reached 50 mSv/year.

After 21 years of living in a contaminated apartment, the average total cumulative dose ranged from 0.4 to 6.0 Sv (Chen et al., 2004). These doses were higher than the average doses received by the survivors after the bombing in Japan and higher than the doses received by the liquidators 2 years after the Chernobyl accident.

This unintentional human experiment proved that chronic exposure to humans at specific dose rates could effectively reduce the rate of human cancer. Indeed, according to the linear thresholdless concept, doses received by residents in Taiwan could induce 36 increased deaths from leukemia and 40 from solid cancers after 21 years of residence. In reality, however, such figures have not been confirmed. On the contrary, in several publications (Luan et al., 2006; Chen et al., 2006), the authors noted that cancer mortality among adults living in these buildings for 9–20 years was only 3% of the total cancer deaths of adult population (Figure 7.18). According to Taiwan government statistics, 200 spontaneous cancer deaths should be expected in the population of people living in ^{60}Co-contaminated apartments, in fact, there were 5 such cases, or 2.5% of the average recorded in Taiwan. Analysis of chromosomal aberrations among these patients did not reveal significant changes.

More than 2,000 children have been born in contaminated apartments since 1983, among them, according to statistics, there should be about 46 children with congenital inherited defects (Down syndrome, hemophilia, adrenoleukodystrophy, mucopolysaccharidosis, cerebral palsy, etc.). In fact, only 3 children have been identified with congenital heart disease, or only 6% of similar diseases in the general population of Taiwan (Luan et al., 2006). On this basis, it was suggested that we live in conditions of a deficit of ionizing radiation.

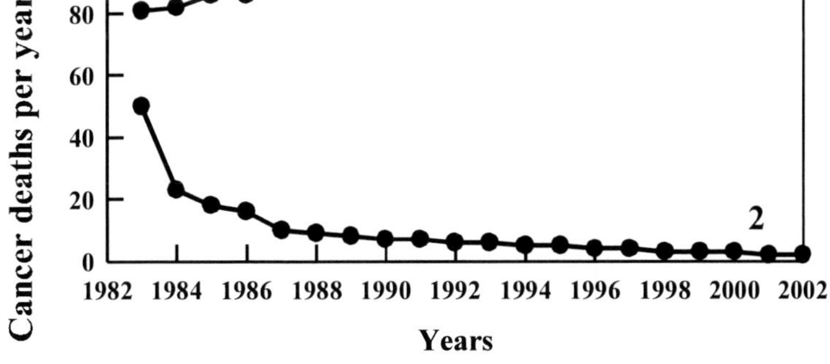

Figure 7.18. Cancer mortality in the adult population of Taiwan (curve 1) and ^{60}Co irradiated residents (curve 2) (Chen et al., 2006).

Since the radiation doses from the natural background and from the peaceful use of nuclear energy, as well as the medical and diagnostic use of ionizing radiation have never been higher than the doses accumulated by residents living in contaminated apartments in Taiwan, it is possible that the increased chronic exposure of the population and nuclear energy professionals might be helpful. There are many reviews and monographs devoted to the positive effect of ionizing radiation (Luckey, 1980 a, 1991; Kuzin, 1991, 1995; Calabrese, Baldwin, 2002; Buldakov, Kalistratova, 2003, 2005; Vaiserman et al., 2010; Koterov, 2010; UNSCEAR, 2010; Petin, Pronkevich, 2011). The main conclusion that follows from these works is the need for a natural background radiation for the existence of the biosphere. It has been shown that a decrease in NBR leads to various deviations from normal development. An increase in the NBR by 10 and even 100 times in many cases had a positive effect on the life expectancy of humans and animals, immunity, fertility and other indicators of vital activity. Moreover, in some of the publications cited above, the positive effect of higher doses of radiation was noted.

***

In this section of the work, a review of the literature data on the effect of low doses of ionizing radiation on life expectancy and the yield of oncological diseases was carried out. A large set of data was analyzed for various cohorts of people exposed to high doses of ionizing radiation. In accordance with these data, people living in conditions of increased natural background radiation (5 mSv/year and above) do not suffer from increased cancer mortality or reduced life expectancy compared to adequately matched controls. Moreover, in some areas with increased NBR, even lower rates of cancer and cancer mortality are observed compared to the national average (Eckhoff et al., 1974; Jagger, 1998; Nair et al., 2009; Wei et al., 1990). These observations, if they do not prove the protective function of irradiation in small doses, then, in any case, refute the thesis of an obligatory increase in carcinogenesis with an increase in the natural background.

The combination of the above examinations of people living with an elevated radon background shows that there really is some optimal concentration of radon, at which the minimum release of lung and laryngeal oncological diseases is recorded (Cohen, 1993, 1995). Both an increase and a decrease in radon concentration in comparison with the optimal value leads to an increase in the incidence of cancer. Thus, contrary to the idea of the predominantly damaging effect of radon, its possible beneficial effect is shown, which reduces the yield of lung cancer. This contradicts the recommendations of the US Environmental Protection Agency (1991), which believes that any amount of radon inevitably leads to poor health and increased incidence of lung cancer.

The atomic bombings of Hiroshima and Nagasaki are the only examples of the use of nuclear weapons in the history of mankind. The total death toll was at least 140,000 in Hiroshima and about 80,000 n Nagasaki. Despite these tragic consequences, it was noted that the population living outside the cities at the time of the bombing and surveyed as "unirradiated control" had a lower mortality rate relative to that of the entire Japanese population. In fact, this was the population exposed to low doses. The ratio of observed mortality to expected mortality is 0.77 (Beebe et al., 1962). A dose range (0.5–20 cSv) was identified, within which mortality from leukemia was lower than in controls (Shimizu et al., 1992).

Other examples of significant severe consequences of exposure to ionizing radiation on people were radiation accidents at the “Mayak” enterprise in 1957 and at the Chernobyl nuclear power plant in 1986. In these cases, despite the tragic consequences, analysis of literature data showed that there were revealed the positive effects of radiation. According to the results of the implementation of the International Chernobyl Project, 150 additional cases of leukemia were predicted, but in fact only 48 such events were verified. The authors of many analyzed works conclude that with chronic irradiation of liquidators, no increase in overall mortality from cancer was recorded (Ivanov, Tsyb, 2002; Luan et al., 2006; Ivanov, 2007).

The positive effect of radiation on the yield of oncological diseases has been published in many papers and monographs for people exposed to radiation due to professional activities. Averaging the results obtained in 15 countries at nuclear facilities did not support these conclusions (Cardis et al., 2007; Thierry-Chef et al., 2007; Vrijheid et al., 2007). However, further analysis of these results showed that protracted exposure to low doses of ionizing radiation is not associated with an increased risk of cancer and mortality (Luckey, 2007; Fornalski, Dobrzyński, 2010).

The most convincing evidence of the positive effect of ionizing radiation has been demonstrated after 25 years of residence of 10 thousand people in ^{60}Co-contaminated apartments in Taiwan. According to Taiwan government statistics, 200 spontaneous cancer deaths should be expected in this population, in fact there were 5 such cases, or 2.5% of the average value reported in Taiwan (Luan et al., 2006; Chen et al., 2006). No genetic consequences of chronic exposure to these residents have also been reported. On this basis, it is suggested that we live in conditions of a deficit of ionizing radiation. Moreover, there are publications suggesting that chronic exposure to ionizing radiation can be used as a means of preventing cancer (Chen et al., 2006; Luckey, 2007).

According to many researchers (Bloom et al., 1983; Luckey, 1991; Kondo, 1993), the basis for the beneficial effect of low doses of radiation is their stimulating effect on cells. The mechanism of health improvement after exposure to low doses of radiation may be associated with an increase in immune activity (Bloom et al., 1983, 1987; Liu et al., 1987; Makinodan, 1992; Hattori, 1994 b; Liu, 2004), with an increase in cellular and body adaptation (Smith-Sonneborn, 1992; Hattori, 1994 a; Liu et al., 1992, 1996), with activation of metabolic reactions (Luckey et al., 1978; Luckey, 1980 b). The process of lymphocyte repair is stimulated by low doses of ionizing radiation (Tuschl et al., 1980). Another point of view is related to the launch of cell DNA repair systems under the influence of low doses of ultraviolet radiation. This assumption was put forward and widely discussed in the works (Morozov, 1996 b; Petin, Morozov, 2015). In accordance with this hypothesis, the Čerenkov radiation, within which ultraviolet light is present, can trigger the cell repair system. This radiation occurs when charged particles move at a speed exceeding the speed of light in a biological environment. Indeed, the stimulation of the restoration of the reproductive capacity of cells under the influence of ultraviolet radiation in low doses has been demonstrated earlier (Abbaszade, Mamedov, 1988).

According to (Makinodan, James, 1990), the immune response to radiation is fundamentally different at low and high levels of radiation. Figure 7.19 show that the number of spleen cells in mice treated with the antigen (sheep erythrocytes), in response to irradiation at a dose of up to 0.5 Gy, increases up to 180% during *in vitro* treatment. The biological effect of radiation is modeled by the stimulating effect of low doses of radiation on the antimutagenic biological system. The DNA damage control biosystem is also stimulated. This leads to a decrease in the level of metabolic mutations.

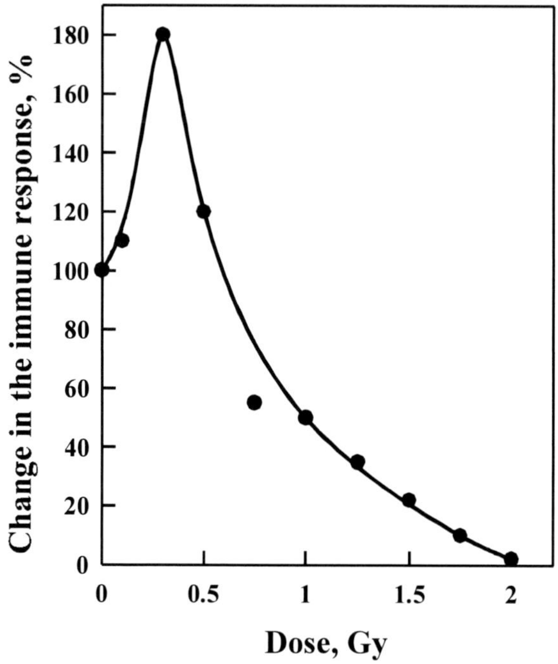

Figure 7.19. The immune response of spleen cells of mice treated with the antigen – sheep erythrocytes, to irradiation with sparsely ionizing radiation (Makinodan, James, 1990).

Our review would not be complete without mentioning the opposite point of view – about the harmfulness of small doses of ionizing radiation (Baryakhtar, 1995; Burlakova et al., 1996; Yablokov, 1997, 1998). The fallacy of the views of these authors has been refuted by life itself and the data presented in this work. Scientific criticism of such views has been presented in sufficient detail in the literature (Ilyin, 1994; Korogodin, Korogodina, 1997; Filyushkin, Petoyan, 1988).

The fear of radiation (radiophobia) is associated with radiation accidents, testing of atomic weapons and even with the operation of nuclear power plants. It is known that radiophobia and ionizing radiation can interact synergistically, i.e., strengthen each other's action. It is not excluded that the consequences of the accident at the Chernobyl nuclear power plant could be intensified by the fear of radiation. That is why it seems relevant to achieve a reduction in radiophobia, especially when exposed to low doses of ionizing radiation. The totality of the presented data shows the possible positive effect of low doses of radiation on the life span and oncological morbidity of a person and thereby demonstrates the potential for reducing radiophobia. In addition, these data indicate the fundamental importance of radiation hormesis and the need for detailed and extensive research in this direction. It should be noted that on the basis of the presented results, one should not draw any conclusions related to changes in radiation safety standards. This requires further long-term studies of the effect of various modes of chronic irradiation on viability, cancer incidence, reproductive capacity, immunity, genetic stability and other indicators of biological objects of varying degrees of complexity.

Chapter 8

Involvement of Molecule Excitations in the Biological Action of Ionizing Radiation

Abstract

The involvement of molecule excitations in the biological action of ionizing radiation is demonstrated for UV-hypersensitive pro- and eukaryotic cells. Photoreactivation after combined action of ionizing radiation and visible light is observed. New experimental data on the antagonistic interaction of ionizing radiation and visible light are presented. Parameters of bacterial photoreactivation after exposure to UV light and ionizing radiation are calculated and compared. The results obtained demonstrate that the action of high-energy ionizing radiation is accompanied by irradiation with UV light that occurs when charged particles move at a speed exceeding the speed of light in a given medium. The dose of UV light accompanying the irradiation of biological objects with high energy of sparsely ionizing radiation has been estimated. The observed plateau of the UV dose with an increase in both the energy of ionizing radiation and in the volume of the exposed cell suspension is shown to be determined by physical properties of the Čerenkov emission.

8.1. Photoreactivation after the Combined Action of Ionizing Radiation and Visible Light

For a long time, it had been considered that the significance of excitation was negligible and the biological effect of ionizing radiation is totally due to ionizations. The discovery of the phenomenon of photoreactivation after exposure of some bacterial and yeast cells to ionizing radiation showed that a part of damage produced by ionizing radiation is UV-like and due to molecule excitation (Myasnik, Morozov, 1977; Redpath, Tortorello, 1977; Redpath, Zabilansky, 1979; Myasnik et al., 1980; Petin, Komarov, 1982, 1985). Such UV-like damage can be detected only by the photoreactivation of UV hypersensitive cells. There are indications that the photoreactivable part of damage is similar whether induced by ionizing radiations or UV light and this damage was thought to arise from production of an excited state (Wang, Smith, 1978; Moss, Smith, 1980). Several lines of experimental data and theoretical considerations are in favor of the idea that the photoreactivable damage induced by ionizing radiation might be UV-like damage produced by Čerenkov light emitted by charged particles when their velocity exceeds the phase velocity of the light in the medium (Morozov, Myasnik, 1980; Michael et al., 1981; Munson, 1981; Redpath et al., 1981, 1982; Morgan et al., 1982; Petin, Komarov, 1985).

Already many years ago, the photoreactivability of bacterial cells of certain genotype was discovered after ionizing radiation and some features of the phenomenon have been identified (Myasnik, Morozov, 1977; Redpath, Tortorerllo, 1977; Redpath, Zabilansky, 1979; Myasnik et al., 1980; Munson, 1981). There are evidences that the photoreactivable damage is the same whether induced by UV light or ionizing radiations and this damage was thought to arise from excited state production (Moss, Smith, 1980; Skvortzov et al., 1981). Some experimental data and theoretical considerations are in favor of the idea that the photoreactivable damage induced by ionizing radiation might be UV-like damage produced by Čerenkov light emitted by charged particles when their velocity exceeds the phase velocity of light in the medium (Morozov, Myasnik, 1980; Moss, Smith, 1980; Michael et al., 1981; Munson, 1981; Redpath et al., 1981, 1982; Morgan et al., 1982).

Last year's interest in the phenomenon of photoreactivation resumed. There appeared some new evidence that the ability of cells to photoreactivation may reduce solar injury, prevents or reduces the risk of tumors induced by ultraviolet light (Ley et al., 1991). The decrease in the effectiveness of UV and γ-disinfection of liquids may be also related with photoreactivation of microorganisms (Hijnen et al., 2006; Chatzisymeon et al., 2011). It was demonstrated that photoreactivation after ionizing radiation exhibits some properties identical to those after UV-irradiation (Vinicombe et al., 1978; Wang, Smith, 1978; Redpath, Zabilansky, 1979; Moss, Smith, 1980; Skvotzov et al., 1981). In this section, we deal with our new data concerning the dependence of photoreactivation of bacterial and yeast cells on both the energy of sparsely ionizing radiation and the volume of the irradiated cell suspension.

Bacteria *Escherichia coli* (strain $B_{s\text{-}1}$), yeast *Saccharomyces cerevisiae* (strain T4) and *Pichia pinus* (strain MN4-D) were used for experiments. The culture conditions, irradiation and dosimetry of sparsely and densely ionizing radiation and the illumination with photoreactivating light were described elsewhere (Petin, Morozov, 2015). The X-ray sources were machines RUM-7 and RUM-17 operated at 40 kV and 150–200 kV, electron accelerator (25 MeV) produced pulsed X-ray beam. Gamma irradiation was performed using a ^{60}Co source (Gammacell 220, Atomic Energy of Canada, Ltd.) and a ^{137}Cs source ('Kolos'). Ferrous sulfate dosimetry, ionization chambers, and LiF dosimeters were used for measuring dose rates. For photoreactivation, a part of irradiated suspension was illuminated for 20 minutes by a visible light. All other manipulations were performed under dark or red ambient light. Colony-forming ability was assessed by plating of irradiated cells on nutrient agar and counting macrocolonies after 24 hours incubation of bacterial cells at 37°C and 3-5 days of yeast cell growth of at 30°C. The magnitude of the photoreactivation was evaluated by the dose-modifying factor (DMF) which was defined as the ratio of radiation doses producing the same effect in survival curves with and without photoreactivation. The data obtained for *E. coli* are presented in Figure 8.1. As can see, the photoreactivation phenomenon takes place after exposure to sparsely ionizing radiation (X-rays of different energy and γ-rays) beginning from certain energy. The DMF is increased from 1.00 for low voltage X-rays up to 1.6 for high-energy X-rays. A statistically significant effect was observed for ^{137}Cs-irradiation (670 keV).

The cells of *E. coli* K-12 (*uvr*A*rec*A) and *E. coli* $B_{s\text{-}1}$ (*uvr*A*lexr*A) for which the photoreactivation phenomenon was demonstrated are hypersensitive to UV-irradiation. It is of interest that biochemical investigations have shown that 100 Gy of ^{137}Cs γ-radiation is equivalent to 0.07 J/m^2 of 254 nm-radiations in the production of pyrimidine dimmers that can be photoreactivated (Wang, Smith, 1978). Since the photoreactivation effect of visible light was observed after exposure to low-LET ionizing radiations in strains, which are defective in some major dark repair systems, one may suppose that photoreactivable damage is produced in all cells irradiated but it is not manifested for cells capable of recovery from this damage due to thymine dimmer excision and postreplication repair.

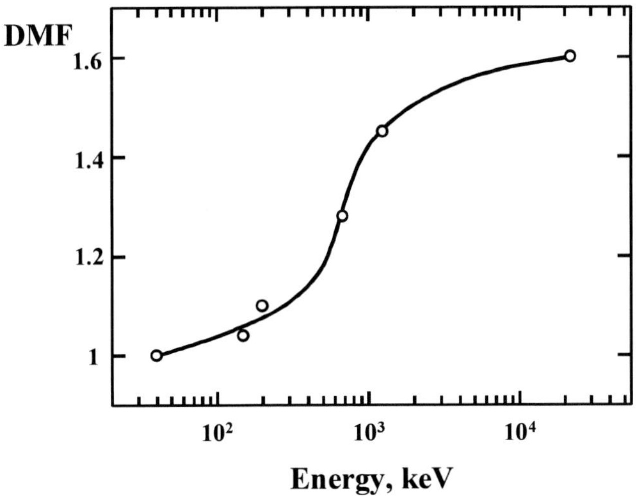

Figure 8.1. The dependence of the magnitude of photoreactivation (DMF) of *E. coli* B_{s-1} on the energy of low-LET ionizing radiation.

A simple mathematical model was suggested (Petin, Komarov, 1982) to estimate the relative contribution of excitations and ionizations in lethal effect of ionizing radiation. The proposed model indicates also an explanation for the difficulties encountered in observing photoreactivation after ionizing radiation in eukaryotic cells. In particular, yeast cells are characterized by a considerably smaller amount of photoreactivation after UV light exposure (DMF = 2.5) in comparison with bacterial cells (DMF = 5–7). The cited model predicts that for equal contribution of excitations to the lethal effect of yeast and bacterial cells the extent of photoreactivation after ionizing radiation will also be much smaller for eukaryotic than for prokaryotic cells.

If the photoreactivated damages are mainly due to Čerenkov light, the increase in both volume of irradiated suspension and energy of ionizing radiation would favor the greater yield of such damages since the yield of Čerenkov light is known to be increased with increasing parameters mentioned (Jelly, 1958). This, indeed, was found to be the case for bacterial cells (Morozov, Myasnik, 1980; Myasnik et al., 1980). The oxygen effect is known to be exerted predominantly on the damages inflicted by ionization rather than excitation events, therefore hypoxic conditions allow to increase the total dose of ionizing radiation and results in the relatively larger contribution of Čerenkov light to the lethal effect (Petin, Komarov, 1982; Skvortzov et al., 1981).

Hence, to observe photoreactivation after ionizing radiation exposure in eukaryotic cells, one needs optimum conditions for the induction of photoreactivable damage: a large volume of irradiated suspensions, a high energy of sparsely ionizing radiation and hypoxia. Many strains of wild type and various rad-mutants of yeast cells were tested in these experiments. However, an appreciable photoreactivation was observed only for two diploid yeast strains that were the most UV-sensitive among other investigated. One of these was a double mutant (*rad2rad54*/*rad2rad54*) of *S. cerevisiae*, strain T4, defective in both excision and double strand break repair, and a wild strain (MN4-D) of *P. pinus*.

To verify the dependence of photoreactivation on the volume of exposed suspension, a 10 and 1,000-ml yeast cell suspensions were irradiated. To check the dependence of photoreactivation on energy of sparsely ionizing radiation, the cells were exposed by 25 MeV photons from 25 MeV pulsed electron accelerator (dose rate 50 Gy/min) and a ^{60}Co γ-ray source (dose rate 10 Gy/min). To test the dependence of photoreactivation on the degree of hypoxia, we have used two different concentration of yeast suspension (10^6 and 10^7 cells/ml). It is known that at high cell concentrations the oxygen consumption by maintenance energy metabolism of cell leads to anoxia under certain conditions. It was shown (Pohlit, 1973) that irradiation (without stirring) of 3 ml cell suspension in small cylindrical plastic dishes (3.2 cm in diameter) results in hypoxic or anoxic conditions for the cell concentration about of $3\cdot10^7$ and $6\cdot10^7$ cells/ml, respectively. In our experiments (without stirring), the area of contact of yeast suspension with air was smaller (2 and 5 cm^2) and the volume of irradiated suspension was larger (10 and 1,000 ml). It causes a partial hypoxia at the cell concentration about of 10^7 cells/ml. Concentrations in excess of 10^7 cells/ml result in increasing of optic density and therefore in decreasing of the value of photoreactivation (Moss, Smith, 1980).

DMF-values obtained for yeast cells are listed in Table 8.1 for yeast *P. pinus* (strain MN4-D) and *S. cerevisiae* (strain T4) exposed to ionizing radiation at various conditions. The errors involved in the DMF-values have been calculated from interexperimental variations. The results revealed the highest photoreactivation was observed for both strains at the following optimum conditions: 25 MeV X-rays, 1,000 ml suspension and cell concentration of 10^7 cells/ml.

Table 8.1. The dose-modifying factor (DMF) after photoreactivation of yeast cells exposed to ionizing radiations at various experimental conditions

Strain	Irradiation	Experimental conditions			
		10^6 cells/ml		10^7 cells/ml	
		10 ml	1,000 ml	10 ml	1,000 ml
Saacharomyces cerevisiae, T4	^{60}Co γ-rays	1.00 ± 0.03	1.04 ± 0.03	1.01 ± 0.02	1.07 ± 0.04
	25 MeV X-rays	1.02 ± 0.04	1.08 ± 0.03	1.05 ± 0.03	1.15 ± 0.03
Pichia pinus, MN4-D	^{60}Co γ-rays	1.01 ± 0.03	1.04 ± 0.03	0.99 ± 0.03	1.10 ± 0.05
	25 MeV X-rays	1.02 ± 0.02	1.12 ± 0.04	1.07 ± 0.04	1.31 ± 0.04

The presented results demonstrated that postirradiation illumination by visible light of yeast cells of certain genotypes exposed to ionizing radiation leads to an increase of the cell survival; the increment over the controls is not very large, but consistent throughout replica experiments, and highly significant for the optimum condition producing the maximum yield of Čerenkov light. From the foregoing it would appear that the effect of photoreactivation of yeast cells exposed to ionizing radiation may depend on the energy of sparsely ionizing radiation as well as on the volume of irradiated suspensions and hypoxic condition during irradiation. Hence, these data support the viewpoint that Čerenkov emission is mainly responsible for the induction of photoreactivable damage in both bacterial and yeast cells irradiated with high energy sparsely ionizing radiation. The hypothesis that Čerenkov light produces the photoreactivable damage in cells irradiated with ionizing radiation has been advanced by several authors (Morozov, Myasnik, 1980; Michael et al., 1981; Moss, Smith, 1980; Redpath et al., 1981). Some other observations are in favor of it. If the hypothesis was correct, then no photoreactivable damage should be produced by densely ionizing slow particles. Precisely this has been observed experimentally (Myasnik et al., 1980). Another confirmation of the hypothesis comes from the fact that the survival fraction of ^{60}Co γ-irradiated *E. coli* B_{s-1} decreases whereas the photoreactivation increases with increasing volume of irradiated suspension in the range from 0.5–500 ml (Morozov, Myasnik, 1980). Our calculations (Petin, Komarov, 1982) have shown that the excitation contribution in these experiments increased from 12% up to 50% with increasing suspension volume. Further confirmation comes from experiments measuring the effect of photoreactivation in bacterial cells exposed to ^{137}Cs γ-rays when using glass or quartz irradiation vessels (0.8 ml) with or without 800 ml of distilled water in the surrounding vessel (Moss, Smith, 1980). The surrounding water results in an increased inactivation and in an increase for photoreactivation, however, only for quartz and not for glass tubes. It was inferred that these effects were due to UV radiation produced by Čerenkov radiation in the surrounding water (Moss, Smith, 1980). For these conditions, the highest excitation contribution in this experiment was found to be of 56% (Petin, Komarov, 1982).

Michael et al. (1981) have also shown that a large part of the photoreactivation that can be achieved after ionizing radiation involves repair of damage induced by UV Čerenkov light. This was demonstrated by irradiation of suspensions either with electrons and with their concomitant Čerenkov light, or with a similar flux of net Čerenkov light that was produced by stopping the electrons in high purity fused silica placed immediately in front of the suspension. In the latter case, the dose-modifying factor of photoreactivable light was about 5, as was estimated from their published graphical data (Michael et al., 1981). It worth noting that this value obtained for net Čerenkov light are in the above-mentioned range (DMF = 5–7) observed for the dose-modifying effect of photoreactivatable light after UV-exposure alone.

It should be noted that our earlier experiments with photoreactivation after ionizing radiation described above were performed under conditions where photoreactivation is not always realized at optimal conditions: the intensity of the visible light used photoreactivation, duration of exposure, the temperature, at which the photoreactivation took place. These factors may lead to an underestimation of photoreactivation after ionizing radiation (Kim et al., 2005 b; Krivova et al., 2012). The data presented in Figure 8.2 have been obtained recently. Irradiation with ^{60}Co γ-rays was occurred at dose rate 27.7 Gy/min. Pulsed beam of bremsstrahlung formed by 18 MeV electron accelerator (SL-20, Phillips) at an average dose rate 2.9 Gy/min. For dosimetry measurements, calibrated ionization chambers (Siemens and Keithleg) have been used. For photoreactivation, bacterial cells irradiated in the dark condition with ionizing radiation were illuminated with visible light (5 W/m^2) at 37°C for 30 min.

It is evident that the value of photoreactivation of bacterial cells irradiated with ^{60}Co γ-rays at 500 ml is larger (DMF = 1.9) compared to the irradiation at 0.5 ml (DMF = 1.3). A comparison of curves 1 and 2 shows that the optimization of photoreactivation resulted in an increase in the effect observed. A similar series of experiments was carried out with bremsstrahlung. Comparing curves 2 and 3, one can draw a conclusion that the magnitude of cell photoreactivation is significantly larger after irradiation with high-energy of ionizing radiation for all volumes of cell suspension. These dependencies can be explained by the fact that electron track increases with the increase in both the energy of ionizing radiation and the irradiated volume of the suspension and, consequently, UV light flux, which is a part of Čerenkov radiation accompanying the action of ionizing radiation, is also increased. These data are in agreement with the idea that the photoreactivable damage induced by ionizing radiation might be UV-like damage produced by Čerenkov light emitted by charged particles when their velocity exceeds the phase velocity of light in the medium.

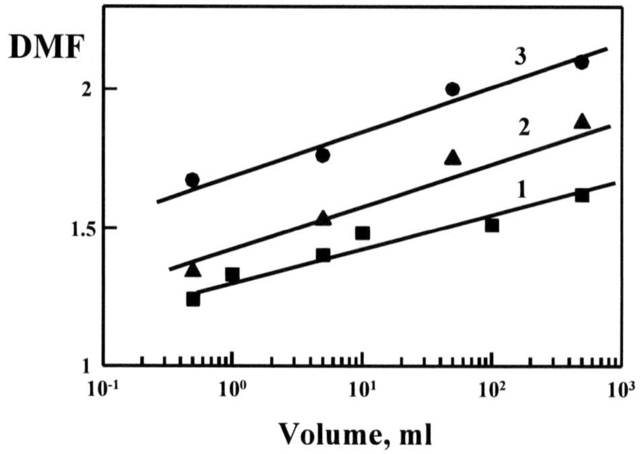

Figure 8.2. The dependence of the magnitude of photoreactivation (DMF) of *E. coli* B_{s-1} on the volume of irradiated cell suspension and energy of low-LET ionizing radiation. Curve 1, recalculated data after cell irradiation with ^{60}Co γ-rays (Morozov, Myasnik, 1980); curve 2, results of our new experiments with ^{60}Co γ-rays performed at the optimal conditions of photoreactivation; curve 3, cells were exposed to X-ray from 18 MeV electron accelerator.

In summary, the most basic conclusion justified by the above findings concerns the general nature of primary radiation damages inflicted by ionizing radiation. One can conclude that irradiation both prokaryotic and eukaryotic cells with high-energy sparsely ionizing radiations result in the production not only single- and double-strand breaks, base damage and cross-links between DNA molecules, but also pyrimidine dimers similar to those induced by UV light, which may be susceptible of photoreactivation. These results show that excitations in addition to ionizations may be considered as important physical interaction processes, which can give rise to a variety of physicochemical, biochemical, and biological reactions that finally produce observable effects in biological objects.

8.2. Parameters of Bacterial Photoreactivation after Exposure to UV Light and Ionizing Radiation

It was demonstrated that the extent and the rate of photoreactivation in *E. coli* following ionizing radiation decreased when compared with UV light (Myasnik, Morozov, 1977). The extent and the rate of bacterial and yeast cell photoreactivation were shown to be modified by a number of chemical and physical factors (Redpath, Tortorello, 1977; Redpath, Zabilansky, 1979, Morozov, Myasnik, 1980; Petin, Komarov, 1985). After exposure to ionizing radiation, the extent of photoreactivation was found to increase with the energy of damage-inducing ionizing radiation in both the bacterial (Redpath, Zabilansky, 1979; Myasnik et al., 1980; Moss, Smith, 1980) and yeast cells (Petin, Komarov, 1985). However, in none of the cited papers the probability of photoreactivation and the irreversible portion of radiation damage have been evaluated quantitatively. The decrease in both the extent and the rate of recovery may proceed through the impairment of the process of photoreactivation itself, i.e., through the decrease in the probability of recovery, or through changes of the repairable and irreversible portions of the damage inflicted. In the latter case, the probability of recovery may either decrease or stay unchanged. However, the data distinguishing these possibilities are lacking in the literature.

Thus, the here we will compare quantitatively the parameters of photoreactivation of a UV hypersensitive strain of *E. coli* B_{s-1} irradiated with UV light and ionizing radiations of different energies. On this basis, one may elucidate whether the change in the extent and the rate of recovery is proceeding through the impairment of the recovery process itself or through the alteration of the repairable and irreversible portions of the damage inflicted. We will also evaluate the influence of the different physical and chemical factors, which modify the cell radiosensitivity, on the parameters of photoreactivation kinetics of the bacterial cells exposed to ionizing radiation.

Figure 8.3A shows the survival curves for the bacterial cells irradiated with graded doses of UV light, pulsed X-ray beam, γ-ray of ^{60}Co and ^{137}Cs. The shape of the survival curve was purely exponential after UV light irradiation and there was a little deviation from the exponential shape after irradiation with ionizing radiations.

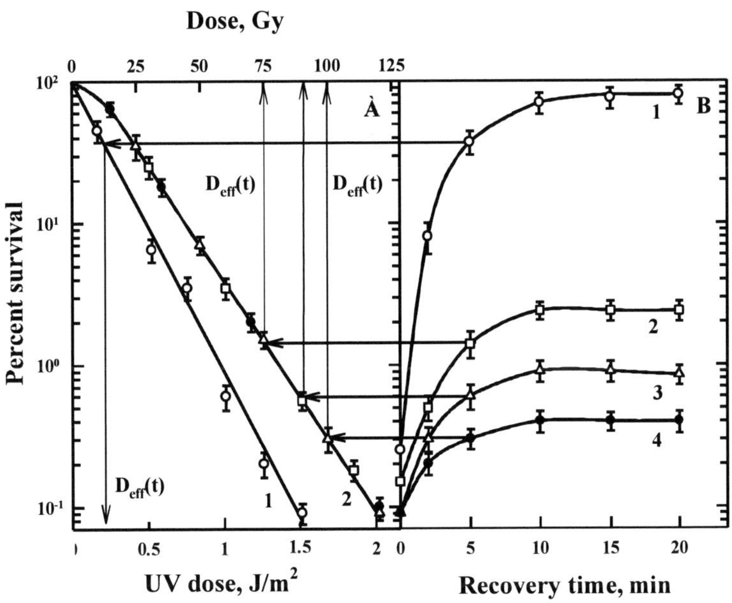

Figure 8.3. Survival curves (A) and photoreactivation kinetics (B) for bacterial cells *E. coli*, strain B_{s-1}. A – cells were exposed to graded doses of UV light (curve 1), γ-ray of ^{137}Cs (curve 2, closed circles), γ-ray of ^{60}Co (curve 2, triangles) and pulsed X-ray beam (curve 2, squares); cells were plated on nutrient agar immediately after irradiation. B – cells were exposed to UV light (curve 1, circles), pulsed X-ray (curve 2, squares), γ-ray of ^{60}Co (curve 3, triangles), and γ-ray of ^{137}Cs (curve 4, closed circles). Cells were plated on nutrient agar after various photoreactivation times. The arrows indicate the examples of the determination of the effective dose $D_{eff}(t)$, after 5 min of photoreactivation.

Figure 8.3B exhibits the photoreactivation kinetics for the bacterial cells exposed to the same kinds of radiation. The recovery kinetics were studied at approximately equal initial levels of survival. The arrows indicate the examples of the determination of the effective dose $D_{eff}(t)$ (Eq. 2.4), after 5 min of photoreactivation. The number of viable cells increased as a function of time, reached a plateau after about 15 min. The present data suggest that both the rate and the extent of photorecovery decreased in the following order of the inactivating agents: UV light pulsed X-ray beam, γ-ray of ^{60}Co and γ-ray of ^{137}Cs. The tangent of angle (the slope ratio) of curves describing photorecovery kinetics can judge the rate of photoreactivation. It is also apparent that this rate gradually decreased with increasing duration of photorecovery. But the question whether or not this decrease is due to the increase in the portion of the irreversible damage or to the impairment of photorecovery itself, i.e., to the change in the probability of recovery or to both these effects is still unclear. Using the results presented in Figrre. 8.3, we estimated (Figure 8.4A) the decrease in the relative part of the unrecovered radiation damage, $K(t)$ (Eq. 2.2), as a function of the recovery time for the cells exposed to UV light, the pulsed X-ray beam, γ-ray of ^{60}Co and ^{137}Cs.

These data demonstrate that $K(t)$ decreased, reaching a plateau after 15–20 min of photoreactivation. It also appears from these data that the ultimate quantities of $K(t)$, i.e., the values of the irreversible component of radiation damage K (Eq. 2.3), increased in the following order: UV light, pulsed X-ray beam, γ-ray of ^{60}Co and γ-ray of ^{137}Cs (Table 8.2).

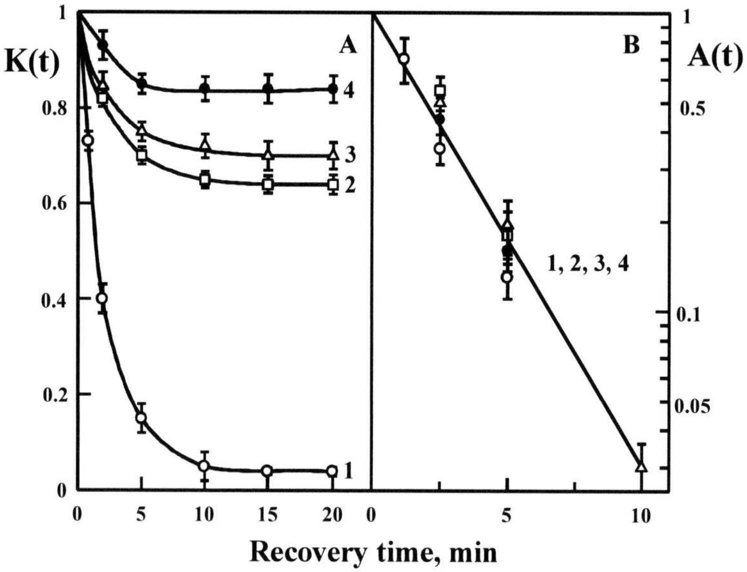

Figure 8.4. Decrease in the portion of the relative part of unrecovered, $K(t)$, (A) and recovered, $A(t)$, (B) radiation damage as a function of photoreactivation time for bacterial cells *E. coli*, strain B_{s-1}. Cells were exposed to UV light (curve 1, circles), pulsed X-ray beam (curve 2, squares), γ-ray of ^{60}Co (curve 3, triangles) and γ-ray of ^{137}Cs (curve 4, closed circles). Cells were plated on nutrient agar after various photoreactivation times. Points are the means from at least three independent experiments. Error bars indicate the standard error of the mean. Curves were fitted to the data points by eye (A) and by the least squares method (B).

The results show the expected trend in the proportion of irreversible damage as a function of photon energy with the highest proportion occurring with the lowest photon energy. This is in agreement with previous results that indicate that the proportion of UV-like damage reduces in such an order (Myasnik et al., 1980; Petin, Komarov, 1985). The experimental data presented make it possible to calculate the relative part of the reparable damage $A(t)$, defined by Eq. 2.6. The outcome is shown in Figure 8.4B. It is evident that this function decreases exponentially with the photorecovery time and does not depend on the quality of the radiation. Using Eq.

(2.7) and the results shown in Figure. 8.4B, we obtained the recovery constant $\beta = 0.37\ min^{-1}$ for all the four kinds of radiation used here (Table 8.2). This means that about 37% of the residual reparable damage is recovered every 1 min independently of radiation quality.

Table 8.2. Radiobiological parameters of the photoreactivation of the *E. coli* B_{s-1} irradiated with UV light (254 nm), 25 MeV photons, γ-ray ^{60}Co and γ-ray ^{137}Cs

Treatment	Irreversible component $K(plateau)$	Recovery constant β, min^{-1}
UV light (254 nm)	0.04 ± 0.01	0.37 ± 0.04
25 MeV photons	0.63 ± 0.03	0.37 ± 0.04
γ-ray ^{60}Co	0.70 ± 0.03	0.37 ± 0.04
γ-ray ^{137}Cs	0.84 ± 0.03	0.37 ± 0.04

Thus, the data presented show that the irreversible component and the probability of recovery vary in different ways with the kind of radiation employed in this study. While the probability of photorecovery appears to be identical for the four kinds of radiation tested, the irreversible component of radiation damage was much smaller after UV light exposure as compared with that for ionizing radiation treatment. For ionizing radiation, this component seems to decrease with the increase in photon energy. Hence, it may be concluded that the change in the extent and the rate of photoreactivation may be entirely explained by the formation of irreversible damage rather than by the impairment of recovery process itself. The fact that the probability (not the rate) of photoreactivation is the same for UV light and whatever photon energy was used indicates a commonality of the photoreactivation process both for UV and for ionizing radiations. This fact supports the idea that a part of damage induced by ionizing radiation is photoreactivatable and this part of damage may be similar to that produced by UV light.

It would be of interest to analyse treatments that selectively modify the cell sensitivity through the change in the relative yield of damage inflicted by ionization and excitation (UV-like damage). For example, some chemical radioprotectors and radiosensitizers, which are effective for ionizing radiation, are known to exert their effect predominantly on events inflicted by ionizations. On the other hand, illumination with visible light during the post-irradiation period is more effective in modifying UV effects called forth by excitations. It would seem feasible that the increase in the relative yield of excitations due to radioprotective action should result in an increase of the extent of photoreactivation (a decrease in the irreversible component) after exposure to ionizing radiation. An inverse situation would be expected after radiosensitizing action of some chemicals. Thus, it would be of interest to evaluate the influence of different physical and chemical factors, which modify the cell radiosensitivity, to the parameters of the photoreactivation kinetics of the bacterial cells exposed to ionizing radiation. One example exhibits in Figure 8.5.

Extensive studies of this matter have been published (Redpath, Tortorello, 1977; Redpath, Zabilansky, 1979). The authors came to the following three major conclusions: (1) the extent of photoreactivation of ionizing radiation-induced damage was independent of the presence of oxygen during irradiation; (2) the presence of caffeine during the photoreactivation process reduced the rate of the recovery by a factor of about 4. This was comparable with the effect of caffeine, a known inhibitor of photoreactivation enzyme, on the kinetics of the photoreactivation of the UV-induced damage (Harm, 1978); (3) the presence of the radioprotective agent dithiothreitol (DTT) during irradiation does not suppress the production of the photoreactivable damage. It is of interest to apply mathematical approach describing the recovery process as a decreasing of the effective dose to their data to evaluate quantitatively the influence of these factors on the irreversible component and the probability of recovery per unit time. The output is presented in Figure 8.5. The analysis was based on the work of Redpath and Zabilansky (1979).

Here, the decrease in the portion of the relative parts of the unrecovered, $K(t)$ (Figure 8.5A), and recovered, $A(t)$ (Figure 8.5B), radiation damage as a function of the photoreactivation time for the bacterial cells *E. coli* are depicted. Cells were exposed to 6 MeV photons in the presence and absence of oxygen; in the presence of 10 mM DTT during irradiation under oxic and anoxic conditions; in the presence of 16 mg/ml caffeine during photoreactivation after irradiation under oxic and anoxic conditions. It is of important to note that during these experiments (Redpath, Zabilansky, 1979) post-irradiation illumination was performed with low intensity light for the following conditions of irradiation: O_2, N_2, O_2 + caffeine, N_2 + caffeine. In contrast, a high intensity of post-irradiation illumination was used for N_2 + DDT and O_2 + DDT conditions. Thus, the conditions of photorecovery were quite different for the two series of experiments, which undoubtedly resulted in the different rates, and probabilities of photorecovery. Therefore, it is not possible to derive any conclusions based on the comparison between these two series of experiments. However, comparisons within each of these series are valid. The parameters of the photoreactivation calculated based on the curves presented in Figure 8.5 are listed in Table 8.3.

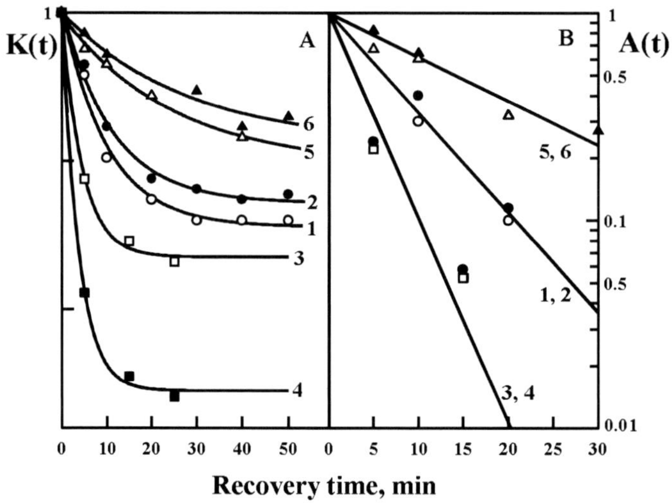

Figure 8.5. The decrease in the portion of the relative parts of unrecovered $K(t)$ (A) and recovered $A(t)$ (B) radiation damage as a function of photoreactivation time for bacterial cells *E. coli*, strain B_{s-1}. Cells were exposed to 6 MeV photons in the presence (open circles, curve 1) and absence (closed circles, curve 2) of oxygen; in the presence of 10 mM DTT during irradiation under oxic (open squares, curve 3) and anoxic (closed squares, curve 4) conditions; in the presence of 16 mg/ml caffeine during photoreactivation after irradiation under oxic (open triangles, curve 5) and anoxic (closed triangles, curve 6) conditions The calculations were based on the work of Redpath and Zabilansky (1979).

Table 8.3. Radiobiological parameters of photoreactivation of *E. coli* B_{s-1} irradiated with 6 MeV photons

Condition of irradiation	Irreversible component K	Recovery constant β, min^{-1}
N_2 + DTT	0.65 ± 0.03	0.20 ± 0.03
O_2 + DTT	0.78 ± 0.03	0.20 ± 0.03
O_2	0.80 ± 0.03	0.11 ± 0.02
N_2	0.82 ± 0.03	0.11 ± 0.02
O_2 + caffeine	0.88 ± 0.04	0.05 ± 0.01
N_2 + caffeine	0.90 ± 0.04	0.05 ± 0.01

The following conclusions can be drawn from these data.

1. Both the extent and the rate of photoreactivation of ionizing radiation-induced damage was dependent of the oxygen presence during irradiation; it was most noticeably seen for the (O_2 + DTT) and (N_2 + DTT) conditions of irradiation. However, the probability of the photoreactivation of ionizing radiation-induced damage was independent of the oxygen presence during irradiation for all the versions of the experiment. It indicates a similar nature and mechanism of recovery of photoreactivatable damage induced at these conditions.
2. The presence of caffeine during the photoreactivation affected both the parameters of photoreactivation. It resulted in the decrease in both the rate and probability of photoreactivation. The proportion of irreversible damage was higher in the presence than in the absence of caffeine and this makes sense in that caffeine is known to sensitize cells to the effects of ionizing radiation-induced damage. If ionizing radiation really produces a part of UV-like lesions, this would be consistent with the data that caffeine inhibits photoreactivation of UV-induced damage by competing with the photoreactivation enzyme for its substrate (Harm, 1978).
3. The presence of the radioprotective agent (DTT) during irradiation does not suppress the production of the photoreactivatable damage. Moreover, the portion of the irreversible damage decreased while photoreactivatable damage increased in the presence of the radioprotective agent DTT during irradiation. The effect was more clearly revealed in the absence of oxygen during irradiation. Both of these facts may be associated with a decrease in the relative yield of lesions inflicted by ionizations and a corresponding increase in the part of UV-like lesions inflicted by excitations due to the protective action of the chemical agent and anoxia. This is in agreement with the known effects of this thiol acting as a radioprotector through free radical scavenging. This also suggests that the photoreactivatable damage was not induced by free radicals, the main conclusion of the original article (Redpath and Zabilansky, 1979).

8.3. Estimation of UV Light Dose Concomitant to Irradiation with Ionizing Radiation

In the previous sections of this chapter, our experimental data and the results published by others concerning the phenomenon of photoreactivation after cell exposure to ionizing radiation have been summarized and it was concluded that the UV-like damage is produced by Čerenkov light emitted by charged particles when their velocity exceeds the phase velocity of the light in the medium. Therefore, it is of interest to estimate a UV light dose concomitant to irradiation with ionizing radiation. We will describe a simple biological method determining the UV light dose concomitant the ionizing radiation exposure and present the results of our calculations of the dependency of UV light dose accompanying 100 Gy of ionizing radiation on the energy of ionizing radiation and on the volume of the exposed cell suspension.

The mathematical model that will be used here has been partly published (Petin, Komarov, 1982). It is worth mentioning in short, the main points of the model. The survival curves of UV-hypersensitive bacterial and yeast cells exposed to ionizing radiation can be simply presented as

$$S = e^{-kD}, \tag{8.1}$$

where S is the cell survival fraction, D the irradiation dose and k is the overall cell sensitivity, defined as the inverse of the mean lethal dose $\left(k = \frac{1}{D_0}\right)$. It is reasonable to assume that excitation and ionization events act stochastically and independently of each other for sparsely ionizing radiation. Then Eq. (8.1) can be rewritten as follows

$$S = e^{-(k_1+k_2)D}, \tag{8.2}$$

where k_1 and k_2 are the cell sensitivities to ionizations and excitations, respectively. We can normalize the overall radiosensitivity and consider that

$$k = k_1 + k_2 = 1. \tag{8.3}$$

Then k_1 and k_2 represent the relative contribution of ionization and excitation to cell inactivation. It is not difficult to choose some factors, which can modify cell radiosensitivity through a selective influence to the consequences produced by ionization and excitation. For example, the oxygen effect is known to be predominantly exerted on cell inactivation produced by ionization. On the contrary, illumination by visible light during a postirradiation period is more effective in modifying the effects caused by excitation.

Let us suppose that the initial survival curve (Eq. 8.1) was obtained under oxic and dark conditions. Then the survival under hypoxia will be given by

$$S = e^{\frac{-kD}{F_1}} = e^{-\left(\frac{k_1}{f_1}+k_2\right)D}, \tag{8.4}$$

where F_1 the DMF of the experimental survival curve obtained under hypoxic and dark conditions and f_1 is the DMF for the ionization component only.

If the initial survival curve (Eq. 8.1) was changed by the photoreactivating light after irradiation, the corresponding survival curve may be described by

$$S = e^{\frac{-kD}{F_2}} = e^{-\left(k_1 + \frac{k_2}{f_2}\right)D}, \tag{8.5}$$

where F_2 the DMF of the experimental survival curve obtained after photoreactivation and f_2 is the DMF for the excitation component only.

Taking into account equations (8.3)–(8.5), we attain the following set of equations

$$\begin{cases} \dfrac{k_1}{f_1} + k_2 = \dfrac{1}{F_1} \\ k_1 + k_2 = 1 \\ k_1 + \dfrac{k_2}{f_2} = \dfrac{1}{F_2} \end{cases}, \tag{8.6}$$

where k_1, k_2, f_1, and f_2 are unknown values, while F_1 and F_2 are known from experiments. The DMF for the excitation component (f_2 = 4.5) was estimated from the experiments on photoreactivation of cells exposed to UV light (254 nm) only. Then the solution for the set in Eq. (8.6) yields

$$k_1 = \frac{f_2 - F_2}{F_2(f_2 - 1)} \tag{8.7}$$

$$k_2 = \left(1 - \frac{1}{F_2}\right)\left(1 + \frac{1}{f_2 - 1}\right). \tag{8.8}$$

Figure 8.6 shows a typical example of the basic experimental data observed in this study. Figure 8.6A exhibits the survival curves for the bacterial cells irradiated with graded doses of pulsed X-rays (25 MeV) and plated on nutrient media without and with post irradiation photoreactivation. The results revealed that postirradiation illumination by visible light of bacterial cells exposed to ionizing radiation leads to an increase of cell survival; the increment over the control cells is consistent throughout replica experiments and the dose modifying factor F_2 = 1.60.

Using equations (8.7) and (8.8) we obtained those values of $k_1 = 0.57$ and $k_2 = 0.43$. This means that the contribution of the ionizations to the total lethal effect induced by pulsed X-rays is 57%, while the contribution of the excitations is 43%. Subdividing the initial survival curve after ionizing radiation action (Figure 8.6B, curve 1) into two components caused by ionizations (57%, Figure 8.6B, curve 2) and excitations (43%, Figure 8.6B, curve 3), i.e., mainly by UV light, we found that for the dose of 100 Gy the value of the cell survival for the UV component was 7%. The survival for the action of 254 nm UV light only (Kim et al., 2005 a) corresponds to the UV dose of 0.56 J/m^2. Thus, the irradiation of bacterial cells (strain B_{s-1}) with pulsed X-rays (100 Gy) was simultaneously accompanied by cell exposure to the UV dose of 0.56 J/m^2. It is important to note that for the sake of simplification we compared the whole spectra of the Čerenkov emission with the most effective equivalent fluence for the lethal and genetic effects of 254 nm wavelength of UV light.

As was mentioned above, Čerenkov emission arises in water solutions irradiated with high-energy ionizing radiation and an increase in energy of ionizing radiation results in an increase in Čerenkov emission intensity (Jelley, 1958). Therefore, it would be of interest to estimate the dependence of UV light dose concomitant to exposure with ionizing radiation on energy of ionizing radiation. Extensive experimental data regarding the dependence of the extent of photoreactivation of bacterial cells of *E. coli* (strain B_{s-1}) on energy of ionizing radiation have been published (Myasnik et al., 1980).

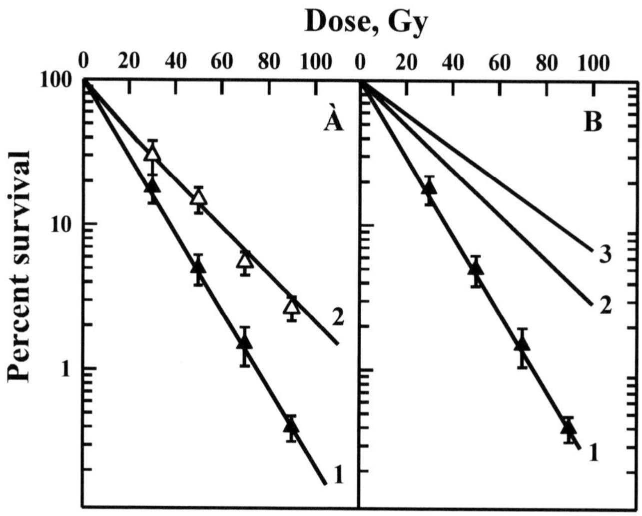

Figure 8.6. Survival of *E. coli* B_{s-1} after pulsed X-irradiation (25 MeV): A, without (curve 1) and with (curve 2) maximum photoreactivation and B, cell inactivation induced by both ionizations and UV light damage without photoreactivation (curve 1), only by ionizations (curve 2) and excitations (curve 3) after pulsed X-irradiation (25 MeV).

Using these data as well as the above-described example of the calculations of the UV dose, we calculated the dependence on the energy of ionizing radiation the relative contribution of excitation to the lethal effect (k_2

Gy of ionizing radiation. The results are presented in Figure 8.7. One can see that the excitation contribution (UV light damage) to the total lethal effect of ionizing radiation increases from

3% for 0.04 MeV X-rays to 43% for 25 MeV pulsed X-rays while the dose of the equivalent UV light accompanied by 100 Gy of ionizing radiation increases from 0.04 to 0.56 J/m^2.

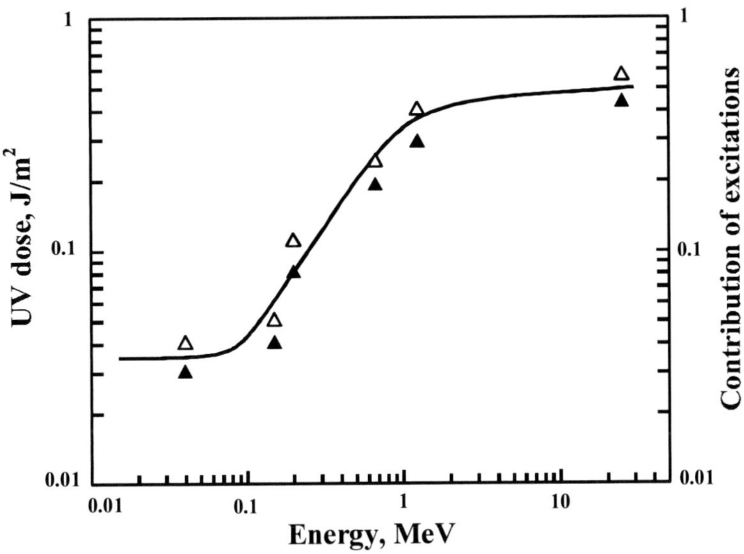

Figure 8.7. UV light dose (open triangles) and the relative contribution of UV damage to the lethal effect of ionizing radiation (closed triangles) as a function of the energy of sparsely ionizing radiation.

As was shown above, the survival fraction of γ-irradiated bacterial cells of *E. coli* B_{s-1} decreased, whereas the value of the photoreactivation increased with an increase in the volume of irradiated suspension. This effect can be explained by the corresponding increase in the UV light dose and the excitation contribution to the total lethal effect of ionizing radiation. Therefore, it would be of interest to use the approach described above to estimate the dependence of UV light dose concomitant to exposure with ionizing radiation on the volume of irradiated cell suspension.

Using the results published before (Morozov, Myasnik, 1980), we calculated the dependence of the relative contribution of excitation in the lethal effect (k_2) and the dose of the equivalent UV light accompanied by 100 Gy of ionizing radiation on the volume of irradiated suspension of *E. coli* B_{s-1}. The results are depicted in Figure 8.8. One can see that the excitation contribution (UV light damage) to the total lethal effect of ionizing radiation increases from 26% for 0.5 ml to 49% for 500 ml of irradiated suspensions while the dose of the equivalent UV light accompanied by 100 Gy of ionizing radiation increases from 0.33 to 0.64 J/m^2.

Thus, the main conclusions of these results can be formulated as follows. First, the relative excitation contribution to the total lethal effect and the value of the UV dose for the same dose of ionizing radiation (100 Gy) is considerably increased with an increase in energy of ionizing radiation. Second, these parameters were increased with an increase in the volume of irradiated cell suspension. Both these observations are in agreement with the supposition that the Čerenkov emission is responsible for the production of UV light damage and the phenomenon of photoreactivation might be observed after ionizing exposure for bacterial and yeast cells hypersensitive to UV light. This hypothesis has been put forward and discussed in various papers (Morozov, Myasnik, 1980; Moss, Smith, 1980; Michael et al., 1981; Redpath et al., 1981; Petin, Komarov, 1985). The plateau of the UV dose with an increase in the energy of ionizing radiation is determined by physical properties of the Čerenkov emission (Jelley, 1958). The plateau of the UV dose with an increase in volume of the exposed suspension is determined by the highest penetration of Čerenkov emission photon, which is about 12 cm in a water medium (Pennarun et al., 1977).

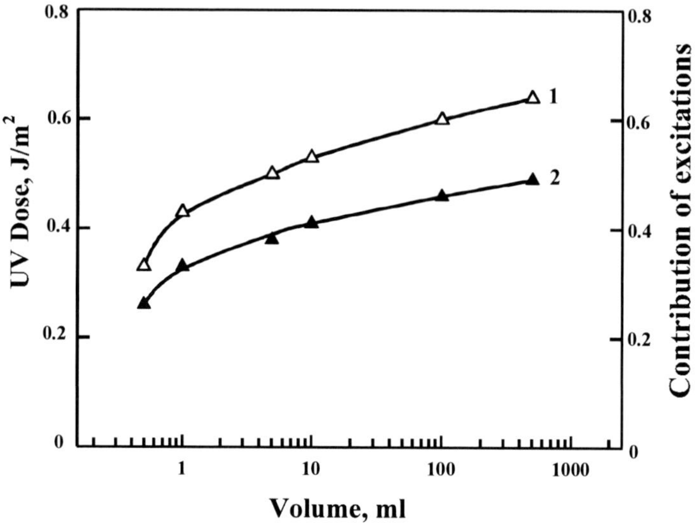

Figure 8.8. UV light dose (open triangles) and the relative contribution of UV damage to the lethal effect of ionizing radiation (closed triangles) as a function of the volume of the exposed cell suspension.

Conclusion

The considerable part of this work provides an overview of patterns of genetic instability display and mechanisms of radiation-induced genetic instability. Genetic instability concerns such important phenomena as radiation mutagenesis, carcinogenesis, and aging – the main long-term consequences of ionizing radiation (Saenko, Zamulaeva, 2007; Vorobtsova, 2008). Genome stability is a priority for the survival and functioning of all organisms (Inge-Vechtomov, 2015). It should be noted that the preservation of inherited damage in distant offspring surviving after the exposure to various physical factors and drugs used in medicine is the subject of intensive modern research in terms of the participation of this effect in aging and the formation of oncological diseases (Davydov et al., 2013; Pankratov et al., 2014). Manifestations of genetic instability, along with destabilization of chromosomes, somatic mutations, changes in sensitivity to the action of ionizing radiation, include the delayed reproductive cell death and the delayed appearance of clones by cells surviving after exposure to ionizing radiation (Chang, Little, 1992 a, b; Mazurik, Mikhailov, 2001; Skoneczna et al., 2015; Sun et al., 2016). The studying of genetic instability became the main part of this work.

Various damage occurs after cell exposure to ionizing and UV radiation in the chemical structure of DNA, which is the most radiosensitive cell target. In response to DNA damage, cells initiate a number of processes that repair this damage to ensure survival and preserve the integrity of the genome. The repair systems of all currently existing living organisms were formed at the early stages of evolution and are represented in cells from bacteria to humans in the form of highly homologous enzymatic systems. The degree of homology of these systems is especially high in eukaryotic cells. In this regard, yeast, being the simplest of eukaryotic organisms, is an excellent model for studying the relationship of repair systems with various radiobiological effects, including genetic instability (Korogodin et al., 2007; Petin et al., 2008; Korolev, 2009; Korogodin, 2010; Stirling et al., 2011; Feldman, 2012; Zhu et al., 2012; Donigan et al., 2015; Bell, Labib, 2016; Duffy et al., 2016; Tombline et al., 2017). The short life cycle of yeast and the ability to produce a large number of cell generations allow even very rare phenomena to be investigated. Most of the radiobiological responses of yeast cells, such as synergistic interactions, the shape of the survival curve, the dependence of the RBE on LET radiation, the oxygen effect, the action of radioprotectors and radiosensitizers, are qualitatively similar to those observed in mammalian cells (Critchlow, Jackson, 1998; Resnick, Cox, 2000; Botstein, Fink, 2011; Petin et al., 2012, 2020; Petin, Kapultcevich, 2014; Petin, Kim, 2016). The use of yeast cells in the experiment provides a clear reproducibility of the results.

The first part of this work is mainly related to the description of the new patterns of genetic instability revealed in experiments with haploid and diploid yeast cells of *Saccharomyces cerevisiae*. The test of genetic instability was the delayed formation of colonies by cells surviving after irradiation with γ-rays, α-particles, and UV light, as compared to the control. Such a test allowed to take a fresh look at the patterns of manifestation of the phenomenon of genetic instability including to assess the dependence of genetic instability on the dose of the agent used, as well as to quantitatively compare the genetic instability of haploid and diploid cells. It was shown that in the distant offspring of cells surviving after irradiation there is an increased yield of various changes in cells from late formed clones – the content of respiratory mutants, the existence of recombinant cells, a number of morphologically altered colonies, a significant content of non-viable cells. These effects serve as evidences of genetic instability and were expressed largely for clones, the formation of which was delayed for several days in comparison with the control. In addition to radiobiological illustrations of genetic instability, the authors' data on re-seeding of late formed colonies are provided. It was shown that cells from different colonies that grew later than the control again formed colonies that were heterogeneous in size and morphology.

It was demonstrated for the first time that genetic instability caused by both sparsely and densely ionizing radiation, in contrast to the long-existing concepts, is largely determined by cell ploidy and not by the shape of the dose-effect curve and the cell ability to recover from radiation damage. This point of view is supported by data demonstrating that both resistant and radiosensitive diploid yeast strains, regardless of the shape of the survival curves and cell ability to recover from radiation damage, show a significant delay in clone formation by cells survived after irradiation (100%), in contrast to haploid strains (20%). This conclusion is consistent with data on genetic instability induced by UV radiation. Note that after UV irradiation the survival curves of both diploid and haploid cells have a sigmoid shape of the dose-effect curves. In this case, diploid cells, in contrast to haploid ones, show a significant delay in the formation of clones by cells surviving after irradiation (100% vs. 30–45%) regardless of the shape of the survival curves, UV sensitivity, and the ability of cells to recover from UV-induced damage. These data show that the effect of the delayed appearance of colonies by diploid yeast cells that survived after irradiation is not unambiguously associated with the shape of the survival curve and cell ability to recover from radiation damage. In other words, the presence of diploid-specific recovery (Saeki et al., 1980) is not the main determinant of the delayed colony formation by irradiated cells. Thus, the data presented here show that the delayed formation of colonies by cells that survived after irradiation with ionizing radiation of various qualities and UV light is mainly determined by cell ploidy and not by the survival curves forms, the cell radiosensitivity and cell ability to recover from radiation damage.

The RBE of α-particles for the survival of wild-type diploid yeast cells capable of recover from radiation damage was shown to be of 4.6 ± 0.2, which does not differ from the RBE for the genetic instability determined by the delayed appearance of colonies (4.7 ± 0.4). The RBE for cells with significantly reduced ability to recover from radiation damage was in a great extent lower for both survival and genetic instability (1.6–2.7). The highest RBE value was observed for wild-type diploid cells. Many authors (Frankenberg et al., 1981; Petin, Kabakova, 1981; Frankenberg-Schwager et al., 1984; Blöcher, 1988; Petin, Kim, 2005; Petin, Kapultcevich, 2014; Petin, Kim, 2016) have published similar RBE values for cell survival, while the RBE values for genetic instability were obtained in this work for the first time. The increased RBE values of densely ionizing radiation may be due to the fact that radiation with high LET creates clusters of ionizations and excitations in DNA (Goodhead, 1994; Zadnepryanets et al., 2018 a, b), which lead to more severe and complex damage than control radiation with low LET. This is also consistent with our previous results (Petin, Kim, 2005, 2016) that densely ionizing radiation inhibits cell repair due to the formation of a greater proportion of irreversible radiation damage, while the recovery constant, characterizing the cell probability to recover for unit time, does not depend on the quality of radiation. For wild-type haploid yeast cells and all mutant cells, regardless of their ploidy, RBE values randomly varied for survival and genetic instability in the range 1.6–2.7 without any systematic pattern. These data clearly indicate that RBE of densely ionizing radiation is determined not only by the increased probability of the formation of primary radiation damage at the physicochemical stage, but also by the ability of cells to recover from radiation damage at the biochemical stage of fixation of such damage. This conclusion is not new for cell survival, while it is fundamentally new for genetic instability. The identity of RBE values for both tests may be associated with the greater efficiency of densely ionizing radiation to produce lethal radiation damage, which are accompanied by sub-damage responsible for the delayed appearance of colonies by cells surviving after irradiation. These sub-damages persist in the distant descendants of cells and thereby cause destabilization of the genome, the consequences of which are recorded by many tests of genetic instability, including the delayed appearance of colonies after irradiation.

The independence of genetic instability from cell survival was displayed for haploid and diploid yeast cells after their irradiation with γ-rays, α-particles and UV light. These data indicate that isoeffective cell exposure to radiation of different quality induces the same number of effective (lethal) radiation damage and the proportional number of sub-damage responsible for genetic instability. These sub-damages cause destabilization of cell genome and persist in the distant offspring of cells surviving after irradiation.

Based on the data presented, it was concluded that the mechanism of different manifestations of genetic instability by haploid and diploid yeast cells surviving after irradiation may be associated with some chromosomal damage, for example, deletions or even loss of chromosomes, which are more fatal for haploid, and not for diploid cells. That is why the genetic instability of haploid cells is manifested to a much lesser extent than diploid cells, regardless of their radiosensitivity or cell ability to recover from radiation damage.

Among a number of unexpected and insufficiently studied radiobiological effects analyzed in this work, considerable attention was paid to synergistic and antagonistic interactions of various agents under their separately and combined application. With a synergistic interaction, the total biological effect of the combined use of two agents exceeds the independent sum of the effects of each agent. The inverse situation is observed for antagonistic interaction.

The assessment of the potential significance of the synergistic and antagonistic interactions of ionizing radiation with hyperthermia or chemical agents in medical radiology and oncology remains an urgent task of our time. The monograph summarizes the universal patterns of the manifestation of combined application of various factors, which do not depend on the agents used, biological objects and tests. In particular, both the synergistic and antagonistic interaction of the simultaneous application of hyperthermia with ionizing radiation or chemical agents is observed only in a certain temperature range or chemical agent's concentration, within which there is an optimal value, at which the greatest synergism or antagonism is recorded. Another example is related to the dependence of both kind of interaction on the intensity of the agents used. It is known that biological effects depend on the dose rate of ionizing radiation unidirectionally, with low dose rates being in the overwhelming majority of cases less effective. The data presented in the monograph show that the synergistic enhancement ratio first increases with an increase in the dose rate or concentration of a chemical agent, then reaches its highest value, followed by a decrease. Analogous pattern was displayed for antagonism. These unexpected dependences were observed for various biological objects, interacting agents, the effect investigated and thereby indicates the existence of optimal dose rates of ionizing radiation, intensities of UV light, microwave radiation, or concentrations of chemical compounds that provide the greatest synergistic or antagonistic interaction.

It was noted that some researchers believed that inhibition of recovery processes might be considered as a reason for the mutual reinforcement of high temperatures with other agents applied together (Zaider, Rossi, 1980; Kurpeshev, Konoplyannikov, 1987; Kuzin, 1983; Kogl, 1986; Joiner, van der Kogel, 2009). In accordance with this approach, hyperthermia disrupts or suppresses the processes of cell recovery from radiation damage and cells become more sensitive to radiation exposure and the synergism of thermoradiation action is observed. A quantitative assessment of recovery parameters after combined effects of sparsely and densely ionizing radiation with chemical agents, many of which are used in medical radiology to increase the radiosensitivity of cancer cells, made it possible to conclude that the recovery constant does not depend on the conditions of combined effects, which is consistent with the previously made conclusions (Komarova, Petin, 2007; Petin et al., 2012, 2020; Petin, Kim, 2016). The totality of these data shows that a decrease in both the rate and the volume of recovery under conditions of combined impacts is due to the formation of irreversible damage, and not to a violation of the recovery process itself. Then the inhibition of recovery, observed by many authors because of combined effects, is not the cause of radiosensitization under combined effects, but a quite expected consequence of the formation of irreversible damage from which the cell is unable to recover.

Comparing the new results obtained in this work with the numerous data related with this problem and published by other authors, it can be stated that our results bring order among the chaos of mutually exclusive facts described in the literature concerning the combined action various agents. In general, the experimental and theoretical material of the monograph gives a fairly complete picture of the features, patterns, mechanisms of action and ways of further development of various manifestations of synergistic and antagonistic effects after the combined actions of various environmental factors on biological objects in the modern world. The results obtained are not only of fundamental importance from the point of view of understanding the mechanisms of interaction and ways to optimize the influencing agents to achieve the desired effect, but are also useful for a number of practical applications in the field of biomedical technologies and environmental research.

The influence of tonicity and chloramphenicol on hyperthermic cytotoxicity and cell permeability under various heating rate was also investigated. The cell lethality and permeability induced in *Escherichia coli* B/r, *Escherichia coli* B_{s-1} and *Zygosaccharomyces bailii* cells by high temperature (52°C) after heating at different rates (mean values – 0.015, 0.25 and 1.5°C per s) and in the media of different tonicity and content (with versus without chloramphenicol 10 mg/ml) have been investigated. Hyperthermia treatment in YEP broth or isotonic 0.01 M phosphate buffer resulted in markedly reduced cytotoxicity with decreasing heat rate. The heating rate effect was larger when the cells were treated in YEP broth. Chloramphenicol, which is known inhibit expression of heat shock proteins in bacteria, in our case did not affect the viability of cells or the development of thermotolerance in cells heated at different heating rates in isotonic phosphate buffer but prevented the development of an additional degree of thermotolerance in cells heated slowly in YEP broth. In contrast, the differential effect of heating rate on cytotoxicity and cell permeability was not demonstrated when the cells were heated in hypertonic solution (1 M NaCl in 0.01 M phosphate buffer, pH 7.0). It is proposed that heat destabilization of the osmotic cell homeostasis, which is more profound after rapid heating, plays a major part in heat induced cellular lethality. The repeated heating and cooling in lethal (2–52°C) and nonlethal (2–37°C) temperature ranges resulted in cell death of *Escherichia coli* B/r and *E. coli* B_{s-1} suspended in 0.01 M phosphate buffer, pH 7.0 at varying osmotic pressure, but not in cow's milk. The lethal effect increased with the rate of heating and cooling and with increasing suspension media tonicity. This effect may be caused by the temperature destabilization of cellular osmotic homeostasis.

New data are presented on the effect of osmolytes (NaCl, glycerin) on the damaging effect of ionizing radiation, hyperthermia, microwave radiation and ultrasound, obtained in experiments with bacterial cells. Some general patterns of the protective effect of the used osmolytes have been revealed. The protective effect of osmolytes against the used damaging factors (antagonistic interaction) can be observed only within a certain range of "doses" of the applied factors, within which there is an optimum providing the maximum effect. It has been shown that to ensure the greatest antagonism with an increase in the intensity of one of the agents, a corresponding change in the intensity of another factor involved in the interaction is required. It was concluded that along with DNA damage, membrane structures and the osmotic homeostasis system can be another critical target of cell damage under the combined action of various environmental factors.

The great attention was given to antagonistic interaction of ionizing radiation and visible light. Some new experimental data of this phenomenon were presented. These data are in line with early suggested hypothesis that the photoreactivable damage induced by ionizing radiation might be UV-like damage produced by Čerenkov light emitted by charged particles when their velocity exceeds the phase velocity of light in the medium. It is concluded that irradiation of both prokaryotic and eukaryotic cells with sparsely ionizing radiation of high-energy results in production not only single- and double-strand breaks, base damage and cross-links between DNA molecules but also UV-like damage. An additional argument supporting these data was obtained in the quantitative evaluation of the parameters of photoreactivation after exposure to ultraviolet light and ionizing radiation of different energy. A mathematical model describing the process of photoreactivation in terms of a decreasing effective dose was applied to the experimental data to evaluate quantitatively the probability of photoreactivation per unit time and the irreversible component, i.e., the proportion of damage from which cells is incapable of recovery. Both the rate and extent of photoreactivation decreased in the following order of inactivating agents: UV light, pulsed X-ray bean, γ-ray of ^{60}Co and γ-ray of ^{137}Cs. The irreversible component increased with the same order of radiations whereas the probability of photoreactivation per unit time was independent of the kind of radiation. Hence, the extent and the rate of photoreactivation following ionizing radiation decreased in comparison with UV light only due to the smaller number of damage susceptible of photoreactivation. On this basis, it can be concluded that the nature of photoreactivable damage produced by UV light and ionizing radiation may be similar to each other.

A simple mathematical model for biological estimation of UV light dose concomitant to ionizing radiation was developed. This approach was applied to determine the dependency of equivalent UV light dose accompanied with 100 Gy of ionizing radiation on the energy of sparsely ionizing radiation and on the volume of the exposed cell suspension. It was shown that the relative excitation contribution to the total lethal effect and the value of UV dose for the same dose of ionizing radiation is increased with an increase both in energy of ionizing radiation and volume of the irradiated cell suspension. Both these observations are in agreement with the supposition that the Čerenkov emission is responsible for the production of UV light damage and the phenomenon of photoreactivation observed after ionizing radiation. In the model presented here we suggested additive summation of damage produced by ionizations and excitations of molecules. However, it is not improbable that they can interact in a synergistic manner.

So, the action of high energy sparsely ionizing radiation is always accompanied by the simultaneous action of UV light. Čerenkov emission can be considered as a means of UV light delivery to deeply located tissues. The estimated UV doses will undoubtedly take place during irradiation of wild-type cells including eukaryotic cells, plants, animals, men etc. However, the induced UV damage may be effectively repaired and not displayed. It is not excluded a probable participation of UV component of ionizing radiation in the mechanism of the manifestation of different low-dose responses: adaptive reaction, radiation hormesis, bystander effect that can be observed after extremely low dose of ionizing radiation when charged particles traversed through only a small part of cell population but the whole population is capable of changing its response to the subsequent action of a higher dose of radiation (Pelevina et al., 1978, 1985; Luckey, 1991, 2007; Luan et al., 2006). Since the number of quanta of Čerenkov emission for the radiation dose of 100 Gy (^{60}Co) is about 10^{13} per 1 cm^3 of irradiated water (Morozov, Myasnik, 1980), there would be a tremendous number of these quantum even for a low dose of radiation – for example 10^9 quanta per 1 cm^3 for the dose of 1 cGy. As many quanta of UV light are more uniformly distributed in an irradiated volume than ionizations, it must not be ruled out that the low dose of ionizing radiation including the biological significance of natural background can trigger reparation processes in cells analogous to the SOS repair by the accompanied UV light, thereby increasing the radioresistance of biological objects to a subsequent irradiation.

The final chapter of the monograph is devoted to a review of works by other authors devoted to the problem of radiation hormesis under the action of low doses of ionizing radiation. This chapter summarizes and analyzes the experimental data on the effect of small doses of ionizing radiation on biological objects of varying degrees of complexity, including humans. Many objects and tests revealed the positive effect of small doses of ionizing radiation. The most impressive results came from a survey of a cohort of people who have lived in Taiwan for more than 30 years in radiation-contaminated apartments. It was found that the incidence of cancer in these people was 40 times less than in the control, although the total doses accumulated by residents exceeded the doses of the Chernobyl accident liquidators. Possible mechanisms of radiation hormesis are discussed. The totality of the presented data shows that exposure to low doses of ionizing radiation in many cases has a beneficial effect, which can serve as a basis for reducing the radiophobia of professionals in the nuclear industry and the population.

This review would not be complete without mentioning the opposite point of view – about the harmfulness of small doses of ionizing radiation (Shevchenko et al., 1992; Baryakhtar, 1995; Burlakova et al., 1996; Yablokov, 1997). The fallacy of the views of these authors has been refuted by life itself and the results presented in this work. Scientific criticism of such views has been analyzed in sufficient detail in the literature (Ilyin, 1994; Korogodin, Korogodina, 1997; Korogodin, 2010; Filyushkin, Petoyan, 1988).

The fear of radiation (radiophobia) is associated with radiation accidents, testing of atomic weapons and even the operation of nuclear power plants. It is not excluded that radiophobia and ionizing radiation can interact synergistically, i.e., reinforce each other's action. It is possible that the consequences of the accident at the Chernobyl nuclear power plant could be intensified by the fear of radiation. That is why it seems relevant to achieve a reduction in radiophobia, especially when exposed to low doses of ionizing radiation. The presented data shows the possible positive effect of low doses of ionizing radiations on life expectancy and on the oncological morbidity of animals and humans, thereby demonstrating the sharp need for radiophobia reducing.

References

Abbaszade I. G., Mamedov T. G. 1988. Features of small-dose UV-light effect of Escherichia coli WP2 cells and their mutants. *Studia Biophysica* 128:89–94.

Abbatt J. D., Hamilton T. R., Weeks J. L. 1983. "Epidemiological studies in three corporations covering the Canadian Nuclear fuel cycle." In *Biological Effects of Low–Level Radiation IAEA Symposium Proceedings*, 351–361.

Adair E. R., editor. 1983. *Microwaves and Thermoregulation*. N.Y.: Academic Press.

Ager D. D., Haynes R. H. 1987. Mathematical description of the interactions between cellular inactivating agents. *Radiat. Res.* 110(2): 129–141.

Aggarwal M., Brosh R. M. 2012. Functional analysis of human DNA repair proteins important for aging and genomic stability using yeast genetics. *DNA Repair (Amst.).* 11(4):335–348.

Akoev I. G., Alekseev S. N., Tyazhelov V. V. 1986. *Biological Effects of Electromagnetic Fields. Questions of Their Use and Standardization.* Pushchino: ONTI NTSBI AN SSSR.

Alexandrov V. Ya. 1977. *Cell Reactivity and Proteins.* Leningrad: Nauka.

Alexandrov V. Ya., Kislyuk I. M. 1994. Cell reaction to heat shock. *Cytology* 36(1):5–59.

Alper T. 1970. "Mechanisms of lethal radiation damage to cells." In *Second Symposium on Microdosimetry*, edited by H. G. Ebert, 5–36. Brussels: Commission of the European Communities.

Alper T. 1971. "Cell death and its modification: the role of primary lesions in membranes and DNA." In *Biophysical Aspects of Radiation Quality,* 171–184. Vienna: IAEA.

Alper T. 1979. *Cellular Radiobiology.* London: Cambridge University Press.

Alper T. 1984. Implications of repair models for LET effects and other radiobiological phenomena. *Br. J. Cancer* 49(VI):137–143.

Alper T., Gillies N. E. 1960. "The influence of metabolic events immediately after irradiation on the difference in sensitivity between *E. coli* B and *E. coli* B/r." In *Immediate and Low-Level Effects of Ionizing Radiations*, 305–310. London.

Anderson R. L. Herman T. S., Van Kersen I., Hahn G. M. 1988. Thermotolerance and heat shock protein induction by slow rates of heating. *Int. J. Radiat. Oncol. Biol. Phys.* 15:717–725.

Anderson R. L., Tao T. W., Betten D. A., Hahn G. M. 1986. Heated shock protein levels are not elevated in heat resistant response B16 melanoma cells. *Radiat. Res.* 105:240–246.

Angleton G. M., Benjamin S. A., Lee A. C. 1988. Health effects of low-level irradiation during development: experimental design and prenatal and early neonatal mortality in beagles exposed to 60Co gamma rays. *Radiat. Res.* 115(1):70–83.

Anisimov V. S., Geraskin S. A., Geshel I. V. 2018. *Radiological Consequences of the Accident at the Chernobyl Nuclear Power Plant: Biological Effects, Migration, Rehabilitation of Contaminated Areas.* Moscow: Science.

Antonelli F., Campa A., Esposito G., Giardullo P., Belli M., Dini V., Meschini S., Simone G., Sorrentino E., Gerardi S., Cirrone G. A. P. and Tabocchini M. A. 2015. Induction and repair of DNA DSB as revealed by H2AX phosphorylation foci in human fibroblasts exposed to low- and high-LET radiation: Relationship with early and delayed reproductive cell death. *Radiat. Res.* 183(4):417–431.

Aoki-Nakano M., Furusawa Y. 2013. Misrepair of DNA double-strand breaks after exposure to heavy-ion beams causes a peak in the LET–RBE relationship with respect to cell killing in DT40 cells. *J. Radiat. Res.* 54:1029–1035.

Aoyama T., Futamura A., Kato H., Nakamura M., Sugahara T. 1987. Mortality study of Japanese radiological technologists. *J. Jpn. Assoc. Radiol. Tech.* 91–96.

Armitage P. 1971. *Statistical Methods in Medical Research.* Oxford and Edinburgh: Blackwell Scientific Publications.

Artandi S. E., DePinho R. A. 2010. Telomeres and telomerase in cancer. *Carcinogenesis* 31(1):9–18.

Asaithamby A., Hu B., Chen D. J. 2011. Unrepaired clustered DNA lesions induce chromosome breakage in human cells. *Proc. Natl. Acad. Sci. USA.* 108(20):8293–98.

Auvinen A., Pukkala E., Hyvonen H., Hakama M., Rytomaa T. 2002. Cancer incidence among Finnish nuclear reactor workers. *J. Occup. Environ. Med.* 44(7):634–638.

Avdeeva L. V., Evstyukhina T. A., Koltover V. K., Korolev V. G., Kutlakhmedov Yu. A. 2018. "Magnetic isotopes: a new approach to the creation of new antiradiation means." In *Problems of Chemical Protection and Reparation under Radiation Exposure*, 78–80. Dubna.

Averbeck D., Moustacchi E. 1975. Methoxypsoralen plus 365 nm light effects and repair in yeast. *Biochim Biophys Acta.* 23(4):393–404.

Babicová A., Havlínová Z., Hroch M., Rezáčová M., Pejchal J., Vávrová J., Chládek J. 2013. In vivo study of radioprotective effect of NO-synthase inhibitors and acetyl-L-carnitine. *Physiol Res.* 62(6):701–10.

Balcer-Kuiczek E. K., Harrison G. H. 1985. Evidence of microwave carcinogenesis in vitro. *Carcinogenesis* 6:859–864.

Bald T., Quast T., Landsberg J., Rogava M., Glodde N., Lopez-Ramos D., Kohlmeyer J., Riesenberg S., van den Boorn-Konijnenberg D., Hömig-Hölzel C., Reuten R., Schadow B., Weighardt H., Wenzel D., Helfrich I., Schadendorf D., Bloch W., Bianchi M. E., Lugassy C., Barnhill R. L., Koch M., Fleischmann B. K., Förster I., Kastenmüller W., Kolanus W., Hölzel M., Gaffal E., Tüting T. 2014. Ultraviolet-radiation-induced inflammation promotes angiotropism and metastasis in melanoma. *Nature* 507(7490):109–13.

Baranski S. 1971. Effect of chronic microwave irradiation on the blood forming system of guinea pigs and rabbits. *Aerospace Med.* 42:1196–99.

Barendsen G. W. 1968. "Relative biological effectiveness as a function of linear energy transfer." In *Proceeding of the Symposium on Microdosimetry*, edited by Ebert H. G. 249–263. Brussels: Commission of the European Communities.

Barendsen G. W. 1974. "Relative biological effectiveness and biological complexity." In *Forth Symposium on Microdosimetry*, edited by Booz J., Ebert H. G., Eickel R. et al., 235–258. Luxemburg: Commission of the European Communities.

Baryakhtar V. E., edit. 1995. *Chernobyl Disaster.* Kiev: Naukova Dumka.

Baverstock K. 2000. Radiation-induced genomic instability: a paradigm-breaking phenomenon and its relevance to environmentally induced cancer. *Mutation Res.* 454(1–2):89–109.

Beebe G. W., Ishida M., Jablon S. 1962. Studies of the mortality of A-bomb survivors. Plan of study and mortality in the medical subsample (Selection I), 1950–1958. *Radiat. Res.* 16(3):253–280.

Beir V. 1990. *National Academy of Sciences. Health Effects of Exposure to Low Levels of Ionizing Radiation.* Washington, DC: National Academy Press.

Belkina S. V., Petin V. G. 2014. Prediction of an increase in the radiosensitivity of mammalian cells with the sequential use of hyperthermia and ionizing radiation. *Russian Biotherapeutic Journal.* 3(1):65–69.

Bell S. P., Labib K. 2016. Chromosome duplication in *Saccharomyces cerevisiae*. *Genetics.* 203(3):1027–67.

Belyaev I. Y., Alipov Y. D., Polunin V. A. and Scheglov V. S. 1993. Evidence for dependence of resonant frequency of millimeter wave interaction with *E. coli* K12 cells on haploid genome length. *Electro-Magnetobiol.* 12:39–49.

Belyakov O. V., Prise K. M., Trott K. R., Michael B. D. 1999. Delayed lethality, apoptosis and micronucleus formation in human fibroblasts irradiated with X-rays or alpha particles. *Int. J. Radiat. Biol.* 75:985–993.

Ben-Hur E. 1976. Mechanisms of the synergistic interaction between hyperthermia and radiation in cultured mammalian cells. *J. Radiat. Res.* 17(2):92–98.

Ben-Hur E., Elkind M. M. 1974. Thermally enhanced radioresponse of cultured Chinese hamster cells: damage and repair of single-stranded DNA and a DNA complex. *Radiat. Res.* 59: 484–495.

Beral V., Fraser P., Carpenter L. Booth M., Brown A., Rose G. 1988. Mortality of employees of the atomic weapons establishment. *B. M. J.* 24(297(6651)):757–70.

Berthoud H. R. 2013. Synergy: a concept in search of a definition. *Endocrinology.* 154(11):3974–77.

Besse H. C., Bos C., Zandvliet M. M. J. M., van der Wurff-Jacobs K., Moonen C. T. W., Deckers R. 2019. Combination of chemotherapy, radiotherapy and hyperthermia *in vitro. Synergy.* 9:100052.

Bezlepkin V. G., Gaziev A. I. 2001. Induced instability of the genome of animal germ cells by mini- and microsatellite sequences. *Radiat. Biology. Radioecology.* 41(5):475–488.

Biological Effects and Exposure Criteria for Radiofrequency Electromagnetic Fields. 1986. Bethesda. NCRP. Report No. 86.

Bishayee A., Rao D. V. Howell R. W. 2000. Radiation protection by cysteamine against the lethal effects of intracellularly localized Auger electron, α- and β-particle emitting radionuclides. *Acta Oncologica.* 39:713–720.

Blackman C. F., Benane S. G., Weil C. M., Ali J. S. 1975. Effects of non-ionizing electromagnetic radiation on single-cell biologic systems. *Annals N.Y. Acad. Sci.* 247:352–366.

Blasco M. A. 2005. Telomeres and human disease: ageing, cancer and beyond. *Nat. Rev. Genet.* 6:611–622.

Blöcher D. 1988. DNA double-strand break repair determines the RBE of α-particles. *Int. J. Radiat. Biol.* 54:761–771.

Bloom E. T., Akiyama M., Kusunoki Y., Makinodan T. 1987. Delayed effects of low-dose radiation on cellular immunity in atomic bomb survivors residing in the United States. *Health Phys.* 52(5):585–591.

Bloom E. T., Korn E. L., Takasugi M., Toji D. S., Onari K., Makinodan T. 1983. Immune function in aging atomic bomb survivors residing in the United States. *Radiat. Res.* 96(2):399–410.

Bochkov N. P. 1971. *Human Chromosomes and Radiation.* Moscow: Atomizdat.

Bond V. P., Feinendegen L. E., Booz J. 1988. What is a 'low dose' of radiation? *Int. J. Radiat. Biol.* 53(1):1–12.

Bonura T., Smith K. C. 1976. The involvement of indirect effects in cell killing and DNA double-strand breakage in γ-irradiated Escherichia coli K-12. *Int. J. Radiat. Biol.* 29:293–296.

Boothman D. A., Trask D. K., Pardee A. B. 1989. Inhibition of potentially lethal DNA damage repair in human tumor cells by β-lapachone, an activator of topoispomerase I. *Cancer Res.* 49:605–612.

Booz J., Feinendegen L. E. 1988. A microdosimetric understanding of low-dose radiation effects. *Int. J. Radiat. Biol.* 53(1):13–21.

Borchers A. H., Kennedy K. A., Straw J. A. 1990. Inhibition of DNA excision repair by methotrexate in Chinese hamster ovary cells following exposure to ultraviolet irradiation or ethylmethanesulfonate. *Cancer Res.* 50(6):1786–89.

Botstein D., Fink G. R. 2011. Yeast: an experimental organism for 21st century. *Genetics.* 189(3):695–704.

Bracalente C., Ibañez I. L., Molinari B., Palmieri M., Kreiner A., Valda A., Davidson J., Durán H. 2013. Induction and persistence of large γH2AX foci by high linear energy transfer radiation in DNA-dependent protein-kinase deficient cells. *Int. J. Radiat. Oncol. Biol. Phys.* 87(4):785–794.

Brendel M., Haynes R. H. 1973. Interaction among genes controlling sensitivity to radiation and alkylation in yeast. *Molec. Gen. Genet.* 125(2):197–216.

Brenner D. J., Ward J. F. 1992. Constraints on energy deposition and target size of multiply damaged sites associated with DNA double-strand breaks. *Int. J. Radiat. Biol.* 61(6): 737–748.

Bresler S. E., Noskin L. A. 1978. Reparative mechanisms and action of radioprotectors on *E. coli* cells. *Radiobiology.* 18(4):548–555.

Bresler S. E., Noskin L. A., Stepanova I. M., Kuzovleva N. A. 1978. Mechanisms of the radioprotecting action of chemical compounds of Escherichia coli cells. *Molec. Gen. Genet.* 163(1):75-85.

Bresler S. E., Noskin L. A., Suslov A. V. 1981. Study of the induction of double-strand breaks in DNA under the action of γ-radiation and their repair in Escherichia coli cells. *Radiobiology.* 21(1):548–555.

Brosh R. M., Bohr V. A. 2007. Human premature aging, DNA repair and RecQ helicases. *Nucleic Acids Res.* 35(22):7527–7544.

Bruskov V. I. 1990. "Model Systems of Protein-Nucleic Acid Recognition." Doctor of Science thesis., Moscow.

Bruskov V. I., Chernikov A. V., Ivanov V. E., Karmanova E. E. 2020. Formation of the reactive species of oxygen, nitrogen, and carbon dioxide in aqueous solutions under physical impacts. *Physics of Wave Phenomena.* 28(2):103–106.

Brustad T. 1962. Heavy ions and some aspects of their use in molecular and cellular radiobiology. *Adv. Biol. & Medical Physics.* 8:161–224.

Buhtoiarov I. N., Sondel P. M., Wigginton J. M., Buhtoiarova T. N., Yanke E. M., Mahvi D. A., Rakhmilevich A. L. 2011. Anti-tumour synergy of cytotoxic chemotherapy and anti-CD40 plus CpG-ODN immunotherapy through repolarization of tumour-associated macrophages. *Immunology*. 132:226–239.

Buldakov L. A., Kalistratova V. S. 2005. *Radiation Effects on the Body – Positive Effects*. Moscow: Inform-Atom.

Burch L. H., Yang Y., Sterling J. F. Roberts S. A., Chao F. G., Xu H., Zhang L., Walsh J., Resnick M. A., Mieczkowski P. A., Gordenin D. A. 2011. Damage-induced localized hypermutability. *Cell Cycle*. 10(7):1073–1085.

Burlakova E. B., Alesenko A. V., Molochkina E. M. 1975. *Bioantioxidants in Radiation Injury and Malignant Growth*. Moscow: Science.

Burlakova E. B., Goloshchapov A. N., Gorbunova N. V. Gurevich S. M., Zhizhyna G. P., Kozachenko A. I., Konradov A. A., Korman D. B., Molochkina E. M., Nagler L. G., Ozerova I. B., Skalatskaya S. I., Smotryaeva M. A., Tarasenko O. A., Treshchenkova Yu. A., Shevchenko V. A. 1996. Features of the biological action of small doses of radiation. *Radiat. Biology. Radioecol.* 36(4):610–631.

Burlakova E. B., Mikhailov V. F., Mazurik V. K. 2001. System of redox homeostasis in radiation-induced instability. *Radiat. Biology. Radioecology*. 41(5):489–499.

Burns C. P., Lambert B. J., Haugstad B. N. 1986. Influence of rate of heating on thermosensitivity of L1210 leukemia membrane lipids and M 20,000 heat shock protein. *Cancer Res.* 46:1882–1887.

Bustad L. K., Gates N. M., Ross A., Carlson L. D. 1965. Effects of prolonged low-level irradiation of mice. *Radiat. Res.* 25(2):318–330.

Bychkovskaya I. B., Ochinskaya G. K. 1964. Study of the "oxygen effect" at different dose rate of radiation. *Radiat. Biology. Radioecol.* 4(1):63–66.

Cai L. 1999. Research of the adaptive response induced by low dose radiation: where have we been and where should we go? *Hum. Exp. Toxicol.* 18(7):419–425.

Calabrese E. J., Baldwin L. A. 2000. Radiation hormesis: origins, history, scientific foundations. *Hum. Exp. Toxicol.* 19(1):41–75.

Calabrese E. J., Baldwin L. A. 2002. Radiation hormesis and cancer. *Hum. Ecol. Risk Assess.* 8(2):327–353.

Calkins J., Ballard R., Gillespie M. 1978. Ultraviolet light-induced reactivation of alpha-irradiated yeast cells. *Radiat. Res.* 73:440–451.

Caratero A., Courtade M., Bonnet L., Planel H., Caratero C. 1998. Effect of a continuous gamma irradiation at a very low dose on the life span of mice. *Gerontology*. 44(5):272–276.

Cardis E., Vrijheid M., Blettner M., Gilbert E., Hakama M., Hill C., Howe G., Kaldor J., Muirhead C. R., Schubauer-Berigan M., Yoshimura T., Bermann F., Cowper G., Fix J., Hacker C., Heinmiller B., Marshall M., Thierry-Chef I., Utterback D., Ahn Y. O., Amoros E., Ashmore P., Auvinen A., Bae J. M., Bernar J., Biau A., Combalot E., Deboodt P., Diez Sacristan A., Eklöf M., Engels H., Engholm G., Gulis G., Habib R. R., Holan K., Hyvonen H., Kerekes A., Kurtinaitis J., Malker H., Martuzzi M., Mastauskas A., Monnet A., Moser M., Pearce M. S., Richardson D. B., Rodriguez-Artalejo F., Rogel A., Tardy H., Telle-Lamberton M., Turai I., Usel M., Veress K. 2007. The 15-Country Collaborative Study of Cancer Risk among Radiation Workers in the Nuclear Industry: estimates of radiation-related cancer risks. *Radiat Res.* 67(4):396–416.

Carlson L. D., Jackson B. H. 1959. The combined effects of ionizing radiation and high temperature on the longevity of Spraque-Dawley rat. *Radiat. Res.* 11(4):509–519.

Carlson L. D., Scheyer W. I., Jackson B. H. 1957. The combined effects of ionizing radiation and low temperature on the metabolism, longevity and soft tissues of the white rat. *Radiat. Res.* 7(2):190–197.

Castellani A. edit. 2012. *Research in Photobiology*. Springer Science & Business Media.

Cebulska-Wasilewska A., Leenhouts H. P., Chadwick K. H. 1981. Synergism between EMS and X-rays for the induction of somatic mutations in Tradescantia. *Int. J. Radiat. Biol.* 40:163–173.

Chadwick K. H., Leenhouts H. P. 1981. *The Molecular Theory of Radiation Biology*. Berlin, Heidelberg, New York: Springer-Verlag.

Chang W. P., Little J. B. 1992 a. Evidence that DNA double-strand break initiate the phenotype of delayed reproductive death in Chinese hamster ovary cells. *Radiat. Res.* 131(1):53–59.

Chang W. P., Little J. B. 1992 b. Persistently elevated frequency of spontaneous mutations in progeny of CHO clones surviving X-irradiation: Association with delayed reproductive death phenotype. *Mutation Res.* 270(2):191–199.

Chapman J. R., Taylor M. R. G., Boulton S. I. 2012. Playing the end game: DNA double-strand break repair pathway choice. *Molecular Cell.* 47(4):497–510.

Chatzisymeon E., Droumpali A., Mantzavinos D. and Venieri D. 2011. Disinfection of water and wastewater by UV-A and UV-C irradiation: application of real-time PCR method. *Photochem. Photobiol. Sci.* 10:389–395.

Chausov V. N., Boreĭko A. V., Krasavin E. A., Mozhaeva A. V., Ravnachka I. I., Tiunchik S. I., Tronov V. A. 2009. The regularities of the induction of DNA double strand breaks and DNA repair in human lymphocytes after irradiation by accelerated heavy ions with different energy. *Radiats Biol Radioecol.* 49(1):72–6.

Chen K. C., Lin C. J. 1978. A system for studying effects of microwaves on cells in culture. *J. Microwave Power.* 13:251–256.

Chen K. C., Samuel A., Hoopingarner R. 1974. Chromosomal aberrations of living cells induced by microwave radiation. *Environ. Letters.* 6:37–46.

Chen W. L., Luan Y. C., Shich M. C., Chen S. T., Kung H. T., Soong K. L., Yeh Y. C., Chou T. S., Mong S. H., Wu J. T., Sun C. P., Deng W. P., Wu M. F., Shen M. L. 2004. Is chronic radiation an effective prophylaxis against cancer? *Journal of American Physicians and Surgeons.* 9(1):6–10.

Chen W. L., Luan Y. C., Shieh M. C., Chen S. T., Kung H. T., Soong K. L., Yeh Y. C., Chou T. S., Mong S. H., Wu J. T., Sun C. P., Deng W. P., Wu M. F., & Shen M. L. 2006. Effects of cobalt-60 exposure on health of Taiwan residents suggest new approach needed in radiation protection. *Dose-response: a publication of International Hormesis Society* 5(1):63–75.

Chen X. F., Meng F. L., Zhou J. Q. 2009. Telomere recombination accelerates cellular aging in Saccharomyces cerevisiae. *PLoS Genet.* 5(6):e1000535.

Cheng K. C., Loeb L. A. 1993. Genomic instability and tumor progression: mechanistic considerations. *Advances in Cancer Research.* 60:121–156.

Chiabrera A., Nicolini C. Schwan H. Editors. 1985. *Interactions between Fields and Cells.* N.Y.-London, Plenum Press.

Choi J., Kang J. O. 2012. Basics of particle therapy. II: Relative biological effectiveness. *Radiat. Oncol. J.* 30(1):1–13.

Chou T. C. 2006. Theoretical basis, experimental design, and computerized simulation of synergism and antagonism in drug combination studies. *Pharmacol. Rev.* 58:621–681.

Clarke R. H. 2001. Control of low-level radiation exposure: what is the problem and how can it be solved? *Health Phys.* 80(4):391–396.

Cohen B. L. 1993. Relationship between exposure to radon and various types of cancer. *Health Phys.* 65(5):529–531.

Cohen B. L. 1995. Test of the linear no-threshold theory of radiation carcinogenesis for inhaled radon decay products. *Health Phys.* 68(2):157–174.

Cohen B. L. 1997. Problems in the radon vs lung cancer test of linear no-threshold theory and a procedure for resolving them. *Health Phys.* 72(4):623–628.

Cohen J. J. 1980. "Natural background as an indicator of radiation-induced cancer." In *Proc. 5th Int. Cong. of IRPA*, 357–360. Yavneh: Israel Health Physics Society.

Costes S. V., Boissiere A., Ravani S., Romano R., Parvin B., Barcellos-Hoff M. H. 2006. Imaging features that discriminate between foci induced by high- and low LET radiation in human fibroblasts. *Radiat. Res.* 165(5):505–515.

Craig L., Seidman H. 1961. Leukemia and lymphoma mortality in relation to cosmic radiation. *Blood.* 17(3):319–327.

Critchlow S. E., Jackson S. P. 1998. DNA end joining: from yeast to man. *Trends Biochem. Sci.* 23(10):394–398.

Croute F., Soleihavoup J. P., Vidal S., Dupouy D., Planel H. 1982. Paramecium tetraurelia growth stimulation under low-level chronic irradiation: investigations on a possible mechanism. *Radiat. Res.* 92(3):560–567.

Dahle J., Kvam E. 2003. Induction of delayed mutations and chromosomal instability in fibroblasts after UVA-, UVB- and X-irradiation. *Cancer Res.* 63:1464–1469.

Dahle J., Kvam E., Stokke T. 2005. Bystander effects in UV-induced genomic instability: antioxidants inhibit delayed mutagenesis induced by ultraviolet A and B radiation. *J. Carcinogenesis.* 4(11):1–9.

Dale R. G., Jones B., Carabe-Fernandez A. 2009. Why more needs to be known about RBE effects in modern radiotherapy. *Appl. Radiat. Isot.* 67:387–392.

Darby S. C., Kendall G. M., Fell T. P., Darby S. C., Doll R., Goodill A. A., Conquest A. J., Jackson D. A., Haylock R. G. 1993. Further follow up of mortality and incidence of cancer in men from the United Kingdom who participated in the United Kingdom's atmospheric nuclear weapon tests and experimental programmes. *British Medical Journal* 307:1530–1535.

Dasu A., Toma-Dasu I. 2008. What is the clinically relevant relative biological effectiveness? A warning for fractionated treatments with high linear energy transfer. *Int. J. Radiat. Oncol. Biol. Phys.* 70:867–874.

Davidson H. O. 1956. *Biological Effects of Whole-body Gamma Radiation on Human Beings.* Operations Research Office. The Johns Hopkins University.

Davoli T., de Lange T. 2011. The causes and consequences of polyploidy in normal development and cancer. *Annu. Rev. Cell Dev. Biol.* 27:585–610.

Davydov M. I., Golanov A. V., Kanaev S. V., Kostylev V. A., Matyakin G. G., Mardynsky Yu. S., Panshin G. A., Tkachev S. I., Khmelevsky E. V., Yurieva T. V. 2013. Analysis of the state and the concept of modernization of radiation oncology and medical physics in Russia. *Problems of Oncology.* 59(5):529–538.

Degtyareva N. P., Chen L., Mieczkowski P., Petes T. D., Doetsch P. W. 2008. Chronic oxidative DNA damage due to DNA repair defects causes chromosomal instability in *Saccharomyces cerevisiae. Molecular and Cellular Biology.* 28(17):5432–5445.

Dergacheva I. P., Morozov I. I., Petin V. G. 1998. Dependence of cell suspension heating in a microwave field on cell concentration. *Biophysics.* 43:487–491.

Dertinger G., Jung H. 1970. *Molecular Radiation Biology.* Springer-Verlag Berlin.

Dethlefsen L. A., Dewey W. C., editors. 1982. Third Intern. Symposium. *Cancer Therapy by Hyperthermia, Drugs, and Radiation.* Bethesda: Nat. Cancer Inst.

Dewey W. C., Holahan E. V. 1984. Hyperthermia – basic biology. *Prog. Exp. Tumor Res.* 28:198–219.

Dietzel F. 1975. *Tumor und Temperature. Actuelle Probleme bei der Anwendung thermischer Verfahren in Oncologie und Strahlentherapie.* Munchen-Berlin-Wien.

Dineva S. B., Abramov V. I., Shevchenko V. A. 1993. Genetic consequences of the action of lead nitrate on seeds of chronically irradiated populations of Arabidopsis thaliana. *Genetics.* 29(11):1914–1920.

Dobrovolsky L. A. 1991. "The principle of maximum" in chronic combined exposure to radionuclides and hyperthermia and criteria for hygienic regulation." In *Problems of Regulation of Ionizing Radiation under the Influence of Modifying Factors*, edited by L. A. Buldakov, V. S. Kalistratova. 250–255. Moscow: Ministry of Health of the Russian Federation.

Donigan K. A., Cerritelli S. M., McDonald J. P., Vaisman A., Crouch R. J., Woodgate R. 2015. Unlocking the steric gate of DNA polymerase η leads to increased genomic instability in *Saccharomyces cerevisiae. DNA Repair (Amst).* 35(1):1–12.

Douglass A. R., Newman P. A., Solomon S. 2014. The Antarctic ozone hole: An update. *Physics Today.* 67(7):42–48.

Duffy S., Fam H. K., Wang Y. K. Styles E. B., Kim J. H., Ang J. S., Singh T., Larionov V., Shah S. P., Andrews B., Boerkoel C. F., Hieter P. 2016. Overexpression screens identify conserved dosage chromosome instability genes in yeast and human cancer. *PNAS.* 113(36):9967–76.

Dugan L. C., Bedford J. S. 2003. Are chromosomal instabilities induced by exposure of cultured normal human cells to low- and high-LET radiation. *Radiat. Res.* 159(3):301–311.

Dugan V. L. 1971. A kinetic analysis of spore inactivation in a composite heat and gamma radiation environment. *Space Life Sciences.* 2(3):498–505.

Durnov L. A., Baikova V. N., Mayakova S. A. 2001. *Children's Health and Radiation: Actual Problems and Solutions.* Edited by Baleva L. S., Tsaregorodtseva A. D. Moscow: Media Sfera.

Dutta S. K., Nelson W. H., Blackman C. F. 1978. Effect of chronic nonthermal exposures of pulsed microwaves on a repair-deficient mutant of Escherichia coli. *Mutation Res.* 53:91–92.

Eccles L. J., O'Neill P., Lomax M. E. 2011. Delayed repair of radiation induced clustered DNA damage: Friend or foe? *Mutation Res.* 711:134–141.

Eckhoff N. D., Shultis J. K., Clack R. W., Ramer E. R. 1974. Correlation of leukemia mortality rates with altitude in the United States. *Health Phys.* 27(4):377–380.

Eichholtz H., Hietel B. 1990. Heat sensitization to cisplatin in two cell lines with different drug sensitivities. *Int. J. Hyperthermia.* 6(1):47–55.

Eidus L. Kh. 1977. *Nonspecific Cell Response and Radiosensitivity*. Moscow: Atomizdat.

Eidus L. Kh. 1984. "General scheme of modification of radiation damage to cells." In *Mechanisms of Radiation Pathology*, edited by Yu. B. Kudryashov. Moscow: Publishing House.

Eidus L. Kh., Korystov Yu. N. 1984. *Oxygen in Radiobiology*. Moscow: Energoatomizdat.

El-Awady R. A., Dikomey E., Dahm-Daphi J. 2001. Heat effects on DNA repair after ionising radiation: hyperthermia commonly increases the number of non-repaired double-strand breaks and structural rearrangements. *Nucleic Acids Res.* 29:1960–1966.

Eliseeva I. M., Iofa E. L., Stoyan E. F. 1994. Analysis of chromosome aberrations and SCO in children from radiation-contaminated regions of Ukraine. *Radiat. Biology. Radioecology.* 34(2):163–171.

Elkind M. M., Beam C. A. 1955. Variation of the biological effectiveness of X-rays and α-particles on haploid *Saccharomyces cerevisiae*. *Radiat. Res.* 3(1):88–101.

Elpiner I. E. 1973. *Ultrasound Biophysics.* Moscow: Nauka.

Engels H., Swaen G. M., Slangen J., van Amersvoort L., Holmstock L., Van Mieghem E., Van Regenmortel I., Wambersie A. 2005. Radiation exposure and cause specific mortality among nuclear workers in Belgium (1969-1994). *Radiat. Prot. Dosimetry.* 17(4):373–381.

Evans H. H., Horng M. F., Ricanati M., Evans H. H., Horng M. F., Ricanati M., Diaz-Insua M., Jordan R., Schwartz J. L. 2003. Induction of genomic instability in TK6 human lymphoblasts exposed to 137Cs γ radiation: comparison to the induction by exposure to accelerated 56Fe particles. *Radiat. Res.* 159:737–747.

Evseeva T. I., Geraskin S. A. 2001. *Combined Effect of Radiation and Non-radiation Factors on Tradescantia.* Yekaterinburg: Ural Branch of the Russian Academy of Sciences.

Evstratova E. S., Kim J. H., Lim Y. K., Kyu K. L., Petin V. G. 2016. Chemical inhibition of cell recovery after irradiation with sparsely and densely ionizing radiation. *J. Radiat. Industry.* 10(4):199–204.

Evstratova E. S., Mishra K. P., Petin V. G. 2018 a. The comparison of genetic instability in haploid and diploid yeast cells exposed to ionizing radiations of different linear energy transfer and ultraviolet light. *J. Radiat. Cancer Res.* 9(2):79–85.

Evstratova E. S., Petin V. G. 2013. Effect of cisplatin on radiosensitivity and yeast cell recovery. *Radiat. Biology. Radioecol.* 53(6):598–603.

Evstratova E. S., Petin V. G. 2015. Increase in radiosensitivity of cells after inhibition of their ability to repair after potential lethal irradiation damage. *Cell and Tissue Biology*. 9(6):462–466.

Evstratova E. S., Petin V. G. 2018 a. "Genetic instability of haploid and diploid yeast cells surviving after exposure to UV light, gamma-rays and alpha-particles." In *Ionizing Radiation: Advances in Research and Applications*, edited by T. Reeve, 111–140. New York: Nova Science Publishers.

Evstratova E. S., Petin V. G. 2018 b. Biophysical Interpretation of the Dependence of Synergy on the Intensity of Applied Agents. *Biophysics.* 63(6):959-966.

Evstratova E. S., Petin V. G. 2018 c. The delayed appearance of haploid and homozygous diploid Saccharomyces cerevisiae yeast cells of wild type and radiosensitive mutants surviving after exposure to gamma rays and alpha particles. *J. Radiat. Res. Appl. Sci.* 11(1):98–103.

Evstratova E. S., Petin V. G. 2021. *Radiobiological Effects and Genetic Instability of Cells.* Moscow: GEOS.

Evstratova E. S., Petin V. G., Zhurakovskaya G. P. 2018 b. Synergistic effects and their potential significance for the influence of natural intensities of environmental factors on cell growth. *Synergy.* 6(1):1–8.

Falcone C., Mazzoni C. 2016. External and internal triggers of cell death in yeast. *Cell. Molec. Life Sci.* 73: 2237–2250.

Fasullo M., Bennett T., Ahching P., Fasullo M., Bennett T., AhChing P., Koudelik J. 1998. The *Saccharomyces cerevisiae* RAD9 checkpoint reduces the DNA damage-associated stimulation of directed translocations. *Molec. Cell. Biol.* 18(3):1190–1200.

Fasullo M., Giallanza P., Dong Z., Cera C., Bennett T. 2001. *Saccharomyces cerevisiae* rad51 mutants are defective in DNA damage-associated sister chromatid exchanges but exhibit increased rates of homology-directed translocations. *Genetics.* 158:959–972.

Federico M. B., Vallerga M. B., Radi A., Paviolo N. S., Bocco J. L., Di Giorgio M., Soria G., Gottifredi V. 2016. Chromosomal integrity after UV irradiation requires FANCD2-mediated repair of double strand breaks. *PLOS Genet.* 12(1):e1005792.

Feldman H., edit. 2012. *Yeast: Molecular and Cell Biology.* New York: John Wiley & Sons.

Feller V. 1969. *Introduction to the Theory of Probability and Its Applications.* Moscow: Mir.

Ferguson L. R., Chen H., Collins A. R., Connell M., Damia G., Dasgupta S., Malhotra M., Meeker A. K., Amedei A., Amin A., Ashraf S. S., Aquilano K., Azmi A. S., Bhakta D., Bilsland A., Boosani C. S., Chen S., Ciriolo M. R., Fujii H., Guha G., Halicka D., Helferich W. G., Keith W. N., Mohammed S. I., Niccolai E., Yang X., Honoki K., Parslow V. R., Prakash S., Rezazadeh S., Shackelford R. E., Sidransky D., Tran P. T., Yang E. S., Maxwell C. A. 2015. Genomic instability in human cancer: Molecular insights for therapeutic attack and prevention through diet and nutrition. *Seminars in Cancer Biology.* 35:S5–S24.

Ferguson L. R., Cox B. S. 1980. The role of dimer excision in liquid-holding recovery of UV-irradiated haploid yeast. *Mutation Res.* 69(1):19–41.

Ferlay J., Soerjomataram I., Dikshit R., Eser S., Mathers C., Rebelo M., Parkin D. M., Forman D., Bray F. 2015. Cancer incidence and mortality worldwide: sources, methods and major patterns in GLOBOCAN 2012. *Int. J. Cancer.* 136(5):359–386.

Filimonova M. V., Evstratova E. S., Makarchuk V. M., Shevchenko L. I., Filimonov A. S., Petin V. G. 2019. Involvement of recovery processes in the effects of radioprotectors at the cellular and organismal Levels. *Biology Bulletin.* 46(2):1619–24.

Filimonova M. V., Proskuryakov S. Ya., Shevchenko L. I. 2012. Radioprotective properties of isothiourea derivatives with NO-inhibiting mechanism of action. *Radiat. Biology. Radioecology.* 52(6):593–601.

Filimonova M. V., Shevchenko L. I., Makarchuk V. M., Chesnakova E. A., Izmest'eva O. S., Korneeva T. S., Filimonov A. S. 2015. Radioprotective properties of the NO-synthase inhibitor E1023: 1. Indicators of antiradiation activity and interaction with other radioprotectors. *Radiat. Biology. Radioecology.* 55(3):250–259.

Filimonova M. V., Shevchenko L. I., Trofimova T. P. 2014. To the question of the mechanism of the radioprotective action of NO-synthase inhibitors. *Radiat. Biology. Radioecology.* 54(5):500–506.

Filyushkin I. V., Petoyan I. M. 1988. *The Theory of the Carcinogenic Risk of Exposure to Ionizing Radiation.* Moscow. Energoatomizdat.

Fomenko L. A., Lomaeva M. G., Bezlepkin V. G. 2006. Genomic instability detected by micronucleus test in the offspring of F-1 generations of mice exposed to ionizing radiation. *Radiat. Biology. Radioecology.* 46(4):431–435.

Fontana L., Partridge L., Longo V. D. 2010. Dietary restriction, growth factors and aging: from yeast to humans. *Science.* 328(5976):321–326.

Fornalski K. W., Dobrzyński L. 2010. The healthy worker effect and nuclear industry workers. *Dose Response.* 8(2):125–147.

Fox J. C., McNally N. J. 1988. Cell survival and DNA double-strand break repair following X-ray or neutron irradiation of V79 cells. *Int. J. Radiat. Biol.* 54:1021–1030.

Franken N. A. P., ten Cate R., Krawczyket P. M., Stap J., Haveman J., Aten J., Barendsen G. W. 2011. Comparison of RBE values of high LET α-particles for the induction of DNA-DSBs, chromosome aberration and cell reproductive death. *Radiation Oncology.* 6(1):64–71.

Frankenberg D. 1984. Interpretations of the dose and LET dependence of RBE values for lethal lesions in yeast cells. *Radiat. Res.* 97:329–340.

Frankenberg D. 1994. Repair of DNA double-strand breaks and its effect on RBE. *Adv. Space Res.* 14:235–248.

Frankenberg D., Frankenberg-Schwager M., Blöcher D., Harbich R. 1981.Evidence for DNA double-strand breaks as the critical lesions in yeast cells irradiated with sparsely or densely ionizing radiation under oxic or anoxic conditions. *Radiat. Res.* 88(3):524–532.

Frankenberg-Schwager M., Frankenberg D., Harbich R. 1984. Repair of DNA double-strand breaks as a determinant of RBE of alpha particles. *Br. J. Cancer.* 49(VI):169–173.

Frankenberg-Schwager M., Frankenberg D., Harbich R. 1987. Potentially lethal damage repair is due to the difference of DNA double-strand break repair under immediate and delayed plating conditions. *Radiat. Res.* 111:192–200.

Frankenberg-Schwager M., Harbich R., Beckonert S., Frankenberg D. 1994. Half-life for DNA double-strand break rejoining in yeast can vary by more than an order of magnitude depending on the irradiation conditions. *Int. J. Radiat. Biol.* 66:543–547.

Friedberg E. C., Walker G. C., Siede W., Wood R. D., Schultz R. A., Ellenberger T. 2006. *DNA Repair and Mutagenesis*, 2nd edition. ASM Press, Washington DC.

Frigerio N. A., Stowe R. S. 1976. *Carcinogenic and Genetic Hazard from Background Radiation. Biological and Environmental Effects of Low-Level Radiation.* Vienna: IAEA.

Fritz-Neagley H. 1961. *Radiobiology, its Foundations and Achievements.* Moscow: Publishing House of Literature in the Field of Science and Technology.

Frumkin J. P., Patra B. N., Sevold A., Ganguly K., Patel C., Yoon S., Schmid M. B., Ray A. 2016. The interplay between chromosome stability and cell cycle control explored through gene–gene interaction and computational simulation. *Nucleic Acids Res.* 44(17):8073–8085.

Fujii Y., Yurkon C. R., Maeda J. 2013. Influence of track directions on the biological consequences in cells irradiated with high LET heavy ions. *Int. J. Radiat. Biol.* 89(6):401–410.

Fujiki H., Sueoka E., Watanabe T., Suganuma M. 2015. Synergistic enhancement of anticancer effects on numerous human cancer cell lines treated with the combination of EGCG, other green tea catechins, and anticancer compounds. *J. Cancer Res. Clin. Oncol.* 141:1511–1522.

Game J. C. 2000. The *Saccharomyces* repair genes at the end of the century. *Mutation Res.* 30(451 (1–2)):277–293.

Game J. C., Cox B. C. 1973. Synergistic interaction between rad mutations in yeast. *Mutation Res.* 20(1):35–44.

Game J. C., Mortimer R. K. 1974. A genetic study of X-ray sensitive mutants in yeast. *Mutation Res.* 24(3)281–292.

Ganem N. J., Godinho S. A., Pellman D. 2009. A mechanism linking extra centrosomes to chromosomal instability. *Nature.* 460:278–282.

Gatti R. A., Boder E., Vinters H. V., Sparkes R. S., Norman A., Lange K. 1991. Ataxia-telangiectasia: An interdisciplinary approach to pathogenesis. *Medicine.* 70:99–117.

George K. 1978. Chromosomal band inducing effect of high frequency electromagnetic fields. *Indian J. Exp. Biol.* 16:390–393.

Geraskin S. A., Sarapultsev B. I. 1995. Stochastic model of induced genome instability. *Radiat. Biology. Radioecology.* 35(4):451–462.

Gilbert E. S., Fry S. A., Wiggs L. D., Voelz G. L., Cragle D. L., Petersen G. R. 1989 a. Analyses of combined mortality data on workers at the Hanford Site, Oak Ridge National Laboratory, and Rocky Flats Nuclear Weapons Plant. *Radiat. Res.* 120(1):19–35.

Gilbert E. S., Petersen G. R., Buchanan J. A. 1989 b. Mortality of workers at the Hanford Site: 1945-1981. *Health Phys.* 56(1):11–25.

Gillies N. E., Alper T. 1959. Reduction in the lethal effects of radiations on *E. coli* by treatment with chloramphenicol. *Nature.* 183:237–238.

Glasunov A. V., Glaser V. M., Kapultsevich Yu. G. 1989. Two pathways of DNA double-strand break repair in Gl cell of *Saccharomyces cerevisiae. Yeast.* 5(2):131–139.

Glazunov A. V. 1989. "Repair of Radiation Injuries and Radiosensitivity." Doctor of Science diss. Dubna.

Glushchenko A. 2010. *Life under the Sign of Nuclear Risk.* Moscow: Litres-Samizdat.

Gmurman V. E. 2014. *Theory of Probability and Mathematical Statistics.* Moscow. Urayt Publishing House.

Golovach I. N., Zhurakovskaya G. P., Komarova L. N. 1998. Prediction and optimization of synergism under the action of ultrasound and hyperthermia. *Acoustic Journal.* 44:354–357.

Goncharova R. I., Kuzhir T. D. Savina N. V., Nikitchenko N. V. 2015. *Genomic Instability and Impaired DNA Repair as Factors of Hereditary and Somatic Human Pathology.* Minsk: Belaruskaya Nauka.

Goodhead D. T. 1994. Initial events in the cellular effects of ionizing radiations: clustering damage in DNA. *Int. J. Radiat. Biol.* 65(1):7–17.

Goodhead D. T. 1999. Mechanisms for the biological effectiveness of high-LET radiations. *J. Radiat. Res.* 40(Suppl.):1–13.

Gribbin M. A., Weeks J. L., Howe G. R. 1993. Cancer mortality (1956-1985) among male employees of atomic energy of Canada Limited with respect to occupational exposure to external low-linear-energy-transfer ionizing radiation. *Radiat. Res.* 133(3):375–380.

Grigoriev Yu. G. 1982. *Space Radiobiology.* Moscow: Energoatomizdat.

Grigoriev Yu. G. 1983. "Effects of long-term chronic gamma irradiation." In *Biological Effects of Small Doses of Radiation*, edited by Yu. I. Moskalev, 11–20. Moscow: Nauka.

Grigoriev Yu. G., Grigoriev O. A. 2016. *Cellular Communications and Health: Electromagnetic Environment, Radiobiological and Hygienic Problems, Hazard Forecast.* 2nd edition, Moscow: Economics.

Grischenko V. L., Merinov S. H., Moiseev V. A. 1995. Effect of temperature and concentration of different substances on erythrocyte microviscosity cytosole. *Biophysica.* 40:106–109.

Grundler W. 1985. Frequency-dependent biological effects of low intensity microwaves. *Interactions between Electromagnetic Fields and Cells. NATO ASI Series, Series A: Life Sciences*. 97:459–481.

Guiliano N. Ya., Malinovsky O. V., Kharlamov A. D. 1981. Radioprotective effect of cysteamine on yeast cells of wild-type and repair mutant strains. *Radiobiology.* 21(3):379–383.

Gulidov I. A., Mardynsky Yu. S. 2010. "Ardon therapy of tumors." In *Therapeutic Radiology*, edited by A. F. Tsyb, Yu. S. Mardynsky. 406–417. Moscow: LLC "MK."

Gulis G. 2003. Cancer occurrence among radiation workers at Jaslovské Bohunice nuclear power plant. *Cent. Eur. J. Public Health.* 11(2):91–97.

Gundem G., Van Loo P., Kremeyer B., Alexandrov L. B., Tubio J. M. C., Papaemmanuil E., Brewer D. S., Kallio H. M. L., Högnäs G., Annala M., Kivinummi K., Goody V., Latimer C., O'Meara S., Dawson K. J., Isaacs W., Emmert-Buck M. R., Nykter M., Foster C., Kote-Jarai Z., Easton D., Whitaker H. C., Neal D. E., Cooper C. S., Eeles R. A., Visakorpi T., Campbell P. J., McDermott U., Wedge D. C., Bova G. S. 2015. The evolutionary history of lethal metastatic prostate cancer. *Nature*. 520:353–357.

Günther K., Schulz W. 1983. *Biophysical Theory of Radiation Action – a Treatise on Relative Biological Effectiveness*. Berlin: Akademie-Verlag.

Habib R. R., Abdallah S. M., Law M., Kaldor J. 2006. Cancer incidence among Australian nuclear industry workers. *J. Occup. Health.* 48(5):358–365.

Hada M., Georgakilas A. G. 2008. Formation of clustered DNA damage after high-LET-irradiation: a review. *J. Radiat. Res.* 49(3):203–210.

Hada M., Sutherland B. M. 2006. Spectrum of complex DNA damages depends on the incident radiation. *Radiat. Res.* 165(2):223–230.

Hahn G. M. 1982. *Hyperthermia and Cancer*. New York: Plenum Press.

Haken G. 1980. *Synergetics*. Moscow: Mir.

Hall B. C. 1983. Yeast thermotolerance does not require protein synthesis. *J. Bacteriology.* 156:1363–1965.

Hall E. J., Giaccia A. J., editors. 2018. *Radiobiology for the Radiologist.* 8th Ed. PA: Wolters, Kluwer.

Hamid M. A. K., Badour S. S. 1973. The effects of microwaves on green algae. *J. Microwave Power.* 8:267–273.

Hammer G. P., Fehringer F., Seitz G., Zeeb H., Dulon M., Langner I., Blettner M. 2008. Exposure and mortality in a cohort of German nuclear power workers. *Radiat. Environ. Biophys.* 47(1):95–99.

Han A., Elkind M. M. 1978. Ultraviolet light and X-ray damage interaction in Chinese hamster cells. *Radiat. Res.* 74(1): 88–100.

Hanahan D., Weinberg R. A. 2000. The hallmarks of cancer. *Cell.* 100(1):57–70.

Handbook for the Montreal Protocol on Substances that Deplete the Ozone Layer. Fourteenth edition. 2020. United Nations Environment Programme. Nairobi. Kenya.

Hanson K. P., Komar V. E. *Molecular Mechanisms of Radiation Cell Death.* Moscow: Energoatomizdat. 1985.

Harder D., Virsik-Peuckert R. P., Bartels E. R. 1994. Theory of intratrack pairwise lesion interaction. *Radiat. Prot. Dosimetry.* 52(1):13–16.

Harm W. 1978. Analysis of photoenzymatic repair of UV lesions in DNA by single light flashes. VIII. Inhibition of photoenzymatic repair of UV lesions in *E. coli* DNA by caffeine. *Mutation Res.* 10:319–333.

Harte C. 1975. Mutagenesis by radiowaves in Antirrhinum majus L. *Mutation Res.* 29:71–75.

Hattori S. 1994a. State of research and perspectives on radiation hormesis in Japan. *Int. J. Occup. Med. Toxicol.* 3:203–217.

Hattori S. 1994b. State of research and perspectives on radiation hormesis. *BELLE Newsletters.* 3:1–4.

Haynes R. H. 1966. The interpretation of microbial inactivation and recovery phenomenon. *Radiat. Res. Suppl.* 6:1–29.

Haynes R. H. 1975. "DNA repair and genetic control of radiosensitivity in yeast." In *Molecular Mechanisms for Repair of DNA*. Part I, edited by P. C. Hanawalt, R. E. Setlow, 529–540. N.Y.: Plenum Publ. Corp.

Haynes R. H., Kunz B. 1981. "DNA repair and mutagenesis in yeast." In *Molecular Biology of the Yeast Saccharomyces: Life Cycle and Inheritance*. 371–413. Cold Spring Harbor Laboratory.

Heilmann J., Rink H., Taucher-Scholz G., Kraft G. 1993. DNA strand break induction and rejoining and cellular recovery in mammalian cells after heavy-ion irradiation. *Radiat. Res.* 135(1):46–55.

Heller J. H., Teixeira-Pinto A. A. 1959. A new physical method of creating chromosome aberrations. *Nature.* 183:905–906.

Herman T. S., Gerner E. W., Magun B. E., Stickney D., Sweets C. C., White D. M. 1981. Rate of heating as a determinant of hyperthermia cytotoxicity. *Cancer Res.* 41:3519–3523.

Hesslewood I. P. 1978. DNA strand breaks in resistant and sensitive murine lymphoma cells detected by the hydroxylapatite chromatographic technique. *Int. J. Radiat. Biol.* 34:461–469.

Hijnen W. A., Beerendonk E. F., Medema G. J. 2006. Inactivation credit of UV radiation for viruses, bacteria and protozoan (oo)cysts in water: a review. *Water Res.* 40(10):3–22.

Hill B. T., Bellamy A. S. 1984. An overview of experimental investigations of interaction between certain antitumor drugs and X-irradiation in vitro. *Adv. Radiat. Biol.* 11:211–267.

Ho K. S. Y., Mortimer R. K. 1973. Induction of dominant lethality by X-rays in radiosensitive strain of yeast. *Mutation Res.* 20(10):45–51.

Hoeijmakers J. H. J. 1993. Nucleotide excision repair II: from yeast to mammals. *Trends in Genetics.* 9(6):211–217.

Hug O., Kellerer A. 1969. *Stochastic Radiobiology.* Moscow: Atomizdat.

Hunnable E., Cox B. S. 1971. The genetic control for dark recombination in yeast. *Mutation Res.* 13(3):297–309.

IARC Monographs on the Evaluation of Carcinogenic Risks to Humans. V. 100D. 2009. Lyon. France.

Ibañez I. L., Bracalente C., Molinari B. L., Palmieri M. A., Policastro L., Kreiner A. J., Burlón A. A., Valda A., Navalesi D., Davidson J., Davidson M., Vázquez M., Ozafrán M., Durán H. 2009. Induction and rejoining of DNA double strand breaks assessed by H2AX phosphorylation in melanoma cells irradiated with proton and lithium beams. *Int. J. Radiat. Oncol. Biol. Phys.* 74(4):1226–1235.

Ignatov V., Panasenko V., Pidenko A. 1978. *The Effect of Microwave Electromagnetic Field on Bacterial Cells.* Saratov: Saratov University.

Iliakis G., Kurtzman S. 1988. Application of non-hypoxic cell sensitizers in radiobiology and radiotherapy: rationale and future prospects. *Int. J. Radiat. Oncol. Biol., Phys.* 15:1235–1241.

Ilyin L. A. 1994. *Realities and Myths of Chernobyl.* Moscow: ALARA Limited.

Inge-Vechtomov S. G. 2015. *Retrospective of Genetics.* St. Peterburg: Publishing House Leningrad.

Ismailov E. Sh. 1987. *Biophysical Action of Microwave Radiation.* Moscow: Energoatomizdat.

Ivanov V. K. 1986. *Mathematical Modeling and Optimization of Tumor Radiation Therapy.* Moscow: Energoatomizdat.

Ivanov V. K. 2007. Late cancer and noncancer risks among Chernobyl emergency workers of Russia. *Health Phys.* 93(5):470–479.

Ivanov V. K., Gorski A. I., Maksioutov M. A., Tsyb A. F., Souchkevitch G. N. 2001. Mortality among the Chernobyl emergency workers: estimation of radiation risks (preliminary analysis). *Health Phys.* 81(5):514–521.

Ivanov V. K., Kaprin A. D., editors. 2015. *Medical Radiological Consequences of Chernobyl: Prognosis and Evidence after 30 Years.* Moscow: GEOS.

Ivanov V. K., Tsyb A. F. 2002. *Medical Radiological Consequences of Chernobyl for the Population of Russia: Assessment of Radiation Risks.* Moscow: Medicine.

Ivanov V. K., Tsyb A. F., Ivanov S. I., Pokrovsky V. 2004. *Medical Radiological Consequences of the Chernobyl Catastrophe in Russia: Estimation of Radiation Risks.* St. Petersburg: Nauka.

Iwasaki T., Murata M., Ohshima S., Miyake T., Kudo S., Inoue Y., Narita M., Yoshimura T., Akiba S., Tango T., Yoshimoto Y., Shimizu Y., Sobue T., Kusumi S., Yamagishi C., Matsudaira H. 2003. Second analysis of mortality of nuclear industry workers in Japan, 1986-1997. *Radiat. Res.* 159(2):228–238.

Jablon S., Boice J. D. Jr. 1993. Mortality among workers at a nuclear power plant in the United States. *Cancer Causes Control.* 4(5):427–430.

Jablon S., Ishida V., Yamasaki M. 1965. Studies of the mortality of A-bomb survivors: 3. Description of the sample and mortality, 1950-1960. *Radiat. Res.* 25(1):25–52.

Jagger J. 1998. Natural background radiation and cancer death in Rocky Mountain States and Gulf Coast States. *Health Phys.* 75(4):428–430.

Jaikrishan G., Andrews V. J., Thampi M. V., Koya P. K., Rajan V. K., Chauhan P. S. 1999. Genetic monitoring of the human population from high-level natural radiation areas of Kerala on the southwest coast of India. I. Prevalence of congenital malformations in newborns. *Radiat. Res.* 152(6):149–153.

Jain V. K., Pohlit W. 1986. *Biocybirnetics of Cancer. Optimizing Cancer Treatment with Ionizing Radiations.* Banglore: INSDOC.

Jakob B., Scholz M., Taucher-Scholz G. 2003. Biological imaging of heavy charged-particle tracks. *Radiat. Res.* 159(5):676–684.

James A. P., Werner M. M. 1965. The radiobiology of yeast. *Radiat. Botany.* 5:359–382.

Jaworowski Z. 1995. Stimulating effects of ionizing radiation: new issue for regulatory policy. *Regul. Toxicol. Pharmacol.* 22(2):172–179.

Jelley J. V. 1958. *Čerenkov Radiation and Its Applications.* London: Pergamon.

Jenner T. J., de Lara C. M., O'Neill P., Stevens D. L. 1993. Induction and rejoining of DNA double-strand breaks in V79–4 mammalian cells following gamma- and alpha irradiation. *Int. J. Radiat. Biol.* 64(3):265–273.

Johnson H. A., Pavelec M. 1973. Thermal enhancement of thio-TEPA cytotoxicity. *J. Natl. Cancer Inst.* 50(4):903–908.

Johnson R. J., Walmsley J. G., Goodin L. J. 1973. Ultraviolet-induced delay of budding in synchronized yeast cells. *Radiat. Res.* 53(1):134–142.

Joiner M., van der Kogel A., editors. *Basic Clinical Radiobiology.* Edward Arnold. 2009.

Jones B., Underwood T. S. A., Dale R. G. 2011. The potential impact of relative biological effectiveness uncertainty on charged particle treatment prescriptions. *Br. J. Radiol.* 84(Spec.Iss.1):S061–S069.

Jozwiak Z., Paleczka D. 1988. Influence of heat treatment on osmotic fragility of carp erythrocyte. *Int. J. Radiat. Biol.* 54:299–303.

Jung H. 1986. A generalized concept for cell killing by heat. *Radiat. Res.* 106(1):56–72.

Kabakov E. N. 1963. "Light and Dark Reactivation of Cells Affected by Ultraviolet Radiation." PhD diss., Medical Radiological Reseach Center, Obninsk.

Kadhim M. A., Lorimore S. A., Hepburn M. D., Goodhead D. T., Buckle V. J., Wright E. G. 1994. Alpha-particle-induced chromosomal instability in human bone marrow cells. *Lancet.* 344(8928):987–988.

Kadhim M. A., Marsden S. J., Wright E. G. 1998. Radiation-induced chromosomal instability in human fibroblasts: temporal effect and the influence of radiation quality. *Int. J. Radiat. Biology.* 73(2):143–148.

Kagramanova V. K. 1995. Inhibition of protein synthesis of archaebacteria. *Progress in Modern Biology.* 115:563–572.

Kalb R., Neveling K., Nanda I., Schindler D., Hoehn H. "Falcone anemia: causes and consequences of genetic instability." In *Genome and Disease*, edited by Volff J. N. 218–242. Basel: Karger Publishers. 2006.

Kaplan M. I., Morgan W. F. 1998. The nucleus is the target for radiation-induced chromosomal instability. *Radiat. Res.* 150(40):382–390.

Kappos A., Pohlit W. 1972. A cybernetic model for radiation reactions in living cells. I. Sparsely ionizing radiation, stationary cells. *Int. J. Radiat. Biol.* 22(1):51–65.

Kaprin A. D., Galkin V. N., Zhavoronkov L. P., Ivanov V. K., Ivanov S. A., Romanko U. S. 2017. Synthesis of fundamental and applied research is the basis for ensuring a high level of scientific results and their implementation in medical practice. *Radiation and Risk.* 26(2):26–40.

Kaprin A. D., Kostin A. A., Andryukhin M. I., Shegai P. V., Leushin V. Y., Agasieva S. V. 2019. Microwave Radiometry in the Diagnosis of Various Urological Diseases. *Biomedical Engineering.* 53(2):87-91.

Kaprin A. D., Mardynsky Yu. S., editors. 2019. *Therapeutic Radiology: A National Guide.* Moscow: GEOTAR-Media.

Kapultcevich Yu. G., Petin V. G. 1977. Probability model for cell responses to irradiation. *Studia Biophysica.* 62:151–165.

Kapultsevich Yu. G. *Quantitative Regularities of Radiation Damage to Cells.* Moscow: Atomizdat. 1978.

Kapultsevich Yu. G., Korogodin V. I., Petin V. G. 1972. Analysis of radiobiological reactions of yeast cells. I. Survival curves and growth effect. *Radiobiology.* 12(2):267–271.

Kapultsevich Yu. G., Petin V. G. 1974. "Determination of possible target shapes and concretization of the concept of hit from the analysis of the RBE dependence on the spatial distribution of absorbed energy." In *Questions of Microdosimetry.* Issue 2. 50–54. Moscow: Atomizdat.

Karpenshif Y., Bernstein K. 2012. From yeast to mammals cells: recent advances in genetic control of homologous recombination. *DNA repair (Amst).* 11:781–788.

Kato H, Schull W. J., Awa A. Akiyama M., Otake M. 1987. Dose-response analyses among atomic bomb survivors exposed to low-level radiation. *Health Phys.* 52(5):645–652.

Kato H., Brown C. C., Hoel D. C. Schull W. J. 1982. Studies of the mortality of A-bomb survivors. Report 7. Mortality, 1950-1978: part II. Mortality from causes other than cancer and mortality in early entrants. *Radiat. Res.* 91(2):243–264.

Keirim-Markus I. B. 2000. More on the regulation of human exposure. *Med. Radiol. and Radiation Security.* 45(3):41–44.

Kellerer A., Rossi H. H. 1972. The theory of dual radiation action. *Current* Topics in R*adiation* Research *Quarterly*. 8:85–158.

Kellerer A. M., Rossi H. H. 1971. RBE and the primary mechanism of radiation action. *Radiat. Res.* 47(1):15–34.

Kellerer A. M., Rossi H. H. 1978. A generalized formulation of dual radiation action. *Radiat. Res.* 75:471–488.

Kendall G. M., Muirhead C. R., MacGibbon B. H., O'Hagan J. A., Conquest A. J., Goodill A. A., Butland B. K., Fell T. P., Jackson D. A., Webb M. A. 1992. Mortality and occupational exposure to radiation: first analysis of the National Registry for Radiation Workers. *British Medical Journal.* 304(6821):220–225.

Kiefer J. 1974. "On the interpretation of the oxygen effect." In *Fourth Symposium on Microdosimetry*, edited by J. Booz, H. G. Ebert, R. Eickel, A. Waker. Luxemburg: Commission of the European Communities. 1:441–461.

Kiefer J. 1975. The effect of caffeine on survival of UV-irradiated diploid yeast strains of different sensitivities. *Mutation Res.* 30(3):317–326.

Kiefer J. 1976. "The oxygen effect: physical, chemical and biological effects." In *Fifth Symposium on Microdosimetry*, edited by J. Booz, H. G. Ebert, B. G. R. Smith. 849–868. Luxemburg: Commission of the European Communities.

Kiefer J., Egenolf R., Ikpeme S. 2002. Heavy ion-induced DNA double-strand breaks in yeast. *Radiat. Res.* 157(2):141–148.

Kielbassa C., Roza L., Epe B. 1997. Wavelength dependence of oxidative DNA damage induced by UV and visible light. *Cancerogenesis.* 18(4):811–816.

Kim J. K., Belkina S. V., Petin V. G. 2007 a. Mathematical description and prognosis of synergistic interaction of radon and smoking. *Iran. J. Radiat. Res.* 4(4):169–174.

Kim J. K., Komarova L. N., Kim S. H., Petin V. G. 2007 b. Inhibition of DNA repair from radiation damages by hyperthermia is mainly realized through the production of irreversible damage. *Ind. J. Radiat. Res.* 4(3–4):188–195.

Kim J. K., Komarova L. N., Tkhabisimova M. D., Petin V. G. 2005 a. Inhibition of recovery from potentially lethal damage by chemicals in Chinese hamster cells is realized through the production of irreversible damage. *Kor. J. Environ. Biology.* 23(4):390–397.

Kim J. K., Komarova L. N., Zhurakovskaya G. P., Petin V. G. 2006. The fluence rate determines the synergistic interaction of UV light and heat for mitotic recombination and cell inactivation in yeasts. *Photochemistry & Photobiology.* 82(4):1053–1057.

Kim J. K., Petin V. G., Morozov I. I. 2005 b. Quantitative evaluation of the parameters of bacterial photoreactivation after exposure to ultraviolet light and ionizing radiation. *Int. J. Radiat. Biol.* 81:55–62.

Kim J. K., Petin V. G., Tkhabisimova M. D. 2004. Survival and recovery of yeast cells after simultaneous treatment of UV light radiation and heat. *Photochem. Photobiol.* 79(4):349–355.

Kim J. K., Petin V. G., Zhurakovskaya G. P. 2001. Exposure rate as a determinant of synergistic interaction of heat combined with ionizing or ultraviolet radiations in cell killing. *J. Radiat. Res.* 42(4):361–369.

Kirshin V. A., Cheremukhin V. N., Verkholetov V. A. 1965. Comparative study of the effect of different doses of X-rays on chickens. *Uchen. Kazan's notes.* 94:47–53.

Klauenberg B. J. Grandolfo M., Erwin D., editors. 1995. *Radiofrequency Radiation Standards Biological Effects, Dosimetry, Epidemiology, and Public Health Policy.* New York and London: Academic Press.

Klein F., Karwan A., Wintersberger U. 1990. Pedigree analysis of yeast cells recovering from DNA damage allow assignment of lethal events to individual post treatment generations. *Genetics.* 124(1):57–65.

Klyuch V. E., Dubrovin S. A. 1967. "Study of the possibility of using semiconductor detectors in biomedical research." In *Materials of the Conference of Young Scientists Dedicated to the 50th Anniversary of Soviet Union.* Obninsk.

Knyazeva E. H., editor. 2013. *Synergetics. Anthology.* Moscow, Saint Petersburg: Center for Humanitarian Initiatives.

Kogl J. 1986. *Biological Effects of Radiation.* Moscow: Energoatomizdat.

Kolganova O. I., Zhavoronkov L. P., Petin V. G., Kim J. K. 2010. Rabbit heating by microwave exposure at various ambient temperatures. *J. Radiat. Protection.* 35:100–105.

Koltovaya N. A. 2017. Kinase CDK/CDC28 and control of DNA integrity in yeast *Saccharomyces cerevisiae. Radiat. Biology Radioecology.* 57(6):573–590.

Komarova L. N., Petin V. G. 2007. *Radiosensitivity Modification: New Horizons and Perspectives.* Obninsk: IATE.

Komarova L. N., Petin V. G., Tkhabisimova M. D. 2002. Recovery of yeast cells after exposure to ionizing radiation and hyperthermia. *Radiat. Biology. Radioecology.* 42(1):54-59.

Kondo S. 1993. *Health Effects of Low-Level Radiation.* Osaka, Japan: Kinki University Press and Madison; Wisconsin: Medical Physics Publishing Co.

Konev S. V., Rudenok A. N. 1973. "The permeability and the heat resistance of the yeast cell cytoplasmic membrane." In *Membrane Biophysics.* 340–343. Kaunas.

Konev S. V., Volotovsky I. D. 1979. *Photobiology.* Minsk: Publishing House BSU.

Korogodin V. I. 1958 a. Some patterns of post-radiation changes in dormant yeast cells. *Biophysics.* 3:703–710.

Korogodin V. I. 1958 b. Forms of inactivation of yeast cells by ionizing radiation. *Biophysics.* 3(2):206–214.

Korogodin V. I. 1966. *Problems of Post-radiation Recovery.* Moscow: Atomizdat.

Korogodin V. I. 1982. Radiotaxons and genome reliability. *Radiobiology.* 22(2):147–154.

Korogodin V. I. 1993. The study of post-irradiation recovery of yeast: the 'premolecular period'. *Mutation Res.* 289(1):17–26.

Korogodin V. I. 2010. *Phenomenon of Life. Selected Works.* V. 1. Moscow: Nauka.

Korogodin V. I., Bliznik K. M., Kapultcevich Yu. G., Korogodina V. L. 2007. "Cascade mutagenesis: regularities and mechanisms." In *Proc. 2-nd Int. N. W. Timoffeev-Ressovsky Conference.* 1:419–447. Dubna.

Korogodin V. I., Bliznik K. M., Kapultsevich Yu. G. 1974. Evaluation of the frequency of mitotic recombination in clones growing from irradiated cells. *Radiobiology.* 14(5):681–685.

Korogodin V. I., Bliznik K. M., Kapultsevich Yu. G. 1977 b. The role of ploidy in the radiosensitivity of cells (experiment on yeast organisms of different species and genotypes). *Radiobiology.* 17(5):700–710.

Korogodin V. I., Bliznik K. M., Kapultsevich Yu. G., Petin V. G. 1977 a. Regularities of the formation of radioraces in yeast organisms. Communication 10. On the non-randomness of the distribution of some hereditary changes in clones growing from irradiated cells. *Radiobiology.* 17(1):7–30.

Korogodin V. I., Egorov A. Ya., Kabakov E. N., Markova L. I. 1962. The comparative study of light and dark reactivation of yeast cells of different ploidy, affected by ultraviolet radiation. *J. General Biology.* 23(4):302–307.

Korogodin V. I., Kapultcevich Yu. G., Myasnik M. N., Mosin A. F., Gridnev V. V. 1968. Cellular repair processes: survival of irradiated yeast, bacteria, and phages under different postradiation conditions. *Adv. Biol. Med. Physics.* 12:253–274.

Korogodin V. I., Kapultsevich Yu. G., Petin V. G., Blisnik K. M. 1996. Radiosensitivity of haplont yeast cells irradiated with sparsely and densely ionizing radiations. *Mutation Res.* 357(1):67–74.

Korogodin V. I., Korogodina V. L. 1997. Oncogenic consequences of human exposure. *Medical Radiology and Radiation Security.* 42(2):26–30.

Korogodin V. I., Korogodina V. L., Fajszi C., Chepurnoy A. I., Mikhova-Tsenova N., Simonyan N. V. 1991. On the dependence of spontaneous mutation rates on the functional state of gene. *Yeast.* 7:105–117.

Korogodin V. I., Krasavin E. A. 1982. Factors determining the differences in the biological effectiveness of ionizing radiation with different physical characteristics. *Radiobiology.* 22(6):727–738.

Korolev V. G. 2009. Genetic effects of radiation and repair mechanisms in yeast as test systems for radiation safety. *Scientific works.* 116(103):9–12.

Kostyuchenko V. A., Krestinina L. Y. 1994. Long-term irradiation effects in the population evacuated from the East-Urals radioactive trace area. *Sci. Total Environ.* 142(1-2):119–125.

Koterov A. N. 2010. *Small Doses of Radiation: Facts and Myths. Book one. Basic Concepts and Genome Instability.* Moscow: A. I. Burnazyan Publishing house.

Kozlov A. A., Dadunashvili K. S., Kakhidze I. G., Tumanishvili T. G., Kharatishvili N. M., Chkhikvashvili M. S. 1987. Some results of exposure to ultra-low doses of penetrating radiation of chicken embryos during incubation. *Radiobiology.* 27(6):783–786.

Krasavin E. A. 1989. *RBE Problems and DNA Repair.* Moscow: Energoatomizdat.

Krasavin E. A. 2016. Radiobiological research at JNR's accelerators. *PHYS-USP.* 59(4):411–418.

Krivova A. A., Petin V. G., Zhurakovskaya G. P. 2012. Formation of ultraviolet damage when exposed to ionizing radiation. *Nuclear Physics and Engineering.* 3(3):239–242.

Kudryashov Yu. B. *Radiation Biophysics (Ionizing Radiation).* Moscow: Fizmatlit. 2004.

Kudryashov Yu. B., Perov Yu. F., Rubin A. B. 2008. *Radiation Biophysics: Radio Frequency and Microwave Electromagnetic Radiation.* Moscow: Fizmatlit.

Kumar A., Kiefer J., Schneider E. 1985. Inhibition of recovery from potentially lethal damage by chemicals in Chinese hamster V79A cells. *Radiat. Environm. Biophys.* 24(2):89-98.

Kumaran R., Yang S. Y., Leu J. Y. 2013. Characterization of chromosome stability in diploid, polyploid and hybrid yeast cells. *PLoS ONE.* 8:e68094.

Kurpeshev O. K., Konoplyannikov A. G. 1987. Radiosensitizing effect of hyperthermia on normal and tumor tissues. *Medical Radiology.* 32(7):16–20.

Kurpeshev O. K., Mardynsky Yu. S. 2010. "Radiomodifiers in radiation therapy of tumors." In *Therapeutic Radiology: Guide for Doctors*, edited by A. F. Tsyb, Yu. S. Mardynsky, 13–26. Moscow: LLC "MK."

Kushner D. J., editor. 1978. *Microbial Life in Extreme Environments.* London, New York, San Francisco: Academic Press.

Kuzin A. M. 1973. *Molecular Radiobiology of the Cell Nucleus.* Moscow: Atomizdat.

Kuzin A. M. 1983. The problem of synergism in radiobiology. *Bulletin of the Academy of Sciences of the USSR, biological series.* 4:485–502.

Kuzin A. M. 1991. *Natural Radioactive Background and Its Significance for the Earth's Biosphere.* Moscow: Nauka.

Kuzin A. M. 1995. *The Ideas of Radiation Hormesis in the Atomic Age.* Moscow: Nauka.

Kuzin A. M., Kaushansky D. A. 1981. *Applied Radiobiology. Theoretical and Technical Foundations.* Moscow: Energoizdat.

Kuzmina N. S., Suskov I. I. 2002. Expression of genomic instability in lymphocytes of children living in conditions of prolonged exposure to radiation factor. *Radiat. Biology. Radioecology.* 42(6):795–739.

Kuznetsov A. N. 1994. *Biophysics of Electromagnetic Influences. Basics of Dosimetry.* Moscow: Energoatomizdat.

Kyryakov P., Gomez-Perez A., Glebov A., Asbah N., Bruno L., Meunier C., Iouk T., Titorenko V. I. 2016. Empirical verification of evolutionary theories of aging. *Aging.* 8(10):2568–2589.

Land C. E. 1980. Estimating cancer risks from low doses of ionizing radiation. *Science.* 209:1197–1203.

Lantz P. M., Mendez D., Philbert M. A. 2013. Radon, smoking, and lung cancer: the need to refocus radon control policy. *Am. J. Public Health.* 103(3):443–447.

Laroche C., Gervais P. 2003. Achievement of rapid osmotic degydration at specific temperature could maintain high Saccharomyces cerevisiae viability. *Appl. Microbiol. Biotechnol.* 60:743–747.

Laskowski W. 1960. [Inactivierungsversuche mit homozygoten Hefestammen verschiedenen Ploidegrades. I. Aufbau homozygoter Stamme und Dosiseffektkurven für ionisierende Strahlen, UV und organische Peroxide.] *Z. Naturforschung.* 15b(8):495–506. German.

Laun P., Bruschi C. V., Dickinson J. R., Rinnerthaler M., Heeren G., Schwimbersky R., Rid R., Breitenbach M. 2007. Yeast mother cell-specific ageing, genetic (in)stability, and the somatic mutation theory of aging. *Nucleic Acids Research.* 35(22):7514–7526.

Lázár I., Tóth E., Marx G., Cziegler I., Koteles G. 2003. Effects of residential radon on cancer incidence. *J. of Radioanalytical and Nuclear Chemistry.* 258(3):519–524.

Lea D. E. 1946. *Action of Radiations on Living Cells.* Cambridge: University Press.

Leatherbarrow E. L., Harper J. V., Cucinotta F. A., O'Neill P. 2006. Induction and quantification of γ-H2AX foci following low and high LET-irradiation. *Int. J. Radiat. Biol.* 82(2):111–118.

Lee M. S., Yu M., Kim K. Y., Park G. H., Kwack K. B., Kim K. P. 2015. Functional validation of rare human genetic variants involved in homologous recombination using Saccharomyces cerevisiae. *PLoS ONE.* 10(5):e0124152.

Lehninger A. L. 1972. *Biochemistry.* New York: Worth Publishers, Inc.

Leith J. T., Miller R. C., Gerner E. W., Boone M. L. 1977. Hyperthermic potentiation. Biological aspects and applications to radiation therapy. *Cancer.* 39: 766–779.

Letavet A. A., Taylor J. H. 1975. "Ultraviolet, visible and infrared rays." In *Fundamentals of Space Biology and Medicine*, 59–77. Moscow.

Lettier G., Feng Q., de Mayolo A. A., Erdeniz N., Reid R. J. D., Lisby M., Mortensen U. H., Rothstein R. 2006. The Role of DNA Double-Strand Breaks in Spontaneous Homologous Recombination in *S. cerevisiae. PLOS Genetics* 2(11): e194.

Lewis L. K., Resnick M. A. 2000. Tying up loose ends: nonhomologous end-joining in *Saccharomyces cerevisiae*. *Mutation Res.* 451(1-2):71–89.

Ley R. D., Applegate L. A., Fry R. J., Sanchez A. B. 1991. Photoreactivation of ultraviolet radiation-induced skin and eye tumors of Monodelphis domestica. *Cancer Res.* 51:6539–6542.

Li G. C. 1985. Elevated levels of 70,000 Dalton heat shock protein in thermotolerant Chinese hamster fibroblasts and their stable heat resistant variants. *Int. J. Oncology, Biology and Physics.* 11:165–177.

Li G. C., Evans R. G., Hahn G. M. 1976. Modification and inhibition of repair of potentially lethal X-ray damage by hyperthermia. *Radiat. Res.* 67(3):491–501.

Limoli C. L., Corcoran J. J., Milligan J. R., Ward J. F., Morgan W. F. 1999. Critical target and dose and dose-rate responses for the induction of chromosomal instability by ionizing radiation. *Radiat. Res.* 151:677–685.

Limoli C. L., Kaplan M. I., Corcoran J., Meyers M., Boothman D. A., Morgan W. F. 1997. Chromosomal instability and its relationship to other end points of genomic instability. *Cancer Res.* 57(24):5557–5563.

Lin P.-S., Kwock L., Hefter K. 1981. Protection of heat-induced cytotoxicity by glycerol. *J. Cell. Physiol.* 108:439–443.

Lindenbraten L. D., Korolyuk I. P. 2000. *Medical Radiology (Fundamentals of Radiation Diagnostics and Radiation Therapy).* Moscow: Medicine.

Ling C. C., Spiro I. J., Mitchell J., Stickler R. 1985. The variation of OER with dose rate. *Int. J. Radiat. Oncol. Biol. Phys.* 11:1367–1373.

Little J. B. 1990. Low-dose radiation effects: interactions and synergism. *Health Phys.* 59(1):49–55.

Little J. B. 1998. Radiation-induced genomic instability. *Int. J. Radiat. Biol.* 74(6):663–671.

Little J. B. 2000. Radiation carcinogenesis. *Carcinogenesis.* 21(3):397–404.

Little J. B. 2007. Non-targeting effects of ionizing radiation: conclusions as applied to low-dose exposures. *Radiat. Biology. Radioecology.* 47(3):262–272.

Little J. B., Nagasawa H., Pfenning T., Vetrovs H. 1997. Radiation-induced genomic instability: delayed mutagenic and cytogenetic effects of X rays and alpha particles. *Radiat. Res.* 148(2):299–307.

Liu S. Z. 2004. Radiation-induced change in lymphocyte proliferation and its neuroendocrine regulation: dose–response relationship and pathophysiological implications. *Nonlinearity Biol. Toxicol. Med.* 2(3):233–243.

Liu S. Z., Cai L., Sun S. Q. 1992. Induction of a cytogenetic adaptive response by exposure of rabbits to very low dose-rate γ-radiation. *Int. J. Radiat. Biol.* 62(2):187–190.

Liu S. Z., Liu W. H., Sun J. B. 1987. Radiation Hormesis: Its Expression in the Immune System. *Health Phys.* 52(5):579–583.

Liu S. Z., Wan H., Chen S. 1996. Molecular changes related to the stimulatory effect of low dose radiation on immunity. *J. Normal Bethune Univ. Med. Sci.* 22(6):559–565.

Liu T., He M. H., Peng J., Duan Y. M., Lu Y. S., Wu Z., Gong T., Li H. T., Zhou J. Q. 2016. Tethering telomerase to telomeres increases genome instability and promotes chronological aging in yeast. *Aging.* 8(11):2827–2640.

Lobrich M. 1998. Induction and repair of DNA double-strand breaks in human fibroblasts after particle irradiation. *Adv. Space Res.* 22(4):551–560.

Lokhmatova S. A. 1994. Influence of long-term pulsed electromagnetic irradiation of the microwave range of low intensity on testes and epididymis of rats. *Radiat. Biology. Radioecology.* 34:279–285.

Lorat Y., Brunner C. U., Schanz S., Jakob B., Taucher-Scholz G., Rübe C. E. 2015. Nanoscale analysis of clustered DNA damage after high-LET irradiation by quantitative electron microscopy – the heavy burden to repair. *DNA Repair.* 28:93–106.

Lorenz E. 1950. Some biological effects of long continued irradiation. *American Journal of Roentgenology, Radium Therapy, and Nuclear Medicine.* 63(2):176–185.

Lorenz E., Hollcroft J., Miller E., Congdon C., Schweisthal R. 1955. Long-term effects of acute and chronic irradiation in mice. I. Survival and tumor incidence following chronic irradiation of 0.11 R per day. *J. Nat. Cancer Inst.* 15(4):1049–1058.

Loshek D. D., Orr J. S., Solomonidis E. 1977. Interaction of hyperthermia and radiation: temperature coefficient of interaction. *Br. J. Radiology.* 50:902–907.

Luan Y. C., Shieh M. C., Chen S. T., Kung H. T., Soong K. L., Yeh Y. C., Chou T. S., Fang W. C., Yao S. L., Pong C. J., Mong S. H., Wu J. T., Wu J. M., Jen H. J., Chen W. L., Deng W. P., Wu M. F., Shen M. L., Sun C. P.2006. Re-examining the health effects of radiation and its protection. *Int. J. Low Radiat.* 3(1): 27–44.

Luchnik A. N., Glaser V. M., Shestakov S. V. 1977. Repair of DNA double-strand breaks requires two homologous DNA duplexes. *Mol. Biol. Repts.* 3(6):437–442.

Luchnik N. V. 1968. *Biophysics of Cytogenetic Lesions and Genetic Code.* Leningrad: Medicine.

Luckey T. D. 1980 a. *Hormesis with Ionizing Radiation.* Tokyo: Boca Raton Publisher, CRC Press.

Luckey T. D. 1980 b. Radiogenic metabolism. *Am. J. Clin. Nutr*. 33: 25–44.

Luckey T. D. 1991. *Radiation Hormesis.* Tokyo: Boca Raton Publisher, CRC Press.

Luckey T. D. 1996. The evidence for radiation hormesis. *21st Century Sci. Tech.* 9:12–20.

Luckey T. D. 2007. Radiation prevents much cancer. *Int. J. Low Radiat.* 4(4):336–344.

Luckey T. D., Johnson W., Krueger S., Tolo D., Vandenboom E. 1978. Ionizing radiation is required for optimum reproduction in Paramecium bursaria. *Bact. Abst.* 94.

Luke G. A., Riches A. C., Bryant P. E. 1997. Genomic instability in haematopoetic cells of F1 generation mice of irradiated male parents. *Mutagenesis.* 12(3):147–152.

Lushnikov E. F., Lantsov S. I. 1999. Mortality of liquidators in the Kaluga region for 10 years after the accident at the Chernobyl nuclear power plant. *Med. Radiol. and Radiation Security.* 44(2):36–44.

Maeda J., Fujii Y., Fujisawa H., Hirakawa H., Cartwright I. M., Uesaka M., Kitamura H., Fujimori A., Kato T. A. 2015. Hyperthermia-induced radiosensitization in CHO wild-type, NHEJ repair mutant and HR repair mutant following proton and carbon-ion exposure. *Oncology Letters.* 10:2828–34.

Maehara H., Iwami Y., Mayanagi H., Takahashi N. 2005. Synergistic inhibition by combination of fluoride and xylitol on glycolysis by mutans streptococci and its biochemical mechanism. *Caries Res.* 39:521–529.

Mager J., Siderius M. 2002. Novel insights into osmotic stress response of yeast. *FEMS Yeast Res.* 2:251–257.

Mairani A., Böhlen T. T., Dokic I., Cabal G., Brons S., Haberer T. 2013. Modelling of cell killing due to sparsely ionizing radiation in normoxic and hypoxic conditions and an extension to high LET radiation. *Int. J. Radiat. Biol.* 89(10):782–793.

Makinodan T. 1992. "Cellular and subcellular alterations in immune cells induced by chronic, intermittent exposure in vivo to very low doses of ionizing radiation and its ameliorating effects on progression of autoimmune disease and mammary tumor growth." In *Low Dose Irradiation and Biological Defense Mechanisms*, edited by T. Sugahara, L. A. Sagan, T. Aoyama, 233–237. Amsterdam: Elsevier Science Publishers.

Makinodan T., James S. J. 1990. T cell potentiation by low dose ionizing radiation: possible mechanisms. *Health Phys.* 59(1):29–34.

Manikowska E., Luciani J. M., Servantie B., Czerski P., Obrenovitch J., Stahl A. 1979. Effects of 9.4 GHz microwave exposure on meiosis in mice. *Experimentia.* 35:388–390.

Mantsygin Yu. A., Nazarova L. F., Kuzin A. M. 1981. On the effect of small doses of γ-radiation on mammalian cells in culture. *Radiobiology.* 21(1):109–113.

Mao Z., Bozzella M., Seluanov A., Gorbunova V. 2008. DNA repair by nonhomologos end joining and homologous recombination during cell cycle in human cells. *Cell Cycle.* 7:2902–2906.

Marder B. A., Morgan W. F. 1993. Delayed chromosomal instability induced by DNA damage. *Mol. Cell. Biol.* 13(11):6667–6677.

Martin N. M. B. 2001. DNA repair inhibition and cancer therapy. *J. Photochem. Photobiol. B.* 63:162–170.

Mason T. J., Miller R. W. 1974. Cosmic radiation at high altitudes and US cancer mortality, 1950–1969. *Radiat. Res.* 60(2):302–306.

Matanoski G. M., Sartwell P., Elliott E., Tonascia J., Sternberg A. 1984. "Cancer risks in radiologists and radiation workers." In *Radiation Carcinogenesis: Epidemiology and Biological Significance*, edited by J. D. Boice and F. Fraumeni, 83–96. New York: Raven Press.

Mattern M. R., Painter R. B. 1979. Dependence of mammalian DNA replication of DNA supercoiling. II. Effects of novobiocin. *Biochim. Biophys. Acta.* 553:306–312.

Maurice B. B., Ferraris J. D. 2008. Intracellular organic osmolytes: function and regulation. *J. Biol. Chem.* 283:7309–7313.

Mayer V. W., Aguilera A. 1990. High levels of chromosome instability in polyploids of Saccharomyces cerevisiae. *Mutation Res.* 231:177–186.

Mazurik V. K., Mikhailov V. F. 2001. Radiation-induced instability of the genome: phenomenon, molecular mechanisms, pathogenetic significance. *Radiats. Biology. Radioecology.* 41(3):272–289.

McEachern M. J., Haber J. E. 2006. Break-induced replication and recombinational telomere elongation in yeast. *Annu. Rev. Biochem.* 75:111–135.

McMurray M. A., Gottschling D. E. 2004. Aging and genetic instability in yeast. *Current Opinion in Microbiology.* 7:673–679.

Meenakshi C., Mohankumar M. N. 2013. Synergistic effect of radon in blood cells of smokers – An *in vitro* study. *Mutation Res.* 757(1):79–82.

Melloni B. B. M. 2014. Lung cancer in never-smokers: radon exposure and environmental tobacco smoke. *Eur. Respir. J.* 44:850–852.

Melloni E., Marchesini R., Emanuelli H., Fava G., Locati L., Pezzoni G., Savi G., Zunino F. 1984. Hyperthermal effects in phototherapy with hematoporphyrin derivative sensitization. *Tumori.* 70(4):321–325.

Merinov S. H., Moiseev V. A. 1991. Influence of temperature on erythrocyte hydration. *Biophysica.* 36:294–297.

Metz-Flamant C., Rogel A., Caër S., Samson E., Laurier D., Acker A., Tirmarche M. 2009. Mortality among workers monitored for radiation exposure at the French nuclear fuel company. *Arch. Environ. Occup. Health.* 64(4):242–250.

Meynell G. G., Meynell E. 1965. *Theory and Practice of Experimental Bacteriology.* London: Cambridge University Press.

Michael B. D., Harrop H. A., Held K. D. 1981. Photoreactivation of *E. coli* after exposure to ionizing radiation: the role of UV damage by concomitant Čerenkov light. *Int. J. Radiat. Biol.* 39:577–583.

Mifune M., Sobue T., Arimoto H., Komoto Y., Kondo S., Tanooka H. 1992. Cancer mortality survey in a spa area (Misasa, Japan) with a high radon background. *Jpn. J. Cancer Res.* 83(1):1–5.

Mikhailov V. F., Mazurik V. K., Burlakova E. B. 2003. Signaling function of reactive oxygen species in regulatory networks of cell response to damaging influences: participation in the realization of radiosensitivity and genome instability. *Radiat. Biology. Radioecology.* 43(1):5–18.

Mikhailov V. F., Mazurik V. K., Burlakova E. B. 2005. Molecular manifestations of radiation-induced genome instability: the possibility of chemical modification. *Radiat. Biology. Radioecology.* 45(5):551–570.

Millar D. B., Christopher J. P., Hunter J., Yeandle S. S. 1984. The effect of exposure of acetylcholinesterase to 2,450-MHz microwave radiation. *Bioelectromagnetics.* 5(2):165-72.

Mille Y., Beney L., Gervais P. 2002. Viability of *E. coli* after combined osmotic and thermal treatment: a plasma membrane implication. *Biochim. Biophys. Acta.* 1567:41–48.

Mitchel R. E. J., Smith B. P., Wheatly N., Chan A., Child S., Paterson M. C. 1985. Sensitivity of hyperthermia-treated human cells to killing by ultraviolet or γ-radiation. *Radiat. Res.* 104:234–241.

Miyakoshi J., Heki S., Kano E. 1984. "Cellular responses to hyperthermia in Chinese hamster cells." In *Modification of Radiosensitivity in Cancer Treatment*, edited by T. Sugahara, 335–350. Tokyo, Academic Press.

Mizzen L. A., Welch W. J. 1988. Characterization of the thermotolerant cell. I. Effects on protein synthesis activity and the regulation of heat shock protein 70 expression. *J. of Cell Biology.* 106:1105–1116.

Moding E. J., Kastan M. B., Kirsch D. G. 2013. Strategies for optimizing the response of cancer and normal tissues to radiation. *Nat. Rev. Drug. Discov.* 12:526–542.

Mohan A. K., Hauptmann M., Freedman D. M., Ron E., Matanoski G. M., Lubin J. H., Alexander B. H., Boice J. D. Jr., Doody M. M., Linet M. S. 2003. Cancer and other causes of mortality among radiologic technologists in the United States. *Int. J. Cancer.* 103(2):259–267.

Mohr U., Dasenbrock C., Tillman T., Kohler M., Kamino K., Hagemann G., Morawietz G., Campo E., Cazorla M., Fernandez P., Hernandez L., Cardesa A., Tomatis L. 1999. Possible carcinogenic effects of X-rays in a transgenerational study with CBA mice. *Carcinogenesis.* 20:325–332.

Monfared A. S., Jalali F., Sedaghat S. Mansoorizade E., Jarrahi A., Hajiahmadi M., Samavat H. 2006. High natural background radiation areas in Ramsar, Iran: can inhabitants feel safe? *Int. J. Low Radiat.* 3(2/3):171–177.

Moody E., McLerran C., Frazer J. W., Segreto V. A. 1979. "Effects of 1.07 GHz fields on microbial systems." In *Abstracts of the Proceedings of the USNC/URSI Annual Meeting and Bioelectromagnetics Symposium*, 438. Seattle. , p..

Moore S., Stanley F. K., Goodarzi A. A. 2014. The repair of environmentally relevant DNA double strand breaks caused by high linear energy transfer irradiation – no simple task. *DNA Repair.* 17:64–73.

Morgan T. L., Redpath J. L., Ward J. F. 1982. Further studies on Čerenkov induced photoreactivable damage in *E. coli*. *Radiat. Res.* 89:217–226.

Morgan W. F. 2003. Non-target and delayed effects of exposure to ionizing radiation. I. Radiation-induced genomic instability and bystander effects in vitro. *Radiat. Res.* 159(5):567–580.

Morgan W. F., Day J. P., Kaplan M .I. 1996. Genomic instability induced by ionizing radiation. *Radiat. Res.* 146:247–258.

Morgan W. F., Murnane J. P. 1995. A role for genomic instability in cellular radioresistance? *Cancer and Metastasis Reviews.* 14(1):49–58.

Morozov I. I. 1983. Method for enumeration of viable microorganisms. *Lab. Delo.* 9:60–62.

Morozov I. I. 1990. "Thermoradiation treatment and cell osmotic homeostasis." In *Actual Problems of Radiation Biology and Genetics*, 90–92.

Morozov I. I. 1992. Thermically cell damage and their osmotic homeostasis. *Med. Radiol.* 37(11−12):39–42.

Morozov I. I. 1996 a. Influence of the state of aggregation, tonicity and nutrient content in the medium on the recovery of *Escherichia coli* bacteria from thermal damage. *Biophysics.* 41:636–641.

Morozov I. I. 1996 b. On the biological role of the Vavilov-Čerenkov glow. *Radiat. Biology. Radioecol.* 36(6):921–925.

Morozov I. I. 2010. Osmotic homeostasis as one of the critical targets of cell damage by some environmental factors. *Radiat. Biology. Radioecol.* 50:514–522.

Morozov I. I., Ansimova N. S., Dergacheva I. P., Petin V. G. 1990. "Osmotic homeostasis and cell thermoradioresistance under combined treatment." In *Problems of Synergism in Radiobiology*, 102–112.

Morozov I. I., Dergacheva I. P., Ansimova N. S. 1986. The effect of an osmotic shock on the viability, optic density and permeability of heated Escherichia coli cells. *Microbiology.* 55:278–281.

Morozov I. I., Dergacheva I. P., Petin V. G. 1996. Features of microwave and thermal heating of samples of cell suspensions and solutions of some compounds. *Radiat. Biology. Radioecol.* 36:671–675.

Morozov I. I., Dubovik B. V., Petin V. G. 1995. Cellular effects of microwaves of thermal intensity. *Radiat. Biology. Radioecol.* 35(1):47–52.

Morozov I. I., Morozova G. V., Petin V. G. 2005. Effect of cooling rate and medium tonicity on survival of Eescherichia coli B/r bacteria after heating at different lethal temperatures. *Cytology.* 47(1):71–76.

Morozov I. I., Myasnik M. N. 1980. The relationship between the phenomenon of photoreactivation in Escherichia coli following ionizing radiation and the Čerenkov emission. *Radiat. Res.* 82:336–341.

Morozov I. I., Petin V. G. 1998. Modification of cell damage caused by heating at different rates by changing the medium osmotic pressure and by chloramphenicol. *Cytology.* 40(2/3):178–184.

Morozov I. I., Petin V. G. 2000. Osmotic modification of thermal damage to bacteria Escherichia coli at different pH values. *Cytology.* 42:244–248.

Morozov I. I., Petin V. G. 2002. On the nature of the increase in the thermal resistance of *E. coli* cells during heating. *Cytology.* 44:450–454.

Morozov I. I., Petin V. G. 2007. Influence of cyclic changes in the heating and cooling regime on the survival of Escherichia coli bacteria. *Microbiology.* 76(2):200–205.

Morozov I. I., Petin V. G. 2008. On the features of the thermal protective effect of osmolytes on bacteria. *Cytology.* 50:182–186.

Morozov I. I., Petin V. G., Dubovick B. V. 1995. Effects of microwave radiation on bacteria *E. coli* B/r and *E. coli* B_{s-1}. *Electromagnetobiol.* 14(2):149–153.

Morozov I. I., Petin V. G., Dubovick B. V. 1997. Influence of tonicity and chloramphenicol on hyperthermic cytotoxicity and cell permeability under various heating rates. *Int. J. Hyperhermia.* 13(1):47–57.

Mortazavi S. M. J., Karam P. A. 2005. "Apparent lack of radiation susceptibility among residents of the high background radiation area in Ramsar, Iran: can we relax our standards?" In *The Natural Radiation Environment VII*, edited by J. P. McLaughlin, S. E. Simopolos, F. Steinhausler, 1141–1147. Amsterdam: Elsevier.

Mortimer R. K. 1955. Evidence for two types of X-ray induced damage in *Saccharomyces cerevisiae*. *Radiat. Res.* 2(4):361–368.

Moskalev Yu. I. 1983. "Dependence of the biological effects of ionizing radiation on the dose rate." In *Biological Effects of Small Doses of Radiation*, edited by Yu. I. Moskalev, 149–160. Moscow: Nauka.

Moskalev Yu. I. 1991. *Long-Term Consequences of Exposure to Ionizing Radiation.* Moscow: Medicine.

Moss S. H., Smith K. C. 1980. Čerenkov ultraviolet radiation (^{137}Cs γ-rays) and direct excitation (^{137}Cs γ-rays and 50 kVp X-rays) produce photoreactivable damage in *Escherichia coli*. *Int. J. Radiat. Biol.* 38:323–334.

Mulla W., Zhu J., Li R. 2014. A simple model system to study complex phenomena of aneuploidy. *FEMS Microbiol. Rev.* 38(2):201–212.

Munson R. J. 1981. Is the major part of the photoreactivable lethal damage induced in Escherichia coli by X- or γ-rays due to Čerenkov ultraviolet radiatioin emitted by secondary electrons? *Int. J. Radiat. Biol.* 40:435–440.

Muramatsu S., Sugahara T., Tsuchiya T., Okazawa Y. 1964. Effect of chronic low-dose irradiation for three successive generations on the breeding behaviour of mice. *Int. J. Radiat. Biol.* 8(6):523–531.

Murthy M. S. S., Deorukhakar V. V., Rao B. S. 1979. Hyperthermic inactivation of diploid yeast and the interaction of damage caused by hyperthermia and ionizing radiation. *Int. J. Radiat. Biol.* 35:333–341.

Myasnik M. N. 1974. *Genetic Control of Bacterial Radiosensitivity.* Moscow: Atomizdat.

Myasnik M. N., Morozov I. I. 1977. The phenomenon of photoreactivation in bacteria *E. coli* irradiated by ionizing radiation. *Int. J. Radiat. Biol.* 31:95-98.

Myasnik M. N., Morozov I. I., Petin V. G. 1980. The value of the photoreactivable component in *E. coli* B_{s-1} cells exposed to densely and sparsely ionizing radiations. *Int. J. Radiat. Biol.* 37(1):81–84.

Myasnik M. N., Skvortsov V. G., Sokolov V. A. 1985. *Photobiological Aspects of Radiation Damage to Cells.* Moscow: Energoatomizdat.

Nair R. R., Rajan B., Akiba S., Jayalekshmi P., Nair M. K., Gangadharan P., Koga T., Morishima H., Nakamura S., Sugahara T. 2009. Background radiation and cancer incidence in Kerala, India-Karanagappally cohort study. *Health Phys.* 96(1):55–66.

Nakajima N. I., Brunton H., Watanabe R., Shrikhande A., Hirayama R., Matsufuji N., Fujimori A., Murakami T., Okayasu R., Jeggo P., Shibata A. 2013. Visualisation of γH2AX foci caused by heavy ion particle traversal; distinction between core track versus non-track damage. *PLoS One.* 8(8):e70107.

Nambi K. S. V., Soman S. D. 1987. Environmental radiation and cancer in India. *Health Phys.* 52(5):653–657.

Neary G. J. 1970. "The interpretation of survival curves in relation with quality and cellular repair mechanisms." In *Second Symposium on Microdosimetry*, edited by H. G. Ebert, 51–74. Brussels: Commission of the European Communities.

Nefedov I. Yu., Nefedova I. Yu., Palyga G. F. 2001. Genetic consequences of irradiation of one and both parents (results of experiments on Wistar rats). Exitus letalis in the offspring of Wistar rats after irradiation of one and both parents. *Radiat. Biology. Radioecology.* 41(2):133–136.

Nelson D. L., Kox M. M. 2021. *Leninger Principles of Biochemistry.* Vol 1–3. 8th Edition. New York: W. H. Freeman and Company.

Nias A. H. M., Gilbert C. M., Lajtha I. G., Lange C. S. 1965. Clone-size analysis in the study of cell growth following single or during continuous irradiation. *Int. J. Radiat. Biol.* 9:275–290.

Nikjoo H., O'Neill P., Terrissol M., Goodhead D. T. 1999. Quantitative modelling of DNA damage using Monte Carlo track structure method. *Radiat. Environ. Biophys.* 38(1):31–38.

Nikjoo H., O'Neill P., Wilson W. E., Goodhead D. T. 2001. Computational approach for determining the spectrum of DNA damage induced by ionizing radiation. *Radiat. Res.* 156: 577–583.

Noskin L. A. 1984. "Molecular Mechanisms of Reparation Balance in Pro- and Eukaryotic Cells." Doctor of Science diss. Universitet of Gatchina, Russia.

Novitsky T. J., Kushner D. J. 1975. Influence of temperature and salt concentration on the growth of facultatively halophile Microbias spore. *Canadian J. Microbiology.* 21:107–110.

Novitsky T. J., Kushner D. J. 1976. A facultatively halophilic coccus Planococcus halophilis sp. *Int. J. Systemic Bacteriology.* 26:53–57.

Nowosielska E. M., Cheda A., Wrembel-Wargocka J., Janiak M. K. 2008. Modulation of the growth of pulmonary tumour colonies in mice after single or fractionated low-level irradiations with X-rays. *Nukleonika.* 53:9–15.

Obaturov G. M. 1987. *Biophysical Models of Radiobiological Effects.* Moscow: Energoatomizdat.

Ogur H., John R., Nagai S. 1957. Tetrazolium overlag technique for population studies of respiration deficiency in yeast. *Science.* 125:928–931.

Okumura K., Kinashi Y., Kubota Y., Kitajima E., Okayasu R., Ono K., Takahashi S. 2013. Relative biological effects of neutron mixed beam irradiation for boron neutron capture therapy on cell survival and DNA double-strand breaks in cultured mammalian cells. *J. Radiat. Res.* 54(1) :70–75.

Olkin I., Gleser L., Derman C. 1994. *Probability Models and Applications.* Pearson College Div.

Omelchenko A. O., Evstratova E. S., Komarova L. N., Ryzhikova E. N. 2014. Manifestation of synergism under simultaneous action of hyperthermia and antitumor drugs on yeast cells. *Cell & Tissue Biol.* 8(3):253–257.

Ormsby R. J., Lawrence M. D., Blyth B. J., Bexis K., Bezak E., Murley J. S., Grdina D. J., Sykes P. J. 2014. Protection from radiation-induced apoptosis by the radioprotector amifostine (WR-2721) is radiation dose dependent. *Cell Biol. Toxicol.* 30:55–66.

Paap B., Wilson D. M., 3rd, Sutherland B. M. 2008. Human abasic endonuclease action on multilesion abasic clusters: Implications for radiation-induced biological damage. *Nucl. Acids Res.* 36:2717–2727.

Pakhomov A. G., Dubovick B. V., Kolupaev V. E., Pronkevich A. N. 1993. Effect of high–peak–power microwave pulsed on isolated nerve function. *Electro-Magnetobiol.* 12:1–15.

Pakhomova O. N., Tsyb T. S. 1994. "Mutagenous effect and the mitotic crossing-over induction in yeasts under combined exposure to alpha-particles and gamma-rays." In *Molecular Mechanisms in Radiation Mutagenesis and Carcinogenesis*, edited by K. H. Chadwick, R. Cox, H. P. Leenhouts et al., 203–207. Brussels: European Commission.

Pankratov A. A., Andreeva T. N., Yakubovskaya R. I., Kogan B. Y., Butenin A. V., Feizulova R. K., Puchnova V. A., Novoseletsky N. V., Khromov A. V., Lukyanets E. A., Vorozhtsov G. N. 2014. Study of photoinduced antitumor activity of phthalocyanine-based nanostructures as pro-photosensitizers in photodynamic therapy of malignant tumors in vivo. *Bull. Experim. Biol. Med.* 157(6):798–803.

Park E. S., Moon K., Kim H. N., Lee W. J., Jin Y. W. 2010. [Radiation exposure and cancer mortality among nuclear power plant workers: a meta-analysis]. *J. Prev. Med. Public Health.* 43(2):185–192. Korean.

Pastwa E., Neumann R. D., Mezhevaya K. I., Winters T. A. 2003. Repair of radiation-induced DNA double-strand breaks is dependent upon radiation quality and the structural complexity of double-strand breaks. *Radiat. Res.* 159(2):251–261.

Patrick M. H., Haynes R. H., Uretz R. B. 1964. Dark recovery phenomenon in yeast. I. Comparative effects with various inactivating agents. *Radiat. Res.* 21:144–163.

Patties I., Kortmann R. D., Menzel F., Glasow A. 2016. Enhanced inhibition of clonogenic survival of human medulloblastoma cells by multimodal treatment with ionizing irradiation, epigenetic modifiers, and differentiation-inducing drugs. *J. Experimental & Clinical Cancer Res.* 35:94–112.

Peak J. G., Peak M. J. 1980. Protection by glycerol against the biological actions of near ultraviolet light. *Radiat. Res.* 83:553–558.

Peak J. G., Peak M. J. 1984. Similar ultraviolet action spectra for the protection by glycerol of transforming DNA against single-strand breaks and inactivation of biological activity. *Radiat. Res.* 97:570–575.

Pelevina I. I., Aleshchenko A. V., Antoshchina M. M. 2001. The level of spontaneous and radiation-induced cytogenetic damage in blood lymphocytes of children depending on age and living conditions. *Radiat. Biology. Radioecology.* 41(5):573–579.

Pelevina I. I., Aleshchenko A. V., Antoshchina M. M. 2003. Responses of a population of cells to irradiation in low doses. *Radiat. Biology. Radioecology.* 43(2):161–166.

Pelevina I. I., Gotlib V. Ya., Kopradov A. A. 2006. 20 years of studying the consequences of the Chernobyl accident – is that a lot or a little for assessing their nature and scale? *Radiation Biology. Radioecology.* 46(2):240–247.

Pelevina I. I., Saenko A. S., Gotlib V. Ya., Sinzinis B. I. 1985. *Survival of Irradiated Mammalian Cells and DNA Repair.* Moscow: Energoatomizdat.

Pennarun A. M., Kovoor A., Heller R. 1977. Čerenkov counting of 24Na in suspension cultures – an economical use of scintillation apparatus for studying ion exchange. *Physiol. Plant.* 39:323–327.

Petin D. V., Petin V. G. 1995. Genetic control of RBE of α-particles for yeast cells irradiated in stationary and exponential phase of growth. *Mutation Res.* 326(2):211–218.

Petin V. G. 1969. Comparative study of the regularities of inactivation and restoration of yeast cells after α- and γ-irradiation. *Radiobiology.* 9(3):421–426.

Petin V. G. 1977. RBE of densely ionizing radiation and cell recovery. *Medical Radiology.* 22(10):8–12.

Petin V. G. 1979. Effect of gamma and alpha irradiation on survival of wild type and sensitive mutants of yeast. *Mutation Res.* 60(1):43–49.

Petin V. G. 1983. Influence of misonidazole on survival of wild type and radiosensitive mutants of yeast. *Mutation Res.* 108(1–3):121–131.

Petin V. G. 1987. *Genetic Control of Cell Radiosensitivity Modifications.* Moscow: Energoatomizdat.

Petin V. G., Anokhin Yu. N. 2014. Synergistic effects of simultaneous action of hyperthermia with physical and chemical agents. *Med. Physics.* 3(63):57–65.

Petin V. G., Belkina S. V., Zhurakovskaya G. P. 2020. *Mathematical Models and Cell Responses to Irradiation with Ionizing Radiation of Various Quality.* Moscow: GEOS.

Petin V. G., Berdnikova I. P. 1979. Effect of elevated temperatures on the radiation sensitivity of yeast cells of different species. *Radiat. Environ. Biophys.* 16(1):49–61.

Petin V. G., Evstratova E. S., Kim J. K. 2014. Radiosensitivity, liquid-holding recovery and relative biological effectiveness of densely-ionizing radiation after repeated irradiation of yeast cells. *Mutation Research - Genetic Toxicology and Environmental Mutagenesis.* 771:37-42.

Petin V. G., Kabakova N. M. 1981. RBE of densely ionizing radiation for wild-type and radiosensitive mutants of yeast. *Mutation Res.* 82(2):285–294.

Petin V. G., Kabakova N. M., Wrigglesworth V. I., Kim J. K. 2006. RBE of α-particles and postradiation recovery in a temperature conditional diploid yeast mutant for various criteria of cell death. *Physics of Elementary Particles and Atomic Nuclei, Letters.* 3(4(133)):241–245.

Petin V. G., Kapultcevich Yu. G. 2014. Radiation quality and the shape of dose–effect curves at low doses of ionizing radiation for eukaryotic cells. *Mathem. Biosci.* 252(1):1–6.

Petin V. G., Kim J. K. 2004. Survival and recovery of yeast cells after combined treatments with ionizing radiation and heat. *Radiat. Res.* 161(1):56–63.

Petin V. G., Kim J. K. 2005. Liquid-holding recovery kinetics in wild-type and radiosensitive mutants of yeast cells exposed to low- and high-let radiations. *Mutation Res.* 570(1):1–8.

Petin V. G., Kim J. K. 2016. *Synergistic Interaction and Cell Responses to Environmental Factors.* New York: Nova Sciences Publisher.

Petin V. G., Kim J. K., Kolganova O. I., Zhavoronkov L. P. 2011. Synergistic interaction of environmental temperature and microwaves: prediction and optimisation. *J. Radiat. Prot.* 36(1):1–7.

Petin V. G., Kim J. K., Kritskiy R. O. 2008. Mathematical description, optimization and prediction of synergistic interaction of fluoride and xylitol on glycolysis. *Chemosphere. Environmental Toxicology and Risk Assessment.* 72:844–849.

Petin V. G., Komarov V. P. 1982. Estimation of the contribution of ionization and excitation to the lethal effect of ionizing radiation. *Radiat. Environ. Biophys.* 20:79–87.

Petin V. G., Komarov V. P. 1985. Photoreactivation of damage induced by ionizing radiation in yeast cells. *Radiat. Environ. Biophys.* 24:281–286.

Petin V. G., Komarov V. P. 1989. *Quantitative Description of Cell Radiosensitivity Modification.* Moscow: Energoatomizdat.

Petin V. G., Komarov V. P. 1997. Mathematical description of synergistic interaction of hyperthermia and ionizing radiation. *Mathem. Biosci.* 146:115–130.

Petin V. G., Morozov I. I. 2015. *Synergetics of Environmental Factors.* Moscow: GEOS.

Petin V. G., Pereklad O. V., Nili M., Kim J. K. 2008. Yeast cells retain a memory of their original radiation-induced insult. *J. Radiat. Industry.* 2(1):59-64.

Petin V. G., Pronkevich M. D. 2013. *Radiation Hormesis under the Action of Small Doses of Ionizing Radiation.* Obninsk: IATE NRNU MEPhI.

Petin V. G., Ryabchenko V. I. 1982. A study of oxygen effect and radiosensitization of hypoxic cells by metronidazole in wild type and radiosensitive mutants of yeast. *Int. J. Radiat. Biol.* 42:491–500.

Petin V. G., Zhurakovskaya G. P. 1995. The peculiarities of the interaction of radiation and hyperthermia in Saccharomyces cerevisiae irradiated with various dose rates. *Yeast.* 11:549–554.

Petin V. G., Zhurakovskaya G. P. 2009. Conceptual bases of synergistic interaction of ionizing radiation and other environmental factors. *Biosphere.* 1(1):58–64.

Petin V. G., Zhurakovskaya G. P. 2014. Some general regularities of manifestation of the greatest synergetic interaction. *Radiat. Biology. Radioecol.* 54(6):589–596.

Petin V. G., Zhurakovskaya G. P. 2015. Influence of the intensity of acting agents on the manifestation of synergistic interaction. *Radiat. Biology. Radioecol.* 55(6):596–606.

Petin V. G., Zhurakovskaya G. P., Kalugina A. V. 2000. "Microwave dosimetry and lethal effects in laboratory animals." In *Radio Frequency Radiation Dosimetry and Its Relationship to the Biological Effects of Electromagnetic Fields*, 375–382. Kluwer Academic Publishers.

Petin V. G., Zhurakovskaya G. P., Komarova L. N. 1999. Mathematical description of combined action of ultrasound and hyperthermia on yeast cells. *Ultrasonics.* 37:79–83.

Petin V. G., Zhurakovskaya G. P., Komarova L. N. 2012. *Radiobiological Basis of Synergistic Interactions in the Biosphere.* Moscow: GEOS.

Petin V. G., Zhurakovskaya G. P., Komarova L. N., Ryabova S. V. 1998. Synergism of environmental factors as a function of their intensity. *Russian Journal of Ecology.* 29(5):338-343.

Petin V. G., Kim K. J., Zhurakovskaya G. P., Rassokhina A. V. 2000. Mathematical description of synergistic interaction of UV light and hyperthermia for yeast cells. *Journal of Photochemistry and Photobiology B: Biology.* 55(1):74-79.

Pirt S. J. 1975. *Principles of Microbe and Cell Cultivation.* Oxford: Blackwell.

Planel H., Soleilhavoup J. P., Liess M. 1968. Evidence of action of natural ionizing radiation of the reproduction of Dr. Melanogaster. *Comptes Rendus de Association des Anatomistes.* 139:1031–1039.

Planel H., Soleilhavoup J. P., Tixador R., Richoilley G., Conter A., Croute F., Caratero C., Gaubin Y. 1987. Influence on cell proliferation of background radiation or exposure to very low, chronic gamma radiation. *Health Phys.* 52(5):571–578.

Pohlit W. 1973. Radiation sensitivity and repair capacity dependent on cell concentration. *Biophysik.* 10:33–37.

Pohlit W. E., Schäfer M. 1974. "Recovery and repair in yeast cells after irradiation with densely ionizing particles." In *Biological Effects of Neutron Irradiation*, 177–184. Vienna: IAEA.

Polyvoda B. I., Konev V. V., Popov G. A. 1990. *Biophysical Aspects of Radiation Damage to Biomembranes.* Moscow: Energoatomizdat.

Ponnaiya B., Cornforth M. N., Ullrich R. L. 1997. Radiation-induced chromosomal instability in BALB/c and C57BL/6 mice: the difference is as clear as black and white. *Radiat. Res.* 147:121–125.

Ponnaiya B., Jenkins-Baker G., Bigelow A., Marino S., Geard C. R. 2004. Detection of chromosomal instability in α-irradiated and bystander human fibroblasts. *Mutation. Res.* 568(1):41–48.

Ponomarev A., Costes S., Cucinotta F. A. 2008. Stochastic properties of radiation-induced DSB: DSB distributions in large scale chromatin loops, the HPRT gene and within the visible volumes of DNA repair foci. *Int. J. Radiat. Biol.* 84(11):916–929.

Prakash S., Prakash L. 2000. Nucleotide excision repair in yeast. *Mutation Res.* 451(1):13–24.

Prise K. M., Pinto M., Newman H. C., Michael B. D. 2001. A review of studies of ionizing radiation-induced double strand break clustering. *Radiat. Res.* 156(5):572–576.

Pronkevich M. D., Evstratova E. S., Belkina S. V., Anokhin U. N., Petin V. G. 2017. Comparison of the effects of combined effects on yeast and mammalian cells. *Medical Radiology and Radiat. Security.* 62(6):21–27.

Putnam C. D., Richard D. K. 2017. Pathways and mechanisms that prevent genome instability in *Saccharomyces cerevisiae*. *Genetics.* 206(3):1187–1225.

Pyatenko V. S., Eidelman Yu. A., Khvostunov I. K. 2013. Radiation-induced instability of chromosomes under conditions of limited growth of the offspring of irradiated cells. *Doklady Akademii Nauk.* 451(2):1–4.

Raaphorst G. P., Frey H. E., Kruuv J. 1977. Effect of salt solutions on radiosensitivity of mammalian cell. III. Treatment with hypertonic solutions. *Int. J. Radiat. Biol.* 32:109–126.

Raaphorst G. P., Kruuv J. 1976. Effect of tonicity on radiosensitivity of mammalian cells. *Int. J. Radiat. Biol.* 29:493–500.

Raaphorst G. P., Kruuv J. 1977. Effect of salt solutions on radiosensitivity of mammalian cells. II. Treatment with hypotonic solutions. *Int. J. Radiat. Biol.* 32:89–101.

Radiation. Doses, Effects, Risks. United Nations Environment Programme. 1985.

Raevsky B. N. 1959. *Doses of Radioactive Radiation and their Effect on the Organism.* Moscow: Medgiz.

Raju M. R. 1980. *Heavy Particle Radiotherapy.* New York: Academic Press.

Redpath J. L., Morgan T. L., Ward J. F. 1982. On the mechanism of induction of photoreactivatable damage in *E. coli* by ionizing radiations. *Int. J. Radiat. Biol.* 42:81–89.

Redpath J. L., Tortorello M. L. 1977. Photoreactivation of *E. coli* irradiated in the absence and presence of oxygen. *Int. J. Radiat. Biol.* 32:505–506.

Redpath J. L., Zabilansky E. 1979. Photoreactivation of ionizing-radiation-induced damage in *E. coli*. Influence of chemical and physical factors. *Int. J. Radiat. Biol.* 35:473–476.

Redpath J. L., Zabilansky E., Morgan T., Ward J. F. 1981. Čerenkov light and the production of photoreactivatable damage in X-irradiated *E. coli*. *Int. J. Radiat. Biol.* 39:569–575.

Reindl J., Girst S., Walsh D. W., Greubel C., Schwarz B., Siebenwirth C., Drexler G. A., Friedl A. A., Dollinger G. 2017. Chromatin organization revealed by nanostructure of irradiation induced γH2AX, 53BP1 and Rad51 foci. *Sci. Rep.* 7:40616.

Rep M., Krantz M., Thevelein J. M., Hohmann S. 2000. The transcriptional response of Saccharomyces cerevisiae to osmotic shock. Hot 1p and Msn 2p/Msn 4p are required for the induction of subsets of high osmolarity glycerol pathway-dependent genes. *J. Biol. Chem.* 275:8290–8300.

Resnick M. A. 1969. Induction of mutations in *Saccharomyces cerevisiae* by ultraviolet light. *Mutation Res.* 7:315–332.

Resnick M. A. 1976. The repair of double-strand breaks in DNA: a model involving recombination. *J. Theor. Biol.* 59(1):97–106.

Resnick M. A., Cox B. S. 2000. Yeast as an honorary mammal. *Mutation Res.* 451(1):1–11.

Resnick M. A., Setlow J. K. 1972. Repair of pyrimidine dimer damage induced in yeast by ultraviolet light. *J. Bacteril.* 109:979–986.

Reynolds M. C., Brannen J. F. 1973. "Thermal enhancement of radiosterilization." In *Rad. Preservation of Food*, 165–170. Vienna: IAEA.

Reynolds M. C., Garst D. M. 1970. Optimizing thermal and radiation effects for bacterial inactivation. *Space Life Sci.* 2(3):394–399.

Ritter M. A., Cleaver J. E., Tobias C. A. 1977. High-LET radiations induce a large proportion of non-rejoining DNA breaks. *Nature.* 266:653–655.

Romano S. L., Raaphorst G. P., Dewey W. C. 1979. The effects of various salt and sucrose solutions on the U. V. L. sensitivity of CHO cells. *Int. J. Radiat. Biol.* 35:401–415.
Rozhdestvensky L. M. 1984. "Oxygen effect and reparation processes." In *Mechanisms of Radiation Pathology*, edited by Yu. B. Kudryashov. Moscow: Publishing House Moscow University.
Rozhdestvensky L. M. 1985. *Mechanism of the Radioprotective Effect and Indication of the Effectiveness of Radioprotectors.* Moscow: Energoatmoizdat.
Rozhdestvensky L. M. 2006. Alternatives of non-threshold and threshold concepts of carcinogenic and mutagenic effects of low LET radiation: the analysis of postulates and arguments. *Int. J. Low Radiat.* 2(3/4):154–171.
Ryabchenko N. I. 1979. *Radiation and DNA.* Moscow: Atomizdat.
Saeki T., Machida I., Nakai S. 1980. Genetic control of diploid recovery after gamma-irradiation in the yeast *Saccharomyces cerevisiae*. *Mutation Res.* 73(2):251–265.
Saenko A. S., Zamulaeva I. A. 2007. "Features of somatic gene mutagenesis in different age groups of persons exposed to low dose radiation." In *Multiple Stressors: a Challenge for the Future*, edited by C. Mothersill, I. Mosse, C. Seymour, 343–349. Springer.
Saenko A. S., Zamulaeva I. A., Smirnova S. G. 1998. Determination of the frequency of mutations at the loci of glycophorin A and T-cell receptor: examination of the liquidators of the Chernobyl accident. *Radiat. Biology. Radioecology.* 38(2):181–185.
Sage E., Shikazono N. 2016. Radiation-induced clustered DNA lesions: Repair and mutagenesis. *Free Radic. Biol. Med.* 107:125–135.
Sagripanti J. L., Swicord M. L., Davis C. C. 1987. Microwave effects on plasmid DNA. *Radiat. Res.* 110:219–231.
Sagripanti L. J., Swicord M. L. 1986. DNA structural changes caused by microwave radiation. *Int. J. Radiat. Biol.* 50:47–50.
Saha G. B. 2006. *Physics and Radiobiology of Nuclear Medicine.* Springer.
Saha J., Wilson P., Thieberger P., Lowenstein D., Wang M., Cucinotta F. A. 2014. Biological characterization of low-energy ions with high-energy deposition on human cells. *Radiat. Res.* 182(3):282–291.
Samoilova K. A. 1967. *The Effect of Ultraviolet Radiation on a Cell.* Leningrad: Nauka.
Sankaranarayanan K., Taleei R., Rahmanian S., Nikjoo H. 2013. Ionizing radiation and genetic risks. XVII. Formation mechanisms underlying naturally occurring DNA deletions in the human genome and their potential relevance for bridging the gap between induced DNA double-strand breaks and deletions in irradiated germ cells. *Mutation Res.* 753:114–130.
Sapareto S. A, Li G. C., White K. A. 1982. Microwave cytotoxicity: lack in vitro evidence for non-thermal effects at high power levels. *Radiat. Res.* 89:124–133.
Sayeg J. A., Birge A. C., Beam C. A., Tobias C. A. 1959. The effect of accelerated carbon nuclei and other radiation on survival of haploid yeast. II. Biological experiments. *Radiat. Res.* 10(4):449–461.
Schlegel G. 1972. *General Microbiology.* Moscow: Mir.
Schneider E., Kiefer J. 1976. Delayed plating recovery in diploid yeast of different sensitivities after X-ray and alpha particle exposure. *Int. J. Radiat. Biol.* 29(1):77–84.
Schwan H. P. 1988. Biological effects of non-ionizing radiations: cellular properties and interactions. *Ann. Biomed. Eng.* 16(3):245–263.
Schwartz J. M., Gaugain C., Nacher J. C., de Daruvar A., Kanehisa M. 2007. Observing metabolic functions at the genome scale. *Genome Biol.* 8(6):R123.
Scott B., Sanders C. L., Mitchel R. E. J., Boreham D. R. 2008. CT scans may reduce rather than increase the risk of cancer. *J. P. and S.* 13:8–11.
Segova-Mendoze M., Jurado R., Mir R., Medina L. A., Prado-Garcia H., Garcia-Lopez P. 2015. Antihormonal agents as a strategy to improve the effect of chemo-radiation in cervical cancer: *in vitro* and *in vivo* study. *BMC Cancer.* 15(1):21–30.
Semenenko V. A., Stewart R. D. 2004. A fast Monte Carlo algorithm to simulate the spectrum of DNA damages formed by ionizing radiation. *Radiat. Res.* 161(4):451–457.
Sevankayev A. V. 1987. *Radiosensitivity of Human Lymphocyte Chromosomes in the Mitotic Cycle.* Moscow: Energoatomizdat.

Seymour C. B., Mothersill C., Alper T. 1986. High yields of lethal mutations in somatic mammalian cells that survive ionizing radiation. *Int. J. Radiat. Biol.* 50:167–179.

Shadley J. D., Whitlock J. L., Rotmensch J., Atcher R. W., Tang J., Schwartz J. L. 1991. The effect of radon daughter α-particle irradiation in K1 and xrs-5 CHO cell lines. *Mutation Res.* 248(1):73–83.

Shay J. W., Wright W. E. 2000. Hayflick, his limit, and cellular ageing. *Nat. Rev. Mol. Cell Bol.* 1(1):72–76.

Sheltzer J. M., Blank H. M., Pfau S. J., Tange Y., George B. M., Humpton T. J., Brito I. L., Hiraoka Y., Niwa O., Amon A. 2011. Aneuploidy drives genomic instability in yeast. *Science.* 33(6045):1026–1030.

Shen R. N., Hornback N. B., Lu L., Young P., Brahmi Z., Broxmeyer H. E. 1988. Curative effect of split low dosage total-body irradiation on mice infected with the polycythemia-inducing strain of the Friend virus complex. *Cancer Res.* 48:2399–2403.

Shen Z. 2011. Genomic instability and cancer: an introduction. *J. Mol. Cell Biol.* 3(1):1–3.

Shevchenko V. A., Snigireva G. P. 2006. The importance of cytogenetic examination for assessing the consequences of the Chernobyl disaster. *Radiat. Biology. Radioecology.* 46(2):133–139.

Shimizu Y., Kato H., Schull W. J. 1990. Studies of the mortality of A-bomb survivors. 9. Mortality, 1950-1985: part 2. Cancer mortality based on the recently revised doses (DS86). *Radiat. Res.* 121(2):120–141.

Shimizu Y., Kato H., Schull W. J. and Mabuchi K. 1992. "Dose-response analysis among atomic-bomb survivors exposed to low-level radiation." In *Low Dose Irradiation and Biologic Defense Mechanisms*, edited by T. Gugahara, L. A. Sagan, T. Aoyama, 71–74. Amsterdam: Excerpta Medica.

Shore R. E. 1990. Overview of radiation-induced skin cancer in humans. *Int. J. Radiat. Biol.* 57(4):809–827.

Siede W., Friedberg E. C. 1990. Influence of DNA repair deficiencies on the UV sensitivity of yeast cells in different cell cycle stages. *Mutation Res.* 245:287–292.

Sinclair W. K. 1964. X-ray-induced heritable damage (small-colony formation) in cultured mammalian cells. *Radiat. Res.* 21:584–611.

Sivintsev Yu. V. 1991. *How Dangerous is Radiation.* Moscow: IzdAT.

Skoneczna A., Kaniak A., Skoneczny M. 2015. Genetic instability in budding and fission yeast – sources and mechanisms. *FEMS Microbiol. Rev.* 39(6):917–967.

Skvortzov V. G., Myasnik M. N., Sokolov V. A. 1981. Action spectrum for photoreactivation of Escherichia coli Bs-1 after γ-irradiation. *Photochem. Photobiol.* 33:187–190.

Smirnova S. S., Zamulaeva I. A., Orlova N. V. 2005. Comparative study of structural and gene somatic mutations in workers of nuclear chemical enterprises. II. The frequency of lymphocytes mutant in the genes of the T-cell receptor. *Radiat. Biology. Radioecology.* 45(2):162–167.

Smith K. C., Hanawalt P. C. 1969. *Molecular Photobiology. First Edition. Inactivation and Recovery.* Academic Press.

Smith L. E., Nagar S., Kim G. J., Morgan W. F. 2003. Radiation-induced genomic instability: Radiation quality and dose response. *Health Physics.* 85(1):23–29.

Smith-Sonneborn J. 1992. "The role of stress protein response in hormesis." In *Low Dose Irradiation and Biological Defense Mechanisms*, edited by T. Sugahara, L. A. Sagan, T. Aoyama, 399–404. Amsterdam: Excerpta Medica, Elsevier Publishers.

Sobue T., Lee V. S., Ye W., Tanooka H., Mifune M., Suyama A., Koga T., Morishima H., Kondo S. 2000. Residential radon exposure and lung cancer risk in Misasa, Japan: a case-control study. *J. Radiat. Res.* 41(2):81–92.

Sohrabi M. 1997. "World high level natural radiation and/or radon-prone areas with special regards to dwellings." In *High levels of natural Radiation*, edited by L. Wei, T. Sugahara, Z. Tao, 57–68. Amsterdam: Elsevier Science Publishers.

Sokurova E. N. 1975. Ploidy of yeast cells. *Advances in Modern Biology.* 79(3):429–443.

Spalding J. F., Brooks M. R., Tietjen G. L. 1981. Comparative litter and reproduction characteristics of mouse populations for 82 generations of X-irradiated male progenitors. *Proc. Soc.* Exp. Biol. *Med.* 166(2):237–240.

Spitz M. R., Johnson C. C. 1985. Neuroblastoma and parental occupation. *Am. J. Epidemiology.* 121:859–864.

Sponsler R., Cameron J. R. 2005. Nuclear shipyard worker study (1980–1988): a large cohort exposed to low-dose-rate gamma radiation. *Int. J. Low Radiat.* 1(4):463–478.

Steel G. G. 2002. "Cell survival as a determinant of tumor response." In *Basic Clinical Radiobiology*, 52–63. New York: E. Arnold Publ.

Stenerlöw B., Höglund E., Carlsson J., Blomquist E. 2000. Rejoining of DNA fragments produced by radiations of different linear energy transfer. *Int. J. Radiat. Biol.* 76:549–557.

Stirling P. C., Bloom M. S., Solanki-Patil T., Smith S., Sipahimalani P., Li Z., Kofoed M., Ben-Aroya S., Myung K., Hieter P. 2011. The complete spectrum of yeast chromosome instability genes identifies candidate CIN cancer genes and functional roles for ASTRA complex components. *PLoS Genetics.* 7(4):e1002057.

Storchova Z. 2014. Ploidy changes and genome stability in yeast. *Yeast.* 31:421–430.

Storchova Z., Kuffer C. 2008. The consequences of tetraploidy and aneuploidy. *J. Cell Science.* 12(23):3859–3866.

Storchova Z., Pellman D. 2004. From polyploidy to aneuploidy, genome instability and cancer. *Nat. Rev. Mol. Cell Biol.* 5(1):45–54.

Streffer C., van Beuningen D., Dietzel F., Rottinger E., Robinson J. E., Scherer E., Seeber S., Trott, K. R., editors. 1978. *Cancer Therapy by Hyperthermia and Radiation.* Baltimore. Munich: Urban and Schwarzenberg.

Streffer C., Vaupel P., Hahn G. 1990. *Biological Basis of Oncology Thermotherapy.* Berlin, Heidelberg, New York, London, Paris, Tokyo, Hong Kong: Springer Verlag.

Sucher N. J. 2014. Searching for synergy *in silico*, *in vitro* and *in vivo*. *Synergy.* 1(1):30–43.

Sugahara T., editor. 1984. *Modification of Radiosensitivity in Cancer Treatment.* Tokyo: Academic Press.

Sugahara T., Saito M., editors. 1988. *Hyperthermic Oncology.* London: Taylor & Francis.

Suganuma M., Okabe S., Sueoka N., Sueoka E., Matsuyama S., Imai K., Nakachi K., Fujiki H. 1999. Green tea and cancer chemoprevention. *Mutat. Res.* 428:339–344.

Sugasawa K. 2008. Xeroderma pigmentosum genes: functions inside and outside DNA repair. *Carcinogenesis.* 29:455–465.

Sun S., Yang F., Tan G., Costanzo M., Oughtred R., Hirschman J., Theesfeld C. L., Bansal P., Sahni N., Yi S., Yu A., Tyagi T., Tie C., Hill D. E., Vidal M., Andrews B. J., Boone C., Dolinski K., Roth F. P. 2016. An extended set of yeast-based functional assays accurately identifies human disease mutations. *Genome Res.* 26(5):670–680.

Suskov I. I., Kuzmina N. S. 2001. The problem of induced genomic instability in a child's body under conditions of long-term exposure to low doses of radiation. *Radiat. Biology. Radioecology.* 41(5):606–614.

Suskov I. I., Kuzmina N. S., Suskova V. S. 2006. The problem of induced genomic instability as the basis of increased morbidity in children exposed to low-intensity exposure to radiation in small doses. *Radiat. Biology. Radioecology.* 46(2):167–177.

Suskov I. I., Kuzmina N. S., Suskova V. S., Agadzhanyan A. V., Rubanovich A. V. 2008. Individual characteristics of transgenerational genomic instability in children of liquidators of the Chernobyl accident consequences (cytogenetic and immunogenetic indicators). *Radiat. Biology. Radioecology.* 48(2):278–286.

Suzuki Y., Honjo S., Kawamura H., Koishi F., Suzuki T., Hirohata T. 1994. Cancer mortality in low radon spa area. *Jpn. J. Cancer Res.* 85(11):1063–1066.

Swicord M. L., Davies C. C. 1982. Microwave absorption of DNA between 8 and 12 GHz. *Biopolymers.* 21:2453–2460.

Szmigielski S., Luczak M., Wiranowska M. 1975. Effects of microwaves on cell function and virus in cell culture irradiated in vitro. *Annals. N. Y. Acad Sci.* 247:305–311.

Tamas M. J., Rep M., Thevelein J. M., Hohmann S. 2000. Stimulation of the yeast high osmolarity glycerol (HOG) pathway: evidence for a signal generated by a change in turgor rather than by water stress. *FEBS Lett.* 472:159–165.

Tao Z., Zha Y., Akiba S., Sun Q., Zou J., Li J., Liu Y., Kato H., Sugahara T., Wei L. 2000. Cancer mortality in the high background radiation areas of Yangjiang, China, during the period between 1979 and 1995. *J. Radiat. Res.* 41:31–41.

Thierry-Chef I., Marshall M., Fix J. J., Bermann F., Gilbert E. S., Hacker C., Heinmiller B., Murray W., Pearce M. S., Utterback D., Bernar K., Deboodt P., Eklof M., Griciene B., Holan K., Hyvonen H., Kerekes A., Lee M. C., Moser M., Pernicka F. and Cardis E. 2007. The 15-country collaborative study of cancer risk among radiation workers in the nuclear industry: study of errors in dosimetry. *Radiation Research.* 167(4):380–395.

Tigranyan R. E., Shorokhov V. V. 1991. *Physical Foundations of the Auditory Microwave Effect.* Pushchino: ONTI PSC AN SSSR.

Timofeeff-Resovsky N. V. 1996. *Selected Works.* Moscow: Medicine.

Timofeeff-Resovsky N. V., Savich A. V., Shalnov M. I. 1981. *Introduction to Molecular Radiobiology.* Moscow: Medicine.

Timofeeff-Resovsky N. V., Zimmer K. G. 1947. [Biophysik. I. *Das Trefferprinzip in der Biologie.*] Leipzig: Hirzel Verlag. German.

Tirmarche M., Rannou A., Mollie A., Sauve A. 1988. Epidemiological study of regional cancer mortality in France and natural radiation. *Radiat. Prot. Dosimetry.* 24(1-4):479–482.

Tobias C. A. 1985. The repair-misrepair model in radiobiology: comparison to other model. *Radiat. Res.* 104(2):77–92.

Tobias C. A., Blakely E. A., Chang P. Y., Lommel L., Roots R. 1984. Response of sensitive human ataxia and resistant T-1 cell lines to accelerated heavy ions. *Br. J. Cancer.* 49(VI):175–185.

Todd P. 1967. Heavy-ion irradiation of cultured human cells. *Radiat. Res. Suppl.* 7:196–207.

Tolstorukov I. I., Benevolensky S. V. 1978. Studying the mechanisms of mating and self-diploidization in the haploid yeast Pichia pinus. Communication 1. Bipolarity of mating. *Genetics.* 140(3):519–522.

Tolstorukov I. I., Bliznik K. M., Korogodin V. I. 1979. Mitotic instability of Pichia pinus diploid yeast cells. Communication 1. Spontaneous segregation. *Genetics.* 15:2140–2147.

Tolstorukov I. I., Bliznik K. M., Korogodin V. I. 1982. Mitotic instability of diploid cells of the yeast Pichia pinus. Communication 2. Cleavage caused by gamma irradiation. *Genetics.* 18(8):1276–1282.

Tolstorukov I. I., Effremov B. D. 1984. Genetic mapping of the yeast Pichia pinus. II. Mapping by tetrad analysis. *Rus. J. Genet.* 20:880–887.

Tomasovic S. P., Steck P. A., Hertzman D. 1983. Heat stress proteins and thermal resistance in rat mammary tumor cells. *Radiat. Res.* 95:399–413.

Tombline G., Millen J. I., Polevoda B., Rapaport M., Baxter B., Van Meter M., Gilbertson M., Madrey J., Piazza G. A., Rasmussen L., Wennerberg K., White E. L., Nitiss J. L., Goldfarb D. S. 2017. Effects of an unusual poison identify a lifespan role for Topoisomerase 2 in *Saccharomyces cerevisiae*. *Aging.* 9(1):68–87.

Tomilin N. V. 1983. *Genetic Stability of the Cell.* Leningrad: Nauka.

Torres-Durán M., Barros-Dios J. M., Fernández-Villar A., Ruano-Ravina A. 2014. Residential radon and lung cancer in never smokers. A systematic review. *Cancer Lett.* 345(1):21–26.

Troshin A. S. 1979. *Photobiology of a Living Cell.* Leningrad: Nauka.

Trujillo R., Dugan V. L. 1972. Synergistic inactivation of viruses by heat and ionizing radiation. *Biophysical J.* 12(2):92–113.

Tsuchido T., Van Bogelen R. A., Neidhardt F. C. 1986. Heat shock response in Escherichia coli influences cell division. *Proc. Natl. Acad. Sci. USA.* 83(18):6959–6963.

Tsyb A. F., editor. 2005. *Radiation and Pathology.* Moscow: Higher school.

Tsyb A. F., Gulidov I. A. 2010. "The current state of radiation therapy for malignant neoplasms." In *Therapeutic Radiology*, 7–12. Moscow: LLC "MK."

Tsyb A. F., Mardynskiy Yu. S., editors. 2010. *Therapeutic Radiology.* Moscow: LLC "MK."

Tuschl H., Altmann H., Kovac R., Topaloglou A., Egg D., Günther R. 1980. Effects of low-dose radiation on repair processes in human lymphocytes. *Radiat. Res.* 81(1):1–9.

Tyrrell R. M., Peak M. J. 1978. Interaction between UV radiation of different energies in inactivation of bacteria. *J. Bacteriol.* 136:437– 440.

UNSCEAR (United Nations Scientific Committee on the Effects of Atomic Radiation). 1994. *Sources and effects of ionizing radiation. Document A/AC 82/R 542*. New York: United Nations.

UNSCEAR (United Nations Scientific Committee on the Effects of Atomic Radiation). 2008. *Report to the General Assembly, with Scientific Annexes. Annex A. Epidemiological studies of radiation and cancer*. New York: United Nations.

UNSCEAR (United Nations Scientific Committee on the Effects of Atomic Radiation). 2010. *Report: "Summary of low-dose radiation effects on health."* New York: United Nations.

Urano M., Kahn J., Majima H., Gerweck L. E. 1990. The cytotoxic effect of cis–diamminedichloro-platinum (II) on cultured Chinese hamster ovary cells at elevated temperatures: Arrhenius plot analysis. *Int. J. Hyperthermia.* 6(3):581–590.

Urushibara A., Kodama S. Yokoya A. 2014. Induction of genetic instability by transfer of a UV-A-irradiated chromosome. *Mutation Res.* 766(1):29–34.

Urushibara A., Shikazono N., O'Neill P., Fujii K., Wada S., Yokoya A. 2008. LET dependence of the yield of single-, double-strand breaks and base lesions in fully hydrated plasmid DNA films by 4He(2+) ion irradiation. *Int. J. Radiat. Biol.* 84(1):23–33.

Vasina M. V., Vasin A. A., Manokhin E. V. 2018. *Probability Theory and Mathematical Statistics.* Moscow: Prometheus.

Vayserman A. M., Mekhova L. V., Koshel N. M., Voitenko V. P. 2010. Morbidity and mortality from cancer in low-dose irradiation: epidemiological aspects. *Radiat. Biology. Radioecol.* 50(6):691–702.

Vilenchik M. M. 1987. *DNA Instability and Long-term Effects of Radiation Exposure.* Moscow: Energoatomizdat.

Vilenchik M. M. 1991. *Radiobiological Effects and the Environment.* Moscow: Energoatomizdat.

Vinicombe D. A., Moss S. H., Davies D. J. G. 1978. Photo-reactivation of γ-radiation damage in Escherichia coli as evidence for the nature of the oxygen-enhancement effect. *Int. J. Radiat. Biol.* 33:483–492.

Volkov E. I., Polezhaev A. A. 1983. Plasma membrane as the target for hyperthermia. *Usp. Sovrem. Biol.* 96(3):363–365.

Volkova O. Yu., Tashinskaya A. D., Kagan M. S. 1961. The impact of radiation and its decay products on the processes of hematopoiesis. *Med. Radiology.* 6(9):54–58.

Vorobtsova I. E. 1988. "Influence of Parental Irradiation on Physiological Usefulness and Risk of Carcinogenesis in the Offspring of the First Generation of Organisms of Different Species." Doctor of Science diss. Leningrad.

Vorobtsova I. E. 2006.Transgenerational transmission of radiation-induced genome instability. *Radiat. Biology. Radioecology.* 46(4):441–446.

Vorobtsova I. E. 2008. Transgenic transmission of radiation-induced genome instability and predisposition to carcinogenesis. *Problems of Oncology.* 54(4):490–493.

Voyutskii, S. S. 1960. *The Solutions of Macromolecular Compounds.* Moscow: Goshimizdat.

Vrijheid M., Cardis E., Ashmore P., Auvinen A., Bae J. M., Engels H., Gilbert E., Gulis G., Habib R., Howe G., Kurtinaitis J., Malker H., Muirhead C., Richardson D., Rodriguez-Artalejo F., Rogel A., Schubauer-Berigan M., Tardy H., Telle-Lamberton M., Usel M., Veress K. 2007. Mortality from diseases other than cancer following low doses of ionizing radiation: results from the 15-country study of nuclear industry workers. *Int. J. Epidemiol.* 36(5):1126–1135.

Waldow S. M., Dougherti T. J. 1984. Interaction of hyperthermia and photoradiation therapy. *Radiat. Res.* 97(2):380–385.

Wang G., Zhao J., Vasquez K. M. 2009. Methods to determine DNA structural alterations and genetic instability. *Methods.* 48(1):54–62.

Wang T. V., Smith K. C. 1978. Enzymatic photoreactivation of Escherichia coli after ionizing irradiation: chemical evidence for the production of pyrimidine dimmers. *Radiat. Res.* 76:540–548.

Wang W., Yang S., Su Y., Xiao Z., Wang C., Li X., Lin L., Fenton B. M., Paoni S. F., Ding I., Keng P., Okunieff P., Zhang L. 2007. Enhanced antitumor effect of combined triptolide and ionizing radiation. *Clin. Cancer Res.* 13(16):4891–4899.

Wang Y., Iliakis G. 1992. Effects of 5′-iododeoxyuridine on the repair of radiation induced potentially lethal damage interphase chromatin breaks and DNA double strand breaks in Chinese hamster ovary cells. *Int. J. Radiat. Oncol. Biol. Phys.* 23:353–360.

Ward J. F. 1981. Some biochemical consequences of the spatial distribution of ionizing radiation-produced free radicals. *Radiat. Res.* 86(2):185–195.

Watanabe R., Wada S., Funayama T., Kobayashi Y., Saito K., Furusawa Y. 2011. Monte Carlo simulation of radial distribution of DNA strand breaks along the C and Ne ion paths. *Radiat. Prot. Dosim.* 143:186–190.

Watson G. E., Lorimore S. A., Wright E. G. 1996. Long-term in vivo transmission of α-particles-induced chromosomal instability in murine haemopoietic cells. *Int. J. Radiat. Biol.* 69(2):175–182.

Wayland J. R., Brannen J. P., Morris M. E. 1977. On the interdependence of thermal and electromagnetic effects in the response of *Bacillus subtilis* spores to microwave exposure. *Radiat. Res.* 71:251–258.

Wei L. X., Zha Y. R., Tao Z. F., He W. H., Chen D. Q., Yuan Y. L. 1990. Epidemiological investigation of radiological effects in high background radiation areas of Yangjiang, China. *J. Radiat. Res.* 31:119–136.

Weichselbaum R. R. 1986. Radioresistant and repair proficient cells may determine radiocurability in human tumors. *Int. J. Radiat. Oncol. Biol. Phys.* 12:637–639.

Weyrather W. K., Ritte S., Scholz M., Kraft G. 1999. RBE for carbon track-segment irradiation in cell lines of differing repair capacity. *Int. J. Radiat. Biol.* 75:1357–1364.

WHO (World Health Organization). World Health Organization & International Programme on Chemical Safety. 1995. Ultraviolet Radiation: an Authoritative Scientific Review of Environmental and Health Effects of UV, with Reference to Global Zone Layer Depletion. Published under the joint sponsorship of the United Nations Environment Programme, the International Commission on Non-Ionizing Radiation Protection and the World Health Organization, 2nd ed. Geneva.

Wiggs L. D., Johnson E. R., Cox-Devore C. A., Voelz G. L. 1994. Mortality through 1990 among white male workers at the Los Alamos National Laboratory: considering exposures to plutonium and external ionizing radiation. *Health Phys.* 67(6):577–588.

Wing S., Shy C. M., Wood J. L., Wolf S., Cragle D. L., Frome E. L. 1991. Mortality Among Workers at Oak Ridge National Laboratory: Evidence of Radiation Effects in Follow-up Through 1984. *JAMA.* 265(11):1397–1402.

Witkin E. M. 1946. Inherited differences in sensitivity to radiation in *E. coli*. *Proc. Nat. Acad. Sci. USA.* 32(1):59–68.

Woodcock J., Griffin J. P., Behrman R. E. 2011. Development of novel combination therapies. *N. Engl. J. Med.* 364:985–987.

Wu A., Lin P. S. 1985. Peak temperature influences on heating rate effect in hyperthermic cytotoxicity. *Int. J. Radiat. Oncol. Biol. Phys.* 11:983–986.

Yalow R. S. 1994. Concerns with low-level ionizing radiation. *Mayo Clin. Proc.* 69(5):436–440.

Yamamori T., Yura T. 1982. Genetic control of heat-shock protein synthesis and its bearing on growth and thermal resistance in Escherichia coli K-12. *Proc. Natl. Acad. Sci. USA.* 79(3):860–864.

Yao K. T. S. 1982. Cytogenetic consequences of microwave irradiation on mammalian cells. *J. Heredity.* 73:133–138.

Yarmonenko S. P., Konoplyannikov A. G., Vainson A. A. 1992. *Clinical Radiobiology.* Moscow: Medicine.

Yoshinaga S., Mabuchi K., Sigurdson A. J., Doody M. M., Ron E. 2004. Cancer risks among radiologists and radiologic technologists: review of epidemiologic studies. *Radiology.* 233(2):313–321.

Yuen K. W., Warren C. D., Chen O., Kwok T., Hieter P., Spencer F. A. 2007. Systematic genome instability screens in yeast and their potential relevance to cancer. *Proc. National Academy of Sciences of the USA.* 104:3925–3930.

Zadnepryanets M. G., Boreyko A. V., Bulanova T. S., Yezhkova L., Krasavin E. A., Kulikova E. A., Smirnova E. V., Falk M., Falkova I. 2018a. Analysis of the structure of complex DNA damage under the action of accelerated 11B ions and 60Co γ-rays. *Radiat. Biology. Radioecology.* 58(3):229–237.

Zadnepryanets M. G., Boreyko A. V., Bulanova T. S., Yezhkova L., Krasavin E. A., Kulikova E. A., Smirnova E. V., Falk M., Falkova I. 2018b. Regularities of the formation and elimination of γh2ax/53vr1 foci under the action of γ-quanta and accelerated heavy ions. *Radiat. Biology. Radioecology.* 58(2):146–156.

Zaider M., Rossi H. H. 1980. The synergistic effects of different radiations. *Radiat. Res.* 83(3):732–739.

Zakharov I. A., Kovaltsova S. V., Kozhina T. N., Fedorova I. V., Yarovoy B. F. 1980. *Mutational Process in Fungi.* Leningrad: Nauka.

Zakharov I. A., Kozhin S. A., Kozhina T. N., Fedorova I. V. 1984. *Collection of Methods on the Genetics of Saccharomycetes Yeast.* Leningrad: Nauka.

Zhang X., Yang H., Gu K., Chen J., Rui M., Jiang G. L. 2011. In vitro and in vivo study of a nanoliposomal cisplatin as a radiosensitizer. *Int. J. Nanomedicine.* 6:437–444.

Zhavoronkov L. P., Petin V. G. 2018. *Quantitative Criteria for Microwave Injury.* Moscow: GEOS.

Zheng D. O., Zhang K., Wu X. C. 2016. Global analysis of genomic instability caused by DNA replication stress in *Saccharomyces cerevisiae*. *PNAS.* 113(50):8114–8121.

Zhestyanikov V. D. 1968. *Recovery and Radioresistance of Cell.* Leningrad: Nauka.

Zhestyanikov V. D. 1979. *DNA Repair and its Biological Significance.* Leningrad: Nauka.
Zhloba A. A., Sevankaev A. V. 1991. Identification of chromosome aberrations reflecting the genome instability of the descendants of irradiated cells. *Dokl. RAS.* 316(5):1239–1244.
Zhu J., Pavelka N., Bradford W. D., Rancati G., Li R. 2012. Karyotypic determinants of chromosome instability in aneuploidy budding yeast. *PLoS Genetics.* 8(5):E1002719.
Zhurakovskaya G. P. 2001. "General Patterns and Prediction of Synergistic Interactions of Environmental Factors." Doctor of Science diss., Obninsk.
Zhurakovskaya G. P., Dergacheva I. P., Pereklad O. V. 2009. Radiobiological properties of the descendants of yeast cells that survived after thermoradiation exposure. *Radiat. Biology. Radioecology.* 49(4):484–489.
Zimmer K. G. 1962. *Problems of Quantitative Radiobiology.* Moscow: Gosatomizdat.

About the Authors

Ekaterina S. Evstratova
Head of Laboratory
National Medical Research Radiological Centre of the
Ministry of Health of the Russian Federation, Russia

PhD in Biological Sciences. She was born in 1989, graduated from the Department of Biology of the Obninsk Institute of Atomic Energy (2011), Russia. From 2011 to 2018 she conducted scientific work in the biophysical laboratory of the Department for the Study of Combined Effects of A.F. Tsyb MRRC, Obninsk, Russia. PhD thesis "Synergism and recovery of cells after the combined use of chemical agents with ionizing radiation or hyperthermia" was defended in 2015. In 2016-2017 she did scientific work in the Department of Stem Cell Biology at the University of Nagasaki, Japan. Currently, she is the head of the department of regenerative technology of the National Medical Research Center for Radiology of the Ministry of Health of Russia. She has over 150 scientific publications, including 1 book. During the entire time of her work at the MRRC, she carried out priority research in the field of radiobiology, biophysics and genetics. She became the Laureate of the Scholarship (2015), she won first place in the International Competition of Scientific Works in the Field of Radioecology (2017), and she was also awarded grants from the Russian Foundation for Basic Research for young scientists (2016, 2018).

E-mail: ekevs7240@mail.ru

Vladislav G. Petin
Head of Laboratory
A. Tsyb Medical Radiological Research Centre –
branch of the National Medical Research Radiological Centre of the
Ministry of Health of the Russian Federation, Russia

Doctor of Biological Sciences, Professor, Head of Biophysical Laboratory in A.F. Tsyb Medical Radiological Research Center, Obninsk, Russia. He was born in Kazakhstan (1939) and graduated from Moscow State University (1962). His postgraduate education was received at Max Planck Institute for Biophysics (Frankfurt-am-Main, 1966-1967), and his doctor degree was awarded in 1984. For more than 10 years, he collaborated with the Korea Institute of Atomic Energy; as a result, the monograph V.G. Petin, J.K. Kim "Synergistic Interaction and Cell Responses to Environmental Factors" was published in New York (2014, 2016). The following priority results have been developed in Biophysical Laboratory: (i) cell photoreactivation after exposure to ionizing radiation; (ii) participation of recovery processes in the RBE of densely ionizing radiation; (iii) mathematical optimization and predictions of the synergistic effects after combined action of harmful environmental agents. He is the author of more than 250 scientific papers and 13 books. He was awarded by N.V. Timofeeff-Ressovsky medal and prize.

E-mail: vgpetin@yahoo.com

Andrey D. Kaprin
CEO, National Medical Research Radiological Centre of the Ministry of Health of the Russian Federation, Russia

Doctor of Medical Sciences, Professor, Academician of the Russian Academy of Sciences, Honored Doctor of the Russian Federation, Director General of the National Medical Research Center for Radiology of the Ministry of Health of Russia, Chief Oncologist of the Ministry of Health of Russia. President of the Association of Oncologists of Russia, President of the Association of Directors of Centers and Institutes of Oncology and X-Ray Radiology of the CIS and Eurasia. Born in 1966. Graduate of the Moscow Medical Dental Institute. Leading specialist in the Russian Federation in the field of diagnosis and treatment of oncological and urological diseases. Leads the development of modern methods of surgical and combined treatment of malignant neoplasms with a wide range of applications of the achievements of nuclear medicine. Actively develops the introduction of Russian radiopharmaceuticals into clinical practice. He is a practicing surgeon, annually performs more than 100 unique and complex surgical interventions.

E-mail: kaprin@mail.ru

Petr V. Shegai
Deputy CEO for Sciences,
National Medical Research Radiological Centre of the Ministry of Health of the Russian Federation, Russia

PhD of Medical Sciences, Head of the Center for Innovative Radiological and Regenerative Technologies, Deputy General Director for Science of the National Medical Research Center for Radiology of the Ministry of Health of Russia. Born in 1980, graduate of the Sechenov University, Moscow. He began his research activity with the study of fundamental issues of carcinogenesis in oncourology and received his PhD degree in 2008. Over the past years, scientific work has covered such areas as molecular biology in oncology, regenerative medicine, the peptide-receptor approach in nuclear medicine, and evolutionary medicine. He is the co-author of more than 150 scientific papers and 6 books. Laureate of various awards, including the state award - the Certificate of Honor of the President of the Russian Federation.

E-mail: dr.shegai@mail.ru

Index

D

E

F

G

H

I

L

M

N

O

P

R

V

W

X

Y